Trouble in the Forest

Trouble in the Forest

California's Redwood Timber Wars

Richard Widick

University of Minnesota Press   Minneapolis / London

Map of Humboldt County, California, created by Richard Widick

Published by the University of Minnesota Press
111 Third Avenue South, Suite 290
Minneapolis, MN 55401-2520
http://www.upress.umn.edu

A Cataloging-in-Publication record for this book is available from the Library of Congress.

ISBN 978-0-8166-5324-9 (hc : alk. paper)
ISBN 978-0-8166-5325-6 (pb : alk. paper)

Printed in the United States of America on acid-free paper

The University of Minnesota is an equal-opportunity educator and employer.

20 19 18 17 16 15 14 13 12 11 10 09   10 9 8 7 6 5 4 3 2 1

Contents

Preface

One ought to read everything, study everything.
In other words, one must have at one's disposal
the general archive of a period at a given moment.
And archaeology is, in a strict sense, the science
of this archive.

—*Michel Foucault*

All production is appropriation of nature on the
part of an individual within and through a specific
form of society.

—*Karl Marx*

Wigi. Later, in a new tongue, even local Wiyot would call it Humboldt
Bay. In 1850 Wiyot tribal lands surrounded the bay, reaching south
beyond the Eel River to Bear River, north beyond the Mad River to
Little River, and inland into the first mountain ranges, and so the
Wiyot took the greatest blow when the men of the Laura Virginia
Company finally discovered the entrance bar and sailed their cutter
into the bay, permanently anchoring the Euro-Anglo-American culture system of capitalism along with their boat right in the center of
the Wiyot universe. So began the colonization that would, over the
next 150 years, first by force and then by way of commodity production to infinity, produce a new place from the old, a place we now call

Humboldt County, California, or simply Humboldt—the place of the redwood timber wars.

The old language of this place was only spoken, not written, and the effect on cultural transmission was tremendous when, on February 26, 1860, pioneer settlers massacred between eighty and one hundred Indians on Humboldt Bay's Indian Island. The killers also struck at Wiyot villages around the mouth of Humboldt Bay and to the south along the Eel River at Eagle Prairie, a spot now occupied by the town of Rio Dell. No final count of the bodies was ever made (estimates range between 150 and 200), but this much we do know: between first contact in 1850 and the massacres of 1860, disease, murder, slavery, enclosure, depletion of wildlife, starvation, war, and state-sanctioned removal and concentration on reservation camps reduced the population of Wiyot from somewhere between fifteen hundred and two thousand to perhaps two hundred. The tribe reports that only one hundred full-blooded Wiyot were living within the tribal area in 1910. Today there are approximately 550 members enrolled, some living on the eighty-eight-acre Table Bluff Reservation overlooking Humboldt Bay from the south. But the tribe has not danced together since that ghastly night 150 years ago. According to Albert James, a descendant of Wiyot tribal leaders and onetime leader of the effort to reclaim Indian Island for the tribe, writing in a report of the Eureka Planning Department in 1971 titled "Far West Indian Historical Center, Indian Island," Wiyot used to perform the Jump Dance and other ceremonial dances using a portion of the island's two massive shell mounds—one was six hundred feet long and fourteen feet high—during weeklong celebrations at which tribal groups from around the region came together for religious rituals and a feast of clams, fowl, venison, and berries. It was at the close of one of these celebrations that the white men struck, killing with club and knife every man, woman, and child they could find.

The Wiyot genocide helps explain why there are no running historical accounts of the seaborne colonization of northern California's Humboldt Bay redwood region handed down to us from Wiyot elders. There is no comparable archive of Wiyot-language counternarratives to challenge the colonizers' own stories of discovery, conquest, appropriation, and settlement of Humboldt. Between 1850 and 1990, more than 96 percent of the ancient forests fell in the name of this cultural onslaught. But the local stories of capital culture and the American nation-state reaching into redwood ecology and Indian territory were told, printed, and archived in English—the New World's hegemonic tongue—and so these are the texts that shape the experience of every new scholar of the place of Humboldt.

Trouble in the Forest: California's Redwood Timber Wars is my search through the textual ruins of capital culture in the bay redwood region, seeking a sense of that colonizing culture's knowledge and power—the motor, in other words, of its institutional dynamism. Fifteen decades of capital in Humboldt structure the place and the contemporary redwood timber wars that animate it, ensuring that the struggle over ancient forests is always about much more than trees. It is a battle for the future, over how this place has been and will be known, over how it has been and will be recognized and represented, and over how its peoples' constitutive memories, energies, and attentions will contribute, or not, to emergent global civil society. The place we discover here today is an outcome of this ongoing struggle over local knowledge, but it also feeds into a larger struggle—perhaps the greatest that humankind has ever faced. We are entering an era of planetary ecological crisis, in which leading establishment environmentalists like James Gustave Speth concur with the Union of Concerned Scientists and socialist environmental theorists like John Bellamy Foster that capitalism as we know it today cannot sustain the environment. For this reason, we critically need studies of specific places, like Humboldt, to help us understand why.

To this end, my project begins from the perspective of place and asks the difficult questions of how historical colonization by the culture system of capitalism made the redwood region what it is today. Inquiring across ecological, political-economic, and historical disciplines in search of answers, I was forced to become a sort of collector. The help I received from archivists at Humboldt State University became more indispensable the further I dug in. Edie Butler and Joan Berman were instrumental in navigating the university's superb Humboldt Room. At the Humboldt Historical Society in Eureka, I was treated with generosity, ensuring my days in the reading room were most pleasurable and rewarding.

At the Noyo Hill House in Mendocino, I found a rare camaraderie and intellectual enthusiasm with local historians and archivists Russell and Sylvia Bartley; their assistance and firsthand knowledge of North Coast history supplemented my energy when I needed it most.

They are now at work on what promises to be an important new resource for regional history, as well as social history in general, especially that of environmental activism and forest defense: the new Bari Collection at the Mendocino County Museum in Willits, California. They treated me to an advance viewing of the as yet to be cataloged collection, which includes the car in which Judi Bari and Darryl Cherney were riding when the bomb exploded, sending chills up and down the spines of forest defenders and environmentalists everywhere and changing the North Coast forever. My great hope is that this collection will soon be open to the public and that my own efforts in *Trouble in the Forest* might compel others to investigate these matters more closely.

For those on all sides of the redwood timber wars who aided my inquiry by granting interviews, my appreciation is limitless. I hope these pages do justice to your generosity. Fellow travelers in social theory and praxis whose guidance was essential to this work include John Foran and especially Martha McCaughey. At the University of California, Santa Barbara, the unflagging support and critical reflections of Roger Friedland, Richard Flacks, Simonetta Falasca-Zamponi, and Elisabeth Weber made this work possible. I would also like to acknowledge financial support received from the Interdisciplinary Humanities Center and the Department of Sociology at the University of California, Santa Barbara.

approx. pre-contact Indian territories
i — Indian Island
H — Holmes-Eureka Lumber Mill
C — David Gypsy Chain death site
B — Luna Tree-Sit/Julia Butterfly Hill
L — town of Stafford, houses hit by landslide
— Simpson Timber Company approx. 300,000 acres
— Maxxam/Pacific Lumber approx. 200,000 acres
0   5 miles
SCALE:

DEL NORTE COUNTY

KLAMATH
RIVER

N

YUROK

SIX

KARUK

REDWOOD
NATIONAL
PARK

CHILULA

HOOPA
VALLEY
Reservation

Trinity
Alps
Wilderness

Simpson Timber

HUPA

Hwy 299

WIYOT

ARCATA

WHILKUT

RIVERS

CHIMARIKO

HUMBOLDT BAY

EUREKA

i

HEADWATERS
FOREST
RESERVE

Table Bluff
Reservation
Wiyot Tribe

NAT.

old growth
fragments

FOREST

NONGATL

TRINITY
RIVER

Highway 36

MAD RIVER

Bear River

SCOTIA

B

MAXXAM/
Pacific Lumber

C

NORTH FORK
VAN DUZEN
RIVER

MATTOLE

HUMBOLDT
REDWOODS
STATE
PARK

VAN DUZEN
RIVER

MATTOLE
RIVER

KING
RANGE
NATIONAL
CONSERVATION
AREA

SINKYONE

LASSIK

NORTH FORK
EEL RIVER

GARBERVILLE

WAILAKI

TRINITY
COUNTY

MENDOCINO
COUNTY
EEL RIVER

U.S. Highway
101

Humboldt County, California, 2000

Entry Point:
Mapping the Humboldt Bay Region

Humboldt Bay, the largest and best deepwater seaport between San Francisco and Puget Sound, lies 231 miles north of that famed gold rush city and sixty miles south of the Oregon border. For the gold- and land-hungry capitalist settlers, the bay meant easy access to the northern heart of the densely forested redwood bioregion that hugs the Pacific coast of North America with increasing density as it stretches north from the Big Sur coast south of Monterey, California, and reaches into southern Oregon. The fifteen-thousand-acre estuary drains about 250 square miles and four major watersheds, forming a key ecosystem of mudflats and eelgrass ideal for salmon, sea otters, and a great many local and migratory birds.[1] It also forms an obvious center for the redwood industry, and Humboldt County, incorporated on May 12, 1853, was at that time 80 percent forested, largely by old-growth temperate rain forests of the towering redwood, *Sequoia sempervirens*.[2]

Whereas the tallest redwoods typically root in saturated creek bottoms where dense coastal fogs protect the trees' short-needled branches from sun-driven evaporation, in 2006 scientists discovered the tallest specimen on record (in fact, at 379 feet, the tallest tree in the world) on a steep slope in northern Humboldt's Redwood National Park. Named Hyperion by the redwood ecologist Stephen Sillett, the tree's immensity challenges simple description and photographic representation. The fibrous bark of a redwood tree grows up to ten inches thick; its wood is colored red by tannins that resist fire and rot and secure its high commercial value. The trunk, up to thirty feet in diameter, teems with lichen and can live for more than two thousand years.

Arboreal canopy soils accumulate several feet deep in trunk fusions and crotches of high-flying branches, creating a barely known aerial sanctuary for salamanders and myriad vascular plants like leather-leaf ferns and spike moss, as well as microscopic crabs and other fauna yet to be discovered. The redwood lives and dies over forest floors blanketed by sword ferns that camouflage fallen trees as they decay over centuries, not years, making the ground beneath as impassable today as it was to wagons and horses in the nineteenth century.[3] These are the world's most massive forests, where a single hectare (2.47 acres) can carry over 10,000 cubic meters (353,000 cubic feet) of biomass.[4]

The redwood belt is approximately one to thirty-five miles wide and 450 miles long, with broad variation in associated species across local climates, landforms, and substrates. Average annual precipitation reaches one hundred inches in Humboldt, and thirty-nine at the Salmon Creek springs that drain the remnant cathedral grove at Headwaters Forest Reserve. From the bluffs in Eureka overlooking Indian Island, redwood coastal mountain peaks dominate every horizon, their sandstone ridges thrust up over tectonic eons of Gorda Plate subduction under North America. Torrential rains cut salmon-bearing streams and river channels that drain predominantly northwest to the sea through Humboldt Bay and the mouths of Redwood Creek and the Mattole, Bear, Van Duzen, Eel, Mad, Little, Trinity, and Klamath rivers. These waters engendered a vast and legendary salmon ecology whose runs of coho, chinook, steelhead, and cutthroat trout sustained Wiyot, Yurok, Karuk, Hupa, Chilula, Whilkut, Nongatl, Mattole, Sinkyone, Lassik, Wintun, Chimariko, New River, Konomihu, and Tolowa peoples for thousands of years—but they could only stand up to world-system markets for less than a century before their collapse. Today these onetime biological powerhouses are devastated, and Indian reservations totaling about 148,000 acres hold the region's First Peoples back from the treasured bay.

On average the great salmon runs reach less than 10 percent of historic levels. More than seven thousand acres of once salmon-rich salt marsh ringing the bay have been reduced to nine hundred acres, the greater portion having been diked off for agricultural use. Miles of additional interwoven channels have been silted in by erosion from logging, a condition exacerbated by road building and development. Upstream dams divert water from the largest rivers, reducing the power of normal rainfall to scour the channels and renew the habitat of clean gravel that salmon need to spawn. Cape Horn dam in Mendocino, for example, diverts up to 90 percent of the flow in the upper main stem of the Eel River, which drains north through the forested mountains of Humboldt to join the Van Duzen and empty into the sea through fertile alluvial flats a few miles south of the bay. The diverted water feeds the hungry grape growers and urban sprawl that Humboldters now see creeping up the coast.[5] Lewiston dam diverted upward of 90 percent of the Trinity River into California's Central Valley irrigation projects in the years after it was completed in 1961, but by 1999 conservation efforts had succeeded in cutting that amount to 75 percent, and debates are ongoing about the ecology and economics of restoring the flow to 50 percent.[6] Six dams on the Klamath River block salmon runs through the Yurok and Hoopa Indian reservations, taking so much water for agribusiness that fish die en masse when the summer sun heats the water.[7] Hundreds of other streams and tributaries run through logged-over industrial timber land, exposing the small numbers of salmon that do hatch to similarly extreme temperatures, aggravating their struggle to survive—they must live for one to two years in the relative safety of these degraded streams to grow large enough to survive in the harsh open seas.

Typical redwood forests are dominated by the so-called ever-living tree, but Douglas fir intrude on north-facing slopes and at higher elevations where the fogs begin to break, and tan oak intermingle on southerly slopes, as do stands of red madrone, grand fir, Sitka spruce, western red cedar, western hemlock, and red alder in certain riparian zones. On the forest floor redwood sorrel thrives on the rich and moist alluvial flats, giving way to sword fern on drier middle and upper slopes or ridges; and everywhere rhododendron, salal, and evergreen huckleberry settle the shady, cool, fog-dripping rain forest bottoms.[8]

Mammalian histories of mule deer, black bear, Roosevelt elk, Humboldt marten, fisher, gray fox, and spotted skunk might each carry a historian's imagination through a volume of social and environmental history—consider, for example, how the grizzly, extirpated in the nineteenth century, could symbolize the total transformation of redwood ecology and the First Peoples' worlds. Alternatively, think of the stories of marten and fisher, commercially valued for their furs and hunted to near extinction in the region. They too incited capitalist imaginings, calling up labor into actions of trapping and hunting and accumulating furs for distant cash markets, so much so, in fact, that taking their bodies was banned in 1946, and today the mere existence of the Humboldt marten is a credible question.[9] They too could serve as emblems of cultural change, more tragic dramas of modernization ecology that future social and environmental historians could write into timber war stories.

Putting these terms of redwood ecology into the service of social history and cultural analysis of the redwood timber wars requires pushing the concept of ecology beyond biological relations to encompass the region's social relations past and present, because social relations have come to govern how ecologies and places are known among the peoples whose institutional actions transform, destroy, and potentially restore them.

My means of telescoping the career of capital culture in this place will not be to study just one or even several ecological constituents of the dynamic whole that now carry its traces but rather to study the collective fixations of its total public culture. In a special sense, that is central to my method, this work is conducted at the intersection of cultural geography and media theory, a junction at which transportation and communication technologies signal their reciprocal constitution and transformation of time and space, and where historical, economic, environmental, and cultural sociologies encounter locals experiencing and classifying things in distinctive ways in the always emergent present moment of long and ongoing developmental processes.[10] For example, consider in what follows how describing the nautical channels and communicative paths by which capital culture entered the places of redwood First Peoples and physical redwood nature will have to blur genres and range across disciplines.

First came the Spaniards Ceremeño, Vizcaíno, and de Hezata, who explored the coast of Alta California between 1595 and the 1790s but failed to enter and map Humboldt Bay or establish a colony anywhere north of San Francisco. An English expedition led by George Vancouver finally reached Trinidad Bay to the north, but it too bypassed Humboldt Bay. Next came joint efforts by Russian and American fur traders using native Alaskan otter hunters. Plying the coast for animal bodies, their first expedition in 1803–5 brought back 1,800 skins. On their second foray, with Captain Jonathan Winship Jr. at the helm, they found passage into Humboldt Bay in 1806. By 1807 the enterprise had taken 4,819 otters and produced the first map we have, pictured here as modified by the ethnogeographer and archaeologist Llewellyn L. Loud, who added the rectangles to indicate the Wiyot villages he dug through on the south spit and eastern shore of

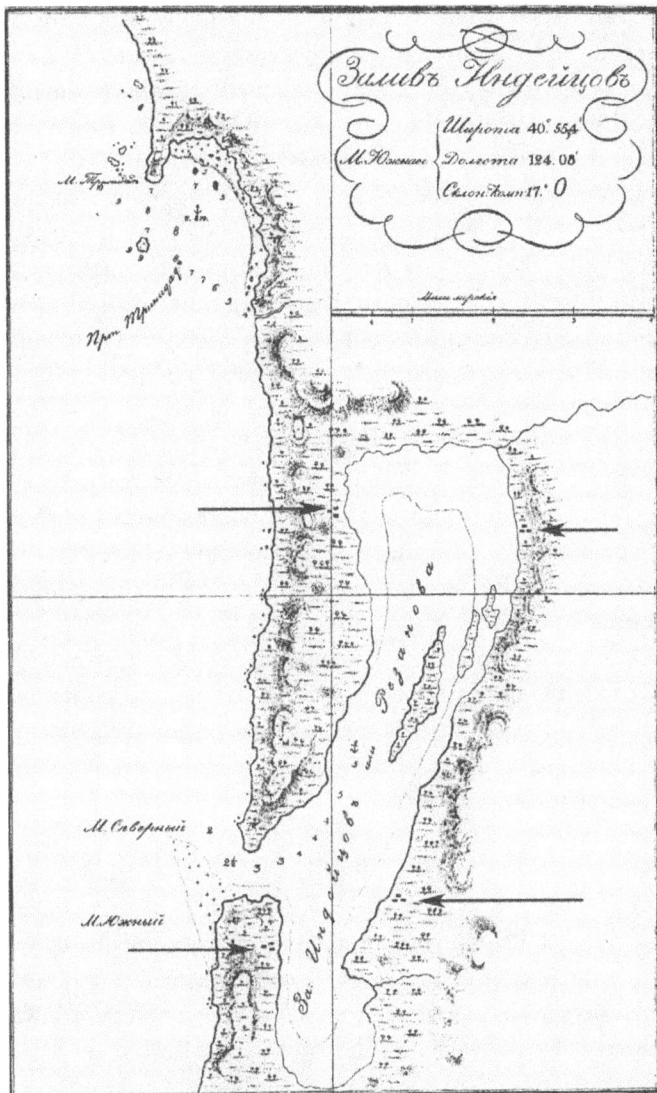

Jonathan Winship's map of Humboldt Bay, 1806. Reprinted from L. L. Loud, *Ethnogeography and Archaeology of the Wiyot Territory* (University of California, 1918), 407, Plate 4. Loud writes: "Explanation of Plate 4: Photographic reproduction of a map of Humboldt Bay sketched in 1806 by Capt. Jonathan Winship, engaged in the fur trade for the Russian-American Company. Published as a subchart to general chart XIII in Atlas of Northwest Coast of America, Aleutian Islands, and other Places in the North Pacific; compiled in 1848 by Captain Tibinkof and printed in 1852 at St. Petersburg. Mad river is not shown upon this map while the portion from Little River northward was probably taken from Vancouver's Chart. Location of four Indian villages are indicated by rectangles." The arrows have been added to highlight the villages, the lower of which were the sites of the 1860 massacres.

the bay in the early twentieth century. But after Winship, the bay was lost again in the coastal fog for another forty-two years.[11]

As far as we know, fur traders returned repeatedly during those years without entering the bay, and an overland route from the California interior was not accomplished by whites until December 1849, when members of the gold-rushing Josiah Gregg expedition arrived after considerable hardship and a lot of help from several Indian communities.[12] The dense coastal mountains and nearly impassable redwood forests formed a natural barrier that ensured passage to Humboldt would be dominated by sea traffic for years to come, if only someone among the Argonaut legions or the land-speculating companies could discover and chart the mouth of the bay.

In the late 1840s, news of gold in the Trinity River stoked a wave of competitive exploration. These ambassadors of capital opened up the bay transportation route that linked white colonial Humboldt to the east coast and the Atlantic world. The same vehicles that shuttled gold and lumber south through San Francisco from Humboldt—at the western limit of the growing republic in 1850—brought knowledge of the nation's market, media, and democratic politics across wide-flung merchant marine and interior railroad routes of newspaper exchange from Capitol Hill, the White House, and the Supreme Court. And this legal, semiotic, and psychical network attached the redwood peoples to the country's founding texts.

In early editions of the region's first newspaper, the *Humboldt Times,* newly published in the fall of 1854, letters addressed to all who would listen illustrate how white colonization explicitly conceived itself as a capitalist adventure from its first public moments; they foretell how deeply the nation would carry its constitutional economic, racial, and gender prescriptions into the making of redwood country. One was signed merely H., a pioneer civic booster for the settlement project:

> The interests of the coast section, of northern California, are the more *inviting for capitalists* and will prove to be the more permanent, from the fact that they are naturally of a treble character—mining, lumbering and agricultural and all of them capable of great extension. . . . The lumber interests of Humboldt County, are certainly unequalled by any other portion of the coast. During the year, terminating on the 30th of June, 1854, there were one hundred and eighty-three arrivals of vessels in Humboldt Bay. The amount of lumber exported in these, was twenty seven million five

hundred and sixty-seven thousand feet. But for the great depression in the lumber market the last half of the year, this amount would have been largely increased. And should the trade be good during the present year, our mills will turn out nearly forty million feet of lumber. And yet, the lumber interest on this bay, is of only two years growth and is capable of multiplication to any extent that a home, or foreign market may demand.[13]

In the following weeks, a series of letters signed "citizen" also spoke favorably of this national project of market revolution to the paper's new reading public, and beyond that to the world:

There is no country on earth, perhaps, which indicates greater fertility of soil than this county, if we may judge from the abundance and luxuriousness of its vegetation; the soil is everywhere covered with trees, shrubs, vines, ferns, grasses, herbs or covers and each to a great size and perfection. . . . Trees claim our attention first and redwood stands per-eminent among them, from its great size and the varied uses to which it is applied; it is found all over the timbered region and is usually 300 ft. high.[14]

Every new country uses certain symbols to represent itself to itself and constitute itself as such, and on one early occasion in Humboldt the national flag served that purpose. Once again the *Times* shows us how—it saturated the symbolic environment with the commentary, history, narrative, and poetry of nationalism:

Our National Flag
"The star spangled banner! Oh long may it wave
O'er land of the free and the home of the brave!"
The following historical Sketch of the origin and progress of the flag of the Union, from the *National Intelligencer,* will be read with interest: Under the head of the "Reminiscences of the Present Century," in the *National Intelligencer,* in September last, we mentioned the fact of that first national flag of the present design, adopted in 1818, was made under the direction of the gallant Captain Reid and made at his house, in New York, by his wife and a number of young ladies. . . . In the Flag, as it now waves, the Past and the present are truly and faithfully remembered. It was a happy idea—simple, republican and comprehensive.[15]

What followed were paragraphs narrating the making of an American symbol—the union flag—but the story itself is a flag of its own, an emblem transmitted and retransmitted, proliferating from newspaper to newspaper and lip to lip, a media spectacle calling up attentions and forming up structures of feeling for nation. The account was intended to bind together the people by wedding paternal values of masculine domination with war, glory, strength, and commitment to revolutionary patriotism—and even family passions, as we learn that Captain Reid designed it, but his unnamed wife and some young ladies actually sewed it. In this way the *Times* editor created, gendered, and racialized public culture with symbolic reminders of a national drama whose originary rhetoric invited in everyone, but whose new institutions of property, press, and polity pointedly excluded women, blacks, and Native Americans. These invitations are well known: *We hold these truths to be self-evident, that all men are created equal. . . . We the People of the United States, in Order to form a more perfect Union. . . . Congress shall make no law . . . abridging the freedom of speech, or of the press; of the right of the people to peaceably assemble and to petition the Government for a redress of grievances.*

The colonial energies of market revolution marshaled in Humboldt cannot be understood apart from the media that the American nation's founding words and charters juridically constituted as such. Labor called up by commercial imagination on continuous display in the mass-mediated public sphere helped make capital and its labor, the bay and its seaport, the roads and their mill towns, the logging crews and their landscapes, the schools and their pupils, and the press and its publics all into archives recording a message that the nation has always broadcast—a message still coming out of Humboldt today, embedded in the voices of the redwood timber war: For better or worse—and especially for profit—we invite you to labor on the land with us and to exercise your rights of free speech, press and property, in the name of the public good. Property and press in free speech public culture are ways we live up to our national calling.

One fact evinces the magnitude of change this colonial labor wrought and links it to the contemporary timber wars; of approximately two million acres of redwood forest before Anglo-European contact, only 4 percent remained uncut in 1990.[16] The staggering scale of this industrial accomplishment has a mythological place in Humboldt, where a thriving industry delivering historical experience doles out the narratives of heroic logging men and their timber baron operators. Today loggers work in an environment permeated with images and narratives memorializing the lives and times of these pioneers—cultural material that saturates schools, parades, museums, and tourist attractions, each one both a vector of literary force and a material structure available for use in assembling provisional self-understandings in the present, perhaps as an ideal on which to model oneself, or maybe as an icon to shatter, positive or negative, either for or against current forestry practice. The celebration of this history is always on display, for example, in the free logging museum at Fort Humboldt State Historic Park in Eureka, where, among a pictorial narrative lining the walls of the museum, one photograph showing white men working at the base of a redwood tree has a caption with this simple message: "Cut'er Down, Boys—There's Plenty More Over The Next Hill! Felling the mighty redwoods was a difficult task. But using his strength, his sweat and his Yankee ingenuity, the American Logger chopped and sawed and hammered and hewed and the big trees came down."

In place of the photo essay at the logging museum, I offer the series of three photos here, taken between the late 1880s and 1930, precise locations within the bay redwood region unknown. What we do know is that they represent two early stages in the development of logging. In the first, men with ax and saw worked with the aid of animal teams, and in the second the aid of steam machines was enlisted, in this case the Dolbeer Spool Donkey. In a third stage, not pictured, tractors and chain saws entered the woods. Today chain saws still reign, but helicopters aid in the yarding of logs, extending the companies' reach to the most remote, steep, and difficult-to-access ridges. Note the near-total destruction of forest habitat. The total symbolic environment to which these images contribute includes the logged-over landscape and its industrial architecture, which together provide a structure of feeling for logging that invites individuals to work and transform themselves and their world. Men toil in the woods with this knowledge on hand, and it helps their labor feel proper to their sex and their place in family and society. It may seem a truism that patriotic working-class masculinity thrives in the industrialized redwood forest, but it has to be mentioned because the ongoing struggle for power over property in redwood ecology runs into this cultural body right where it stands between big corporate owners, whose property value originates largely in the labor this body produces, and the forest defenders, who can only hope to achieve their aims by persuading timber folk and loggers to identify with the cause. And though this dominant culture can be reflected and cast into language, it remains largely tacit, embodied as

Redwood logging between the 1880s and 1930 in the Humboldt Bay region. Courtesy of the Ericson Collection, Humboldt Room, Humboldt State University Library, Arcata, California.

meaning expressed in the practice not of "Chop 'er down, girls!" but of "Cut 'er down, boys!"

How deep this gendered capital culture of man and machine still runs was again on display at a retired loggers' conference I attended at a home for seniors in Eureka. The boys told stories that extend the narratives of these black-and-white photos. They were World War II and Vietnam veterans who had seen enough real war not to be much distracted by timber war talk, which in fact never came up as the camera crew recorded their oral histories for a public television documentary that the producer told me would deal with "past ways of life." But the conversation did turn to the way things have changed, first to the land and the dearth of big trees and then to the logging machinery that extended their hands and thoughts to the world.

Though Humboldt's economy is now in transition toward a lower-paying service economy, with poverty and even hunger on the rise as the timber and fishing industries continue long downward trends, timber extraction remains crucial, and the big timber operators are still the largest generators of the private wage-labor payroll on which most other sectors in the region have always depended.[17] While the search for a sea route to Trinity gold brought capital to the bay, and California Highways 299 and 101 still follow early paths up from the bay into ghost towns and gold camps on coastal range rivers, over time the railroads, highways, and city services were all drawn away to the commodity circuits of emergent lumber culture. Likewise regional architecture can be traced down the corridors that run from the lumber camps along the rivers to the company towns and bayside population centers and finally out the mouth of the bay into markets around the world. These are the spatial, temporal, and semantic coordinates of the redwood timber wars in which *Trouble in the Forest* sets to work, assembling the traces of Indian war, labor trouble, and environmental resistance with which I hope to show what the place of Humboldt can teach us about rights-driven capital culture in its moment of globalization and converging peoples', labor, and environmental movements.

Introduction
The Case of Humboldt: Violence, Archive, and Memory in the Redwood Timber Wars

On the morning of September 17, 1998, in the coastal forest of Humboldt County, a logger working for the Maxxam corporation's Pacific Lumber Company felled a redwood tree that crashed through the skull of David "Gypsy" Chain. The event occurred as Chain and several allies in the North Coast Earth First! movement confronted the loggers directly, interrupting their work and challenging them to stop an illegal harvest of timber in the nesting grounds of an endangered seabird called the marbled murrelet. In a rare moment of apparent reconciliation between history-making individuals usually compelled by social position into hostile confrontation, the fifty-three-year-old logger and the activists knelt together in prayer next to the corpse—a mere instant in conventional time but a veritable lightning strike in the historical space of California's redwood timber wars.[1]

National newspapers carrying the story incited my search for more-detailed accounts. It proliferated as magazines picked up the drama and environmental groups posted Humboldt's local *Times-Standard* reportage to the Internet. The flurry of discourse captured my attention, and I started to archive every trace of the incident. The conditions under which I began this archival and ultimately ethnographic descent into the timber wars are integral to the story: when the new media of cyberspace cultural transmission addressed me from afar, attracting my concern and identifying me first with the struggle to save ancient redwoods and then with new network movements against corporate globalization, I entered a worldwide public that was at that moment just finding its voice. I came to the matter at hand in this

1

way—through the media portal of David Chain's death, through its archive, that is, its medium and its message.

By *archive* I mean something more than a collection of documents and the building that houses them. In what follows, I consider the archive of any event to contain every mark it occasions in the field of cultural production in which it occurs. All recorded images of the event, comments on it, and narrations of it become monuments to it, contributing to the growth of its archive; and the archives of certain events become reservoirs of psychical energies, investments, attachments, and interests that mediate the event in the sense of standing in for it or bearing its impression across time and space to inform the experience of those who come after. Happenings are in this way collectively made into historically freighted affairs. They create public culture. Subsequent attitudes, actions, politics, and modes of memory and historical consciousness are always mediated by such archives, and by the term *mediation* I mean that they enable and constrain future attitudes, actions, political consciousness, and memory in the manner that language does speech—they are, or rather over time they become, a priori meaning-making structures that people use to constitute their lives and identities, their fortunes and politics. The case of Humboldt I introduce here begins in the emergent archive of the killing of David Chain.

"Death in the Forest" read the day-after headline in the *Eureka Times-Standard,* showing a map to the "death site" and a picture of activists circled up, arm in arm, heads down, mourning their loss. Also on the front page: "Activists weep for comrade: Friends lament loss but don't blame logger," and then "PL officials 'saddened.'" Said company spokeswoman Mary Bullwinkel, "Despite all our precautions, a trespasser was apparently killed by a falling tree at one of our logging sites on private property."[2] Her words are telling. *Property* is the corporation's first line of defense. Pacificlumber.com, Maxxam's online corporate bullhorn in Humboldt, posted an immediate news release that reiterated Bullwinkel's reaction and addressed the public in the company's distinctive idiom of property rights, which situates the company as a guardian of the law: Chain was killed by a "falling redwood tree while trespassing on private property."

The twenty-four-year-old Texan had been in California for about two weeks, according to fellow forest defenders, who told reporters that "he was just here because the forest was being cut down and he felt he could make a difference."[3] After a day of public mourning and understanding on both sides, the accusations started to fly. Saturday's headline read "PL blamed for forest death."[4] An activist named Farmer who was with Chain at the time of his death told reporters that the logging crew knew activists were close by and that the logger was purposely aiming trees at them. Farmer, aged sixteen, published his eyewitness account at enviroweb.org on October 1. "I arrived at the drop-off point and started hiking up the hill," he said. "There were nine people including myself and a camera person. . . . When we arrived at the tree the loggers shouted obscenities at us while someone tried to reason with them." Carey Jordan broadcast her account over Berkeley radio's KPFK on September 29. "We went there to talk to the loggers," she said. "We [had] demonstrated the day before at California Department of Forestry to make them aware that we thought PL was logging illegally. . . . A road was punched in before September 15. That's the official end of marbled murrelet nesting season. They're not supposed to do any work before then. Plus they hadn't finished the murrelet surveys before they started and also there was the danger of landslides because the slope they're cutting on is practically straight up and down."[5]

Forest defenders who videotaped the scene extracted audio from the tape and posted it on the Web, publicizing the unforgettable screams of the angry logger. The *Times-Standard* quoted it with discretion, but we tune in here to the uncensored Internet files: "You've got me hot enough now to fuck," screeched Maxxam tree feller A. E. Ammons, shown charging the activists in a copy of the video I later obtained from a North Coast Earth Firster, "I wish I had my fuckin' pistol"; and then "I'm gonna start fallin' into this fuckin' draw!" Chain was dead within the hour, killed by a tree that Ammons cut.

Another Saturday *Times-Standard* headline declared, "PL Workers shocked, not surprised at death." Joe Rogers, an employee at Maxxam/Pacific Lumber for thirty-two years, told reporters he tried to stay out of the controversy: "We need people to pursue causes," he said, but he had hoped everything would end with the so-called Headwaters deal that had recently preserved the largest existing unprotected ancient redwood grove. "This brings it home," he concluded—the timber wars were still on. Less sympathetically, Mark Cobb, a twelve-year employee, told reporters: "PALCO is taking care of me and my family, providing me with insurance, a decent paying job, a great place to raise a family—I am sick and tired of only hearing negative things about PL."[6]

On the following Wednesday, the *Times-Standard* ran an editorial titled "Lessons must be learned after death": "David Chain was a person," wrote the editors; yes, he was trespassing, but "he didn't deserve to die, like some callous people who have grown weary of the protestors have said." On the other hand, they continued, "we can't see many loggers laying down their saws and refusing to cut. They have jobs to do. They have families to feed." And then the lesson—the solution—offered by the editors: "So what we're stuck with is a problem with no solution—unless Earth First! puts an end to the predicament."[7] The question of corporate forestry violations is elided: The problem is dissent, not domination, not hegemony.

This way of framing the timber wars is characteristic of the *Times-Standard,* whose editorial lean is well known and predictable without being monolithic. While letters to the editor representing all positions are regularly printed, Humboldt is timber country, and this is a timber-friendly paper. Stuck in the middle, the *Times-Standard* receives criticism from hardcore timber supporters, who claim it glorifies radical forest defenders and gives them an undeserved stage by reporting their actions against Maxxam as news, as well as certain of the forest defense, among whom one nickname for the paper is the "Slime-Standard." These acute expressions do not divide the field of opinion into opposing camps as much as suggest a finely graded polarization in the timber war public.

What forest defenders do know is that media spectacle is vital to big timber's cultural hegemony.[8] They know that mass media make public consciousness. But because they are unable to compete with the corporations' public relations and advertising budgets—Maxxam's Bullwinkel, for example, was a paid corporate spokeswoman, and pacificlumber.com was maintained for both commercial advertising and corporate image management—they strive to make news instead. Of course, they also build the movement with grassroots organizing; they network, do research, monitor timber harvest plans, raise money, sell T-shirts, direct-mail to members, and dedicate their own time and resources to the cause. But symbols, they know, are powerful things. Forest defenders thus wage a permanent struggle to project their symbols and get their message out, sometimes with great success, as, for example, with the Luna tree-sit campaign.

At the time of Chain's death, Julia "Butterfly" Hill had been living for over nine months two hundred feet up in the canopy of Luna, one of the corporation's ancient redwoods. Speaking through mass media to the world by cell phone from her tree-sit platform, high on a steep ridge several miles south and west from the mountain where Chain was killed, Hill broadcast a message of love for the tree she was protecting *and* for the workers from whom she claimed to protect it, as well as a lesson in global economics and a call for alliance between labor and environmental movements against rapacious corporations. Fifteen months later she made an agreement with Maxxam to purchase Luna for $50,000 and descend from her protest, preserving the tree and herself—two living symbols of peaceful resistance for the redwood forest defense, the environmental movement and its union labor sympathizers, and thousands of others who heard her story on TV and radio or read about it in newspapers and magazines all over the world.

During the tree-sit, Hill spoke out continuously against the so-called Headwaters deal, in which Maxxam was slated to receive cash and land valued at $480 million in return for the highly contested 2,700-acre Headwaters Grove—the sacred center of the forest defense since the grove's discovery by activists in 1986—around which the state would create an old-growth biopreserve. In March 1999, six months after the killing of Chain and just as the Luna tree-sit was building a global public, the federal government, the state, and the corporation completed their deal to transfer the largest remaining unprotected ancient redwood grove out of the market and into the public trust. While many hoped that the deal would end the redwood timber wars, in fact it had the opposite effect. It sparked scrutiny and years of litigation; forest defenders from across the movement network protested its "sacrifice zones," and Earth First! carried on refining its political art of sitting in trees and forming tree villages to defend specific sites, increasingly aggravating both loggers and management. By the end of the year, United Steelworkers, locked out at Maxxam's Kaiser Aluminum plant in Tacoma, Washington, linked arms with Humboldt forest defenders in the Battle of Seattle, helping not just to shut down the World Trade Organization's ministerial conference but to transform the discourse of free trade itself and with it the future of globalization.[9]

On the highway between Grizzly Mountain, where Chain was killed, and the high redwood ridge where Hill sat in Luna, lies a third point of intense collective psychological investment that captured my attention: Scotia, the last authentic company logging town in California and, from the standpoint of forest defense, the symbolic center of profane power in the timber wars. It was the headquarters of Maxxam's Pacific Lumber, Humboldt's second-largest landowner (more than 220,000 acres) and largest private employer in 1998 (more than 1,000 employees). To the chagrin of many workers, community members, and company men, the global extraction conglomerate from Houston had succeeded in its hostile acquisition of the family-controlled and widely respected local timber company in 1986.

Charles Hurwitz, CEO of Maxxam, moved quickly. Whereas he retained the Pacific Lumber name, he set to work changing what mattered most to many citizens, landowners, lumber workers, and forest defenders—he abandoned the company's conservative, selective-cut forestry methods and doubled the harvest rate, intent on converting the ancient forest inventory into cash. While the old Pacific Lumber would leave up to half the trees standing on every acre cut, the new Maxxam plan called for clear-cutting everything fast. It was asset liquidation designed to raise capital and cover high-yield junk-bond debt created to buy the company. But the region's nascent forest defense movement closed ranks rapidly in response, raising the stakes of the conflict and ultimately producing the largest forest rallies and mass arrests in U.S. environmental history, as well as the precedent-setting deal that created Headwaters Forest Reserve.

In 1992, amid the escalating tension, an earthquake struck Scotia, crumbling the town's central shopping complex and consuming it in fire. Maxxam rebuilt with great fanfare and a promise to sustain the logging community, beautifying the town center in an architectural spectacle that doubled as political legitimation for a company beset by environmental critics. Today Scotia's presence looms large over Humboldt, a historical cipher and architectural rebus that both masks and reveals a local transmutation of twentieth-century American capitalism. The town itself is a fluid but material signifier expressing the long struggle of the industry to maintain its position in the cultural occupation of the region. When Maxxam declared Pacific Lumber bankrupt in 2007 and left the county in 2008, Scotia's fate was all but certain. The new owner was the Mendocino Redwood Company, and its business plan included the parceling out and privatization of the last real redwood company town.[10]

Trouble in the Forest is an account of my journey to this place—an ethnographic, historical, and cultural analysis of its redwood timber wars. In the twilight hour of the great lumber culture that made Humboldt, in the ruins of the timber industry, the forest, and the twentieth-century communities that thrived there, I immersed myself in the timber wars and discovered a public struggle between forces of globalizing capital, embodied in Maxxam Inc., and a new social movement against neoliberal corporate globalization and for social justice, embodied in the redwood forest defense. It is a struggle that exemplifies perhaps the biggest challenges facing the twenty-first century: the growing contradictions of capitalism, planetary ecology, and social justice.

All over the world, these contradictions are expressed in place and locale, where concrete particular struggles are waged over land and its remnant communities of labor and environment. In every case, technology and labor, employed by capital, blast latent values out of the environment into commodity circuits that sustain distinct cultures. Local ecologies, by definition self-regulating and self-sustaining, are severely disrupted. Communities of labor suffer doubly where resource-dependent economies short-circuit after boom-time extraction, leaving people underemployed in a depleted environment characterized by increasing difficulties in extracting values. Grievances arise and public struggles ensue as corporations and workers try to hold on to what they have built, while changing conditions associated with declining extraction economies invite worker unrest, state regulatory intervention, environmental activism, and new economies of restoration, tourism, and service. Such developments put capital under increasing cost pressures, producing incentives to seek lower wages and weaker regulations both at home and abroad. In these social spaces and places of conflict, outcomes register as changes in the land that determine linked social and ecological futures. Place by place, community by community, conflict by conflict, ecosystem by ecosystem, and violence by violence, these social struggles determine the planetary ecological future.

Trouble in the Forest addresses this transformation by examining one place that globalization is producing with savage distinction. My initial position as an outsider looking in through mass-media representation and historical investigation presented an opportunity to ask wide-ranging questions. What deep cultural and social forces are driving the long-running timber wars? How did they originate, and how do they work? Why do they continue even after the largest remaining groves have been preserved? How do they embody the twentieth-century rupture of globalization? And what do they say about the United States, not just as a nation with feelings for its own history but as a nation divided by its principal role in making the history of glo-balization and thus our collective ecological future? As the spectacle of American hegemony rises within the global system of modern capi-talism, helping drive that system—by its own inexorable logic of con-tinuous exploitation, reinvestment, and expansion—into ever greater scales of commodity production, it also drives deeper the world con-tradictions of economy, ecology, and social justice and pushes social actors everywhere into increasing conflict. When and where these global forces and actors implode in local conflicts and place-bound events, charges of capitalist *empire* ring out, and new claims are made on the natural-cum-human rights that were constitutionally inscribed in the United States' vaunted self-image of liberty, equality, and jus-tice. People demand that the established power deliver on the liberal promise inscribed in those virtues and pay up for the social and envi-ronmental costs it imposes on communities of labor and nature. And so if emergent forms of new social movements indeed carry messages, as the Italian theorist Alberto Melucci has written, then place-based conflicts like the one Maxxam incited in Humboldt can and should be treated as messaging machines or broadcast devices.[11] This raises the question: just what is the redwood struggle transmitting?

By way of exploring these open-ended questions, I set out for Humboldt with the ethnographer's dream of "going out there" to engage in the struggle and "coming back here" with a story to tell. By means of field research, participant observation, archival study, and wide-ranging interviews, I documented the timber wars, reading them as a symptom of our historical moment. What I learned about Humboldt's property culture, spectacular politics, new movement networks, and violent landscape of social memory—within which all these communicative actions make sense—sent me back through the region's long social history of hard common struggle, back through its archive of conflict in the public sphere. Here I discovered how preced-ing epochs of labor trouble and Indian war had set the conditions for the timber wars and in fact share a deeper cause with them: namely, the performative utterances of the nation's republican constitutional framers, whose nation-building and people-making speech acts and texts institutionalized private property, the press, and democratic pol-ity in the New World and drove the American market revolution to its western frontier.

When I set about tracing the social and environmental conditions of these successive conflicts, I found that each had produced a particular moment of unusual violence around which social memory had crystallized over time, archiving them and creating public culture, just as the killing of Chain did in the closing moments of the Headwaters deal. The massacre of Wiyot on Indian Island in 1860, the killing of redwood strikers in the great lumber strike of 1935, and the car bombing of forest defenders in 1990 are events whose archives similarly inhabit Humboldt's various media and structure its living, symbolic, and built social memory. In three historical chapters, I treat the archives they occasion as reservoirs of valuations, investments, desires, and discourses that carry a signature of social relations in their historical moment. Taken together these horrific events record and map out a social history of place, showing how it became a traumatic structure that structures emergent practices of timber production and oppositional politics. They tell a difficult story of changing contradictions in the capital culture that colonized Humboldt and made it the place it is today. They suggest how integral rights-driven juridical institutions of property and press were, and still are, to the national, racial, gendered, political, and economic—in other words, the *cultural*—formation of place, power, and politics on this capitalist frontier.

This is a work of historical and environmental sociology as well as descriptive cultural and media theory in which my portrayal of social history gives context for theoretical interpretation of the ethnographic present—a present that must be understood as an expression of economic, environmental, and cultural conditions opened up and transformed by emerging events and history-making agents. But the whirling phantasmagoria of this ethnographic *now* can be grasped only by arresting it for contemplation—by blasting it out of the chaotic flow of mundane events, images, and narratives.[12] Dialectical thought must begin like this, theoretically synthesizing the experience of complex totalities like the timber wars, then seeking patterns and using them to ask questions, analyzing their conditions and reflecting on comparable cases and society in general, in view of strengthening the theory that originally shaped the research project.[13] David Chain, Julia "Butterfly" Hill, and the company town of Scotia are psychically charged events of violent death, extraordinary life, and geographic sites where I enter the timber wars and seize them in representation, informing my descriptions with social, environmental, media, and cultural theory. I then use my experience in the archives they occasion to guide my study back through the labor troubles and Indian wars that set the conditions of the present. Immersing myself in the archival, mediated moments of extraordinary violence that defined these preceding conflicts, and similarly using them as points of entry into history, I write the emergent timber war story with a method of arrest. What I produce is a series of snapshots taken at those historical moments when collective cultural colonization by the agents of Euro-American modernity culminated in violence. What was the object of such intense collective attention in those moments? In each case, a struggle over property was at hand.

Property and Place

On one side, timber corporations and their supporters argue that logging is a matter of private property rights protected by law. On the other, forest defenders shout, "Not one more ancient tree!" With less than 4 percent of the original redwood forest left uncut and approximately three-quarters of these remnant acres already protected in parks and other conservation arrangements, forest defenders demand preservation of the final 1 percent and regulated, sustainable industrial forestry on the rest—Humboldt's vast, cutover timber production zone (TPZ). Decades of logging produced this social, symbolic, and environmental landscape, the conditions of which forest defenders use in constructing their demands and building concepts not just of the environment (i.e., physical nature) but also of the capitalist system and its local subculture of redwood commodity production. Saving these forests is a matter not just of biological principle or quasi-spiritual connection; it is also an opportunity to contest the whole reigning social order. What drives the symbolic wedge between these positions is not just trees and their disputed value—biodiversity versus profit or some such schematic—but the long-running contest over character and culture that animates the U.S. tradition of political life and expresses the seminal concepts of Enlightenment rationality inscribed in the juridical engines of national experience. The conflict, in other words, has roots as deep as the nation itself and therefore a history at least as long as American modernity.

When I encountered Humboldt's language of property and protest, I was driven back to the constitutional grounds of the claims being made. My original and far more limited intent of documenting the redwood struggle in the contemporary moment of globalization collapsed. Writing the timber wars entailed writing a history of the place that informs them, the place wherein they make sense as communicative action. The timber wars are embedded in a history of conflicts that are similarly broad in scope and institutional in character. They stand at the present end of capitalism's long career in the North Coast redwoods—a crowning achievement of the cultural colonization that capitalized the redwoods in an extraction economy that simultaneously established the nation in Humboldt.

These two processes—colonization and capitalization—are actually one, and together they form a conceptual umbrella under which I gather all the signal forces that made Humboldt modern and set the conditions for environmental conflict. The timber wars today are an expression of that making, and as such they are haunted by the indigenous First Peoples whom that making devastated, the labor power it channeled, and the bio-zoological landscape it transformed, the character of which, as we will see, necessarily entered into the local formation of capital culture.

Every domain of social memory I encountered in my ethnographic and historical investigation pointed me in the same direction. The imaginative personal remembrances, media chronicles, historical records, cultural museums, local architecture, and landscape each pointed back at a long social history of property conflict.

First came a period of so-called Indian trouble in the 1850s and 1860s, when what had been tribal lands, the indigenous commons, were signified as fungible property, appropriated by whites and enclosed for agriculture, subjected to industrial husbandry, and earmarked for timber production. This was the time of primitive accumulation—accumulation by force prior to and constitutive of capitalist accumulation as such, whereas the latter phase of accumulation proceeds by profit-driven commodity production and exchange in competitive markets and grows by reinvestments in labor and machinery aimed at staying competitive by keeping costs down.[14] With the advent of redwood industrialization, labor trouble was imminent. It culminated in the 1930s and 1940s, when the practical meaning of this redwood property was deeply transformed as workers won rights to collective bargaining. This was the time of internal contradictions, when capital exercised great power over workers in wage labor markets and sparked the revolts that won higher wages and better conditions. But organized redwood labor's collective prosperity, in league with redwood capital, represented the fruit of increasing scales of lumber production, and the resulting acceleration of deforestation laid down conditions for a new round of struggles—the era of environmental conflict. This was and remains the time of external, so-called second contradictions, when the externalized costs of expanding capital are piling up fast, spawning new social movements and regulatory responses that again change the practice of property rights. From their place in the signature archives of violence, these conflicts come to dominate politics in Humboldt, giving them both structure and a lot of material for future political actors to use.[15]

Today the stories of Indian war, labor trouble, and forest defense that circulate constantly through redwood country place working people and communities on a tenuous middle ground between indigenous, corporate, and environmental claims. American Indians make public claims that their struggles are, like those of their ancestors, based on the loss of their lands and autonomy; working men claim that their troubles are still about fairness and unaccountable corporate power; and forest defenders build both colonial and labor stories into their analyses of ecological decline and demands for species protection, habitat preservation, sustainable forestry, economic democracy, human rights, and social justice in general—the big timber corporations stand in the sights of their critical narratives.

In the register of material culture, these conflicts have produced a physical landscape and cultural geography that provides additional structure that people here can and do use in personal and collective self-understanding. The place of Humboldt, in other words—as it has been achieved, as it has been built by this history of struggle—is a condition of possibility for the claims and counterclaims in the ongoing timber wars. The history accumulated in bodies, narratives, traditions, archives, architecture, and landscape gives to the conflict a communicative inertia. From within the timber industry that grew out of the settlers' first struggles with Indians for ownership of their land, an extended struggle developed for control of the labor that industry required. Decades later, from within the environs that organized and controlled labor continuously consumed and transformed, the conditions for environmental conflict emerged. This built history of memory and conflict ensures that social life in the redwoods will always be speaking a language of contested property in land, labor, and ecology—and because it all pivots on property rights, it fosters a cultural politics contesting the institutionalized, philosophical ideas of nation and citizenship that originally gave America its sacred name.

What I found on the ground beneath all this history is a place of contradictions. It is a place of Indian museums and somber monuments to genocide, but also of living reservations and active tribal life; a place of labor halls and an unorganized redwood labor force racked by memories of violence and repression, but also of Labor Day picnics, visions of converting to sustainable forestry, and nascent labor alliances with forest defenders against corporate domination; a place of industrial landscapes, decimated forests, species extinction, and memories of forest defenders' bodies exploded, crushed and bleeding, but also a place of redwood parks, old-growth reserves, restoration economies, and collective struggles to halt the decline.

To grasp what is at stake in revealing the inner connection of these events and archives, I begin with a brief historical sketch of the timber wars.

Maxxam in Humboldt

Globalization came to Humboldt with a vengeance in 1985, when Maxxam announced its takeover bid for the Pacific Lumber Company. By that time there was already a forest defense movement in the works among local residents, a grassroots effort distinct from the long-running work of professional, national groups like the Save the Redwoods League, the Sempervirens Fund, and the Sierra Club. This new activism emerged primarily from southern Humboldt and Mendocino counties, to the south, in a campaign to save the last, largest groves in Mendocino from the Louisiana-Pacific and Georgia-Pacific corporations. But when Maxxam seized control of Pacific Lumber, it became the principal private owner of surviving ancient redwoods. The center of gravity in forest defense quickly shifted north into Humboldt.

Organized as the Environmental Protection Information Center (EPIC) and based in the southern Humboldt hamlet of Garberville, these original forest defenders formed a core around which many other groups and alliances would eventually gather. When Maxxam arrived, EPIC had just won a precedent-setting court decision. *EPIC v. Johnson, 1985* held that the California Department of Forestry must consider the cumulative environmental impacts of timber harvesting each time it approves a timber harvest plan. The ruling established that the timber harvest review process is governed by the California Environmental Quality Act (CEQA) of 1970, according to which every action taken by the state that affects an environment must first be considered within the total field of effects of state action on that environment.[16] The timber harvest plan must act as an environmental impact review. Wildlife surveys and watershed science were thus legally mandated for each timber harvest plan, because assessing the impact of logging requires an understanding of the forest being logged. It thus became law that the forests' species and its full web of life must be documented *before* the chain saws turn on.

With this court decision, the ecology movement came of age in the redwoods, and both the federal and state endangered-species acts became powerful tools of forest defense. The legal grounds on which the battle to save the Headwaters forest would soon be launched were prepared. Maxxam blue-lined Headwaters Grove, physically marking the redwood trunks with the blue line of paint that says to the tree feller that the trees are ready to be cut. Marking Headwaters for total liquidation was the move that called the forest defense into action and emboldened it to assert local control over the social and ecological values at stake. For them, the very balance of life and death was at stake, as that balance was embodied in the number and nature of lumbering jobs and the remnant populations of owls, marbled murrelets, salamanders, and salmon, and in the accumulation of corporate capital and the long-term viability of ecosystems, species, and fresh watersheds. But Maxxam fought with moneyed finesse, all the while cutting old growth quickly in advance of the anticipated regulatory wave. The company took 3.3 billion board feet out of the forest between 1987 and 1996, and approximately $3.6 billion out of the county, but somehow still left Pacific Lumber saddled with more than $700 million in debt in 2007, the year that Hurwitz finally called it quits and declared the company bankrupt. Where did all the money go? Upstream to Maxxam timber note holders. As Hurwitz famously explained when he bought the company: "The function of PL is to throw off cash flow."[17]

As Maxxam ramped up the cut rate in 1986, direct-action forest defenders joined the struggle alongside EPIC, led by an Earth First! group originally calling itself the Redwood Action Team.[18] With the Mattole Restoration Council and the Salmon Group, also formed in the early 1980s, Humboldt's local network of new social movements for environmental defense and sustainable forestry had emerged. Each element had its particular interests and expertise, but the arrival of Maxxam gathered and unified their intentions without effacing their differences. Their collective focus on ancient redwoods occurred within the wider context of a national ancient forest preservation movement that peaked in the late 1980s, when a federal judge yielded to environmental interests and ruled that the northern spotted owl must be listed as an endangered species. Across the Pacific Northwest, traditional lumber communities, steeped in working-class lumber-mill culture and familiar with industrial labor organization, struggled to come to grips with rising public sentiment for forest preservation. When the owl was finally added to the endangered-species list, forcing the government to restrict timber harvest on millions of acres of national forest, many feared that mill towns from Washington to California would be shut down. And that is precisely how the big timber companies tended to publicize the story.[19]

Not surprisingly, mass media followed suit, structuring reports along the same lines. Timber workers and families completed the formation, reproducing the dominant interpretation in their own lived experience: "owls versus jobs" and a spirited defense of "our way of life" characterized their response. Corporate public relations firms were employed by big timber companies to help construct this perspective by forming citizen front groups to organize and fund yellow ribbon campaigns in defense of the lumber communities.[20] Countering these narratives of owls versus people, forest defenders argued that conversion to sustainable forestry methods would preserve environments *and* jobs. They worked to show that the companies' cut-and-run, boom-and-bust extraction forestry was the real threat to jobs and to thriving communities. When the big trees are gone and the forest is converted to monoculture tree farms, mills are shut down, hours curtailed, and workers laid off.

But the situation developing in Humboldt was distinct. Elsewhere in California and the Pacific Northwest, the ancient-forest conflict revolved around timber sales in the publicly owned national forests, sales that had for decades functioned as a subsidy to the private timber industry and a pipeline of economic values into timber culture. While the ancient Douglas fir and mixed conifer forests of Oregon and Washington were being sold by the state and cut by the corporations, the redwoods were almost all privately owned. Thus whereas the movement to preserve ancient forest on public lands required forcing the federal government to do a better job of public land stewardship, its inarguable mandate, the movement to preserve redwoods required forcing private landowners to relinquish property rights over their land. This ensured that the redwood timber wars would be fought in the terms of property and person that occupy the center of American national cultural identity.

When the pivotal year of 1990 arrived, and the decision to list the owl as endangered grew nearer, tensions were flaring across the ancient forest belt. Maxxam was cutting its big trees fast, and the growing alliance of forest defenders was preparing what they hoped would be the largest direct-action protest campaign in history. They called it Redwood Summer—a whole season of rallies, marches, blockades, and nonviolent demonstrations of mass civil disobedience. Adding to the growing social hostility between the forest defenders and the redwood timber industry—by which I mean management, pro-management workers, and their communities of support—was another powerful factor: a California voting initiative on the fall ballot that would permanently preserve all the state's ancient trees. Its supporters named it *Forests Forever,* and if it passed, it would take out of production every acre that contained six or more trees aged over two hundred years. The listing of the owl and the ballot initiative promised to transform the redwood commodity circuit dramatically, rechanneling long-established flows of labor energy and capital accumulation.

Then the signature event of the timber wars occurred. After a period of harassment by redwood logging supporters in early spring, including a series of anonymous but closely linked death threats, Redwood Summer organizers Judi Bari and Daryl Cherney were car-bombed in Oakland.

The 1990s began with that bang, so to speak, and the timber war tumult has not stopped since. The spotted owl was listed, and eventually millions of acres of national forest were set aside by court order; and Redwood Summer proceeded, without the energy of Bari, however, who had been temporarily knocked out of the action. But the corporations defeated the voter initiative with the help of a high-stakes corporate image consulting company that ran a campaign publicly linking the conservation initiative with the falsely accused and not yet exonerated Earth First! "ecoterrorists." This freed Maxxam and others to continue liquidating their ancient trees and left the forest defenders scrambling to protect each isolated grove in any way they could, one timber harvest plan at a time.

Having lost at the state level, grassroots redwood defenders fell back on local nonviolent direct-action and continued to press on the legal front. The grove at Owl Creek, for example, first targeted by Maxxam in 1988, was successfully protected by a combination of EPIC lawsuits, Earth First! direct actions, and ultimately state purchase of the property in 2000 as part of the Headwaters deal, but not before Maxxam surreptitiously entered the grove on Thanksgiving Day in 1992, cutting a million dollars worth of logs before EPIC could work through the California Court of Appeals to gain an emergency stay on Maxxam operations. Forest defenders dubbed it the Thanksgiving Day Massacre. In 1993 EPIC sued for a violation of the Endangered Species Act at Owl Creek, and in 1995 a federal judge issued a permanent injunction.[21] Then, in September, more than two thousand people rallied at the gates of Maxxam's Carlotta sawmill, calling for an end to the company's assault on Owl Creek and the preservation of Headwaters forest. More than two hundred people were arrested for civil disobedience.[22] The Carlotta action was repeated on September 15 in both 1996 and 1997, during which first three thousand and then a record six thousand people gathered, respectively. In 1996 the number arrested reached 897. In 1997 three hundred police officers participated in the arrest of one thousand peaceful protesters.[23] The escalating commitment of forest defenders was transmitting an unmistakable message.

But defenders continued to press on other legal fronts as well. In 1992 the marbled murrelet, a seabird that reproduces by laying a single egg in the branches of old growth, had gained protection under the California Endangered Species Act, winning the forest defense another opportunity to obstruct Maxxam's plan to cut all its remaining old-growth forest, including the majestic Headwaters Grove. Then, when EPIC pressed Maxxam at Owl Creek and Headwaters using the murrelet ruling, Hurwitz responded with a historic Fifth Amendment "takings" lawsuit, charging that enforcement of the murrelet rule had in effect seized all the value of his property without due process or just compensation. With this appeal to the law, Maxxam made the issue of redwood forest defense an explicit constitutional question, driving the redwood timber wars even deeper into the domain of national character and culture.[24]

On September 28, 1996, the takings case was essentially settled out of court when the state of California and the U.S. Department of the Interior agreed in principle to the preservation plan, which would not be completed until 1999. Eventually they paid Maxxam almost half the amount Hurwitz initially put up for the entire company. It was a stunning profit. But local forest defenders fought the deal because it did more than pay for the grove—it also instituted a precedent-setting

Habitat Conservation Plan that, despite its innocuous name, gave the company a fifty-year license to kill endangered species in so-called sacrifice zones on the company's remaining 200,000-plus acres.[25]

By the time the deal was completed and signed by all parties in March 1999, the global media events of David Chain's death in 1998 and Julia Butterfly's occupation of Luna were well under way, promising to make problems for Maxxam indefinitely. But forest defense was not the only trouble brewing for Maxxam in the 1990s. In 1988 the company had acquired the transnational and unionized Kaiser Aluminum Corporation, headquartered in Spokane, Washington. When contract negotiations with the United Steelworkers broke down in the late 1990s, Maxxam locked out the strikers and shipped in laid-off lumber workers from Scotia for use as scabs.[26] In response, United Steelworkers came to Humboldt, climbed Luna to meet with Julia Butterfly Hill, and shared in founding the Alliance for Sustainable Jobs and the Environment, an organization committed to creating a world "where nature is protected, the worker is respected and unrestrained corporate power is rejected through grassroots organizing, education and action."[27] By giving the steelworkers and forest defenders a common target, Maxxam had set the stage for a historic coalition between labor and environmental movements.

At the Seattle protest against the World Trade Organization in 1999, forest defenders and steelworkers linked arms and marched in front of a towering two-story Hurwitz puppet, its global media debut graphically representing the seminal role of Maxxam and the redwoods in the historic rupture of antiglobalization protest. Redwood forest defenders had made the connection between the destruction of local environments and the global unfettering of corporate capitalism, which was manifest, as they saw it, in the rush of so-called free trade agreements and the rise of the WTO. They took this realization to the streets in Seattle and made Maxxam into a global symbol of the corporate destruction of interwoven communities of labor and nature.

The Battle of Seattle was a moment of global identification that revealed once again the crucial role of emergent communications technologies. Redwood forest defenders were able to identify their plight with those of Mexican farmers, French cheese makers, Brazilian Indians, African villagers, and Chinese sweatshop laborers because they could see the faces and landscapes of faraway destruction and read all about each other's regional and local movements, call each other on the telephone, communicate instantaneously and anonymously if

Street puppet depicting Charles Hurwitz, CEO of Maxxam/Pacific Lumber, is carried by redwood forest defender at the Battle of Seattle anti–World Trade Organization protest, November 30, 1999.

necessary via the Internet, and through all these channels effectively plan on marching together. The victory of WTO protesters, who managed to scramble the entire conference, revealed how the movement had already begun changing the character and direction of globalization. The ensuing militarization of security for global trade meetings became one surface indication of how seriously the event was taken by the advocates of free trade globalization. More important, perhaps, was the considerable shift in the actual policy language that globalists themselves began using. Free trade is now increasingly described as a global program for good jobs and sustainable environments. Global industrial associations in the extractive industries, for example, almost uniformly proclaim, especially in their online mission statements, that the primary goal of their organizations is sustainable development. And in the post–Headwaters deal environment, Maxxam redesigned pacificlumber.com in a way that reflected the same transformation. Are these merely co-optations of the movements' messages and a chimerical greening of the same old corporate programs, or something more significant—a signal, perhaps, of a real operational shift in attitudes governing corporate citizenship and world trade policy? One thing is certain: these events put us squarely in the domain of history, and the case of Humboldt has been and still is an exemplary part of its making.

When I began extended field research in Humboldt during September 1999, Julia Butterfly was nearing the end of her second year in Luna, and the World Trade Organization (WTO) ministerial conference at Seattle was on everyone's calendar for November. For the next two years, I lived and worked in this globalizing landscape of capitalist knowledge and power, immersed in Humboldt's ancient forest conflict. I followed it to the Seattle protests, camped in its archives, and interviewed its people. I attended logging conferences, demonstrations, blockades, and protest rallies. I got to know its forests, Indian reservations, manufacturing plants, and logging towns. Everything I encountered sent me into the past on my search for the present. How could it have been otherwise, given the question I asked in the wake of Chain's untimely death and in the light of Butterfly's extraordinary life? That question was the timber wars, their origin, meaning, logic, implications, and message to the future—and the answer, again, was landscape and history.

Landscape and History

Concurring with many in the burgeoning field of culturally tuned environmental history, the historians Richard White and John Findlay see place as collective ideas imposed on time and space; they view the American West as a text written large by a people set in motion by the nineteenth-century market revolution in national culture and society.[28] Pushing that concept, I see Humboldt as a place that a people could make only as subjects of a culture system much larger than they—a system that called their action into specific forms of world historical labor and transformed the so-called frontier into what we see now. Culture enters nature through labor, in the exacting terms of the eco-philosophizing social theorist James O'Connor, where by nature he means the physical environment on whose tremendous riches capitalism ultimately depends for primary inputs of material and energy.[29] But nature enters culture in the same transaction, materializing nature "in historico-geographically contingent and variant ways," as Noel Castree puts it, positioning the work of Bruce Braun and other cultural geographers who grapple with Marxism at the vital theoretical threshold of nature-culture dialectics.[30]

To the insights of O'Connor, Castree, Braun, and the western environmental theorists and historians, I add those of psychoanalytic social and cultural theory, with special emphasis on an idea drawn from Louis Althusser. Where Althusser saw ideological state apparatuses call subjects into actions that reproduce social relations of domination, I see culture systems call place into being through bodies that work by invoking those systems' meaning-making potential. Language is the model for this understanding. It exists before the subject does, embracing it, prefiguring its psychical functions, as do the rituals, traditions, institutional practices, and collective representations that also always already have the subject in their grip, so to speak, even before birth, as Althusser put it, explaining what is simplest to grasp in Freud.[31] We are born, prematurely, into an ordered and ordering universe of systemic social-symbolic practices that is already up and running; for example, in those systems of marriage ties about which the psychoanalyst Jacques Lacan wrote, they "are governed by an order of preferences whose law concerning kinship names is, like language, imperative for the group in its forms, but unconscious in its structures." Unconscious language and social structures constrain the subject to speak and act in particular ways while enabling and inviting it—calling it—to communicate and be social in the first place. "Man thus speaks," said Lacan, "but it is because the symbol[ic order] has made him man."[32] Likewise people work on the land, but it is because the sociosymbolic system has made them working people: In Humboldt, people value and fight over redwood property, but it is because the system of property is already up and running—a juridical culture available to them for making meaningful lives and material gains. Indian, labor, and timber war stories give their politics flesh-and-blood purchase. And landscape, too, from this perspective, is a shared structure for signifying action, one that labor, over time, imposes on the physical world as it builds that world into *place*. The given, built environment—that which every experience in a place must encounter—addresses the subject precisely like a language, both hailing and enforcing its meaningful practice. Landscape is culture, in other words, and it calls on the subject to act.[33]

In this way, we can grasp the native intelligence of a phrase commonly heard in Humboldt: "This is redwood country," people say, expressing a naturalized competence in this language of place and an understanding of the powerful role played by environment in making the symbolic and material conditions of meaningful social life. At one time the region had been physically dominated by the gigantic trees, as it is today by the monoculture redwood tree farms that replaced the ancient forests. Signified as property by the culture system, those ancient forests addressed a massive invitation to labor in redwood commodity production, contributing greatly to the making of timber culture.

That such a system of signification imposed itself on the redwoods from the outside and made the place we call Humboldt is the simple thesis guiding this study. That meaning-making system was nineteenth-century American capital culture exploding through market revolution—an ecology-gobbling, territory-colonizing machine fueled by slave accumulation, genocidal Indian removal, patriarchal family socialization, corporate paternalism, labor exploitation, universal education, and a Protestant calling to Manifest Destiny—an institutional dynamo that crucially took additional energy from emergent print culture and public-sphere media. Without the continuous display of the nation's uniquely enumerated founding speech of natural political and civil rights, which this rising media culture provided, *the people* would have been hard-pressed to identify collectively as such, as Americans, as members of the group, as part of something they saw as big, noble, legitimate, and historic. Adapting Lacan for our purposes here, this institutional order must be seen as imperative for the group in its forms, but unconscious in its structures.

Psychoanalytic social theory emphasizes the importance of the visual and physical-spatial as well as the structural-linguistic registers of these objective institutional unconscious structures.[34] They are the means—the media—of interpersonal experience that subjects use in building the self- and object representations that become the internalized building blocks of self-identification and ultimately complex identity formation. The lovers, for example, form a group of two, but under normal circumstances they must somehow meet eyes, whereupon the image of the other, garbed in all its cultural accoutrements (clothes, eyeglasses, nose rings, circumcisions, and so forth), enters the realm of possible use in self-representation. The nuclear family forms a group of two or three or more, wherein the intimate home establishes proximity and with its furnishings helps mediate and thus channel desire through libidinal investments into social bonds that endure over time. People commit their attention and energies to their lovers and children, their friends and neighbors, using spatial coordinates such as these—they also impose an order that is imperative in its form but unconscious in its structures.

Timber culture, too, must be bound together, but such larger groups set in urban and wide-open spaces need transportation and communication technologies to identify and maintain social-psychical bonds. According to Freud, psychological and emotional ties or bonds involve libidinal investments, where by "libido" he means the passionate life energies that animate the psychical drives for both love (including sexuality) and self-preservation. Ultimately libidinal investments manifest themselves in a universally observable human impulse to combine in the service of pleasure, friendship, procreation, family, collective self-defense, and so forth. This is the energy of desire (*libido* is Latin for desire, longing, wish, and fancy, including sexual appetite, lust, and passion). It is the energy of the investments that form attachments and make collective subjects what they are—emergent, tenuous, fluid, and collectivizing foci of individuals' desirous attentions, labors, and actions into public formations. Media make powerful collective subjects like hegemonies and social movements possible because they put individual subjects in contact, bridging time and space and making possible the common experience of events, leaders, ideas, values, and symbols around and through which collective identifications are built. More than this, and in the same way that languages, landscapes, homes, and family relationships impose an order that greets every new child with life-changing force, mass public-sphere media also impose an order that tends to be imperative for the group in its form but unconscious in its structures.

In this way, public-sphere media can be described as technologies for producing *the common*. They constitute a kind of technological a priori, conditions of possibility, in other words, for collective subjects to take a form that I call identificatory publics. Conceived as aggregated and directed psychical attentions and energies—we can even say *psychical labors*—identificatory publics are constituted by the partial object orientations of individual subjects as they channel their attention into the common when they concretely participate in a psychical collectivity by identifying with a cause or investing their attentions in projects and objects.[35] Only through wide-reaching media can collective subjects self-organize and focus their psychical energies into world historical labors on the world. And as already noted, such media also have storage capacities that play important roles—namely, they accumulate traces of public discourse in archives that are necessary to sustain collective identifications and projects over time.

In concrete terms, consider that a forest and factory can bring workers together, and a town can bring its people together, but only mass media can bring the wide-flung regional timber folk together, spread out as they are across cultural spaces of forest, family, factory, and town. But local publics like this reach further still, upward and outward in scale, to identify local practice with national cultures and ultimately global political and scientific communities, larger collectives whose universalizing concepts of self-identification and inclusion—for example, citizenship and globalization as capitalist world system—are now used with facility in the collective self-identifications of timber workers, Maxxam managers, and forest defenders alike. People now cast themselves in global terms. They project themselves imaginatively into identification with global public cultures, contributing physical and psychical labor to ever greater unities by directing their attentions into world historical events and projects.

In this way psychoanalytic social theory allows us to speak of collective political subjects, for instance, the hegemonic cultural order of capitalism and oppositional social movements like the forest defense, without falsely isolating individuals into discrete categories—that is, without hypostatizing publics into groups of discrete actors that mobilize their bodies in unitary directions. This logic of collective identification helps us better understand a number of complex situations encountered by ethnographic field-workers, for example, a timber worker who consented to work on antiunion shop floors while criticizing the corporation, sympathizing with forest defenders, and attending environmentalist rallies. Is he a forest defender or a logger? His identity, not unitary, flows in both directions and presumably in others. Likewise, a forest defender who supports timber workers could organize against Maxxam while defending the traditional, pre–Maxxam Pacific Lumber's reputation as a good environmental steward. The timber war field of cultural politics is precisely this struggle for power over flows of attention and psychical investment in the sociomental environment.

The concrete expression of this struggle, through the long detour of political processes that ultimately control elections and policy decisions, including the forest practice rules governing redwood production, appears in the channeling of the flows of values that human labor, attending to nature, blasts out of nature into the commodity circuits of capitalism. Capital accumulations appear as compromise formations in the material pattern of values projected by opposing identificatory public and counterpublic forces.

It follows that the timber wars must be viewed as a symbolic politics of subject formation, embedded not just in the social relations of capital to labor but in those of the democratic republican—that is, the liberal—constellation of institutions that define everyday life, especially media. They are a redwood politics of libidinal-economic production. Just as Michel Foucault described the human sciences as power-knowledge complexes—discourses that produce and further subjectify the bodies they represent, technologies through which the European Enlightenment remade the masses that remade the world—so too do new social movements of labor and environmental defense create new public cultures with newfangled powers that reconfigure social actors and redirect their (psychical) labor (energies) into new collective place-making projects that carve out alternative places in alternative futures.

The Deep Culture Drive of Perpetual Conflict

At the energizing core of this constellation of liberal institutions—this colonizing culture of rights—lies the concept of individual property right. The framers of the U.S. Constitution, the pioneer lumbermen of Humboldt, the big redwood timber barons of the twentieth century, the lumber and sawmill union folk, and the CEO of Maxxam all agreed on one basic point: This is America, they repeated, the singular nation of liberty and equality, of which the distinguishing character is a specific program for the collective defense of individual rights, with property, free speech, press, assembly, and religion the most popularly understood.[36] But among these rights, property has historically exercised extraordinary power. According to the prevailing faction of the nation's founders—the Federalists—the Constitution was conceived and written to represent and thus constitute the citizen as a free person, owner of his own body, mind, labor, and products, thereby forging him into a concrete Lockean bulwark against intruding power, governmental or otherwise. It was a necessary mechanism, they argued, for a newly conceived democratic polity in which a propertyless but newly enfranchised majority faction would certainly threaten minority rights sooner or later.[37]

In the words of the historian George Mace, "the major innovation of the American Founding Fathers [was] the conversion of economic social conflict from confrontation based on the amount of property to confrontation based on the kind of property."[38] This understanding of changing class relations can be traced in the words of Publius (Alexander Hamilton, writing in the Federalist Papers), who explained that by combining the democratic institution of direct election with the republican institution of representation, and repeating this structure at the state and federal levels while checking and balancing the powers, "the federal Constitution forms a happy combination." By ensuring the rights of property and setting a course for expansion of the nation's geographic sphere, it guarantees the public good somewhat paradoxically by guaranteeing a proliferation of opposing private property interests anchored in places distant in space and time.[39]

What the authors of the Federalist Papers, Alexander Hamilton, John Jay, and James Madison—and the white owning class they represented—dreaded most was concentrated, unaccountable political power and its possible embodiment in a tyrannous majority. Two methods of preventing majority faction presented themselves: destroy the liberty that allows destructive differences to emerge, or produce "the same opinions, the same passions, the same interests" in everyone (10.4). The first cure would be "worse than the disease," while the second is impractical and unwise because "the reason of man is fallible and he is at liberty to exercise it, [so] different opinions will be formed," with the result being continuous instability and violence (10.6). This is because "as long as the connection subsists between his reason and his self-love, his opinions and his passions will have a reciprocal influence on each other; and the former will be objects to which the latter will attach themselves" (10.6). Citizens, in other words, are driven both by reason and by their passions. Reason will consistently fail if the passions are not contained. This familiar refrain of Enlightenment philosophy is directly embodied in the sacred institution of individual property and must be interpreted as the founders' most concrete solution to majority faction and limited government. Pure democracy could foster tyranny of the masses—a united, impassioned majority—unless, that is, the countervailing institution of a civil right to property is made equally as sacred as the political right of franchise. In property lay the life or death tendency of the national body politic.

One consequence of this program is clear: it helped carve out for the nation a colonial future of perpetual property conflict driven by reciprocally constitutive institutions of free speech, press, and assembly in each new place over which the colonizing culture extended its sphere—institutions that establish a modern public sphere and constitute a social space of media technology for the formation of collective will and public power. Such was the ambivalent nation of public rights and individual liberties imagined and construed in the founding discourse.

From these remarks, we can draw several conclusions. There was a riotous, libidinal, and embodied subject conceptualized in the framers' performative and people-making constitutional utterances. We must therefore see the framers not merely as politicians but as philosophical psychologists as well—their Enlightenment views represented the essential nature of the human being as passionate and driven by impulses beyond its own control. People are ambivalent creatures whose drives, if not contained, overpower their reason. Their energies must be bound in productive institutions. The ambivalent, vindictive, rapacious, conceited, envious, fearful, loving, and ultimately irrational subject must therefore be subjected to rationalization by the rule of law. Only the law can make this creature into a citizen—a rational modern. And the law of property was central to the plan, as was the panoply of civil rights, including free speech, press, pulpit, and assembly, through which the institution of property is continuously made into a public affair. Property is, in fact, a state institution that hails all people into citizenship with spectacular public representations of their national character.[40]

The Declaration of Independence, the Articles of Confederation, the Constitution, the Federalist Papers, and innumerable lesser documents, including a litany of Supreme Court decisions, invoke a disciplinary, psychological discourse that channels its classifications and concepts of essential human nature into the great project of constituting the nation. In so doing, they did more than just recognize a passionate, interested, and conflicted subject desirous of property and fascinated by the law; they called it into being. Revolutionary U.S. nationalism must therefore be viewed as an economic psychology with a normalizing force that energized the colonizing culture, facilitating its privatization of the New World. It established constitutionalism as a deep cultural drive, among whose most profound effects are a constant proliferation of rights-based forms of property and of free-speech public spheres, which together ensure the ongoing production of our modern archive culture and the filling of it with evidence of rights discourse.

The political architects of American modernity understood that it is not possible to extract the psychical character of human organisms from their economic, political, and spiritual livelihoods. To produce and maintain a successful nation, a political constitution must extend its government to the realm of subjectivity, where the liberty of economic, political, religious, and sexual energies inextricably merges in the psychosocial performance of citizenship.

It will help to recall that property is not the *thing* suggested by common sense and much property discourse but rather a social relationship defined by a bundle of enforceable rights that govern the relations between individuals with regard to things.[41] Property rights are made by communities of struggle and institutionalized in laws that establish such relations, relations that are ultimately backed by force of some kind, for example, the state's monopoly on legitimate violence.[42] They are philosophical concepts being put—and again, eventually forced—into action.[43] The right of individual property, for example, puts the philosophical idea of communally defined and publicly limited personal freedom into action.

But over time a problem emerges. The juridical institution of property rights begins to demonstrate its advantage over the coarticulated and reciprocally constitutive political and civil rights of universal franchise, free speech, press, religion, and assembly: being anything legitimately appropriated from nature through labor, property accumulates materially as power over labor under conditions of relatively open competition and freedom of contract. Accumulating power over labor then subverts its own conditions when, deployed in emerging markets, it perverts the operation of rational discourse in the public sphere. The philosophy of freedom, institutionalized as property, provides for, and even invites aspirations to, domination in the public sphere. The basic rights package turns out to be a program for perpetual conflict in the public sphere over property (rights).

But there is something more primary still in the representation carried by these institutions, something now built into this program for social conflict: a deep cultural context of modern European philosophy—the sciences of man!—a new certainty in the knowledge of the human being's intimate connection to nature, its vulnerability to nature, and its rootedness in nature. Knowledge of man as nature suggests that man must be dominated like nature—that it can and must be improved just as surely as wild external nature must be. Modern democratic polity makes these improvements a mandatory state project—through them Enlightenment philosophy addresses and forms a new national public of continuous improvement. When the framers wrote this perspective into the textual engines of national self-identification, it was part of a rational plan to defend against and to improve that alien, wild, natural force—the passionate, erotic, unruly, angry, envious, and greedy nature of actually living people—the unreasoning body.[44] Thus did modern American democracy begin on the psychical defensive. The labor of government was rationally divided against itself, separated into tripartite powers, and set up to be continuously revolutionized with updated technology for the exercise of free speech, press, and assembly. These were conditions for republican democracy that alone made possible the aggregation of popular attentions and sentiments that gave substance to the philosophical concept of a general will embodied in a secular state, a state that was legitimately sovereign for just that reason—a state that had the right to rule because it was the rule not just of right reason but of collective, public reason.

Critical theory and history of American modernity—and by extension its subsidiary conflicts like the redwood timber wars—should begin here, in the juridical culture system that combines the legal authority of property rights with the other core symbols of the revolution (namely, the other civil and political rights) to form an institutional engine that proliferates public struggles and expands geographically as it constitutes the affective performance of American nationalism. The end result is a colonizing knowledge system, among whose chief institutional achievements must be included the collective force of its patriotic worker-citizens' deep and pleasurable feeling of consent to be governed by a perpetual conflict of interests.

The Public-Sphere Spectacle of Rights

We should not be surprised to find that this conflict pervades the permanent record of media spectacle in Humboldt, for reasons intrinsic to the concept of the public sphere. Jürgen Habermas has shown how the natural rights constructed by modern constitutions in effect called the public sphere into its modern configuration, guaranteeing its role as the technology of public address through which nations would call themselves to order. With the rights of free speech, press, assembly, and association, he wrote, "the functions of the public sphere were clearly spelled out in law."[45] These constitutional choices also inaugurated "the [juridical] protection of the intimate sphere (with the freedom of the person and, especially, of religious worship)," in what amounted to an "early expression of the protection of the private sphere in general that became necessary for the reproduction of capitalism in the phase of liberalized markets."[46] Nicholas Garnham lauds this Habermasian model for its "focus upon the indissoluble link between the institutions and practices of mass public communication and the institutions and practices of mass democratic politics," for its "focus on the necessary material resource base for any public sphere," and for its "escape from the simple dichotomy of free market versus state control."[47] My point here is that the so-called free market is a political construction deeply imbued by the state constitution with imbricated rights of free speech, press, and assembly.

On the nation's frontier, where the story of the colonization of Humboldt begins, newspapers were a singular transmission line for the cultural discourse of the nation. They dominated the public sphere with a spectacle of words from the distant capital and eastern population centers, a vital technology connecting Humboldt's local conversation to the continuous address that was forming the nation. They made possible a relatively informed, nominally free, and increasingly heterogeneous discourse in which something called informed public opinion might ostensibly form, something from which an idea of consensus could be derived through electoral process, something like a collective will.

Of course there was much more to the public sphere. There were bars, conversations on the docks and in the streets, citizens groups, voluntary associations, and even Humboldt's genocidal volunteer Indian-hunting militias—these were all places where the conversations took place that boiled down opinion. They were sites for exchange between citizens. But the function of the newspaper system stands out among these collectivizing channels. It projected the culture of rights and perpetual conflict into the redwoods. And while this public-sphere rhetoric claimed universality and spoke as if it had no body at all, it was, as Michael Warner succinctly puts it, "structured from the outset by a logic of abstraction that provides a privilege for unmarked identities: the male, the white, the middle class, the normal."[48]

This was precisely the character of newspaper address that spoke to Humboldt through the region's first local paper, the *Humboldt Times*. From its first issue in 1854, through the period of Indian trouble in which the indigenous lands were enclosed and otherwise appropriated, the *Times* was there, holding up a mirror of universal republican virtue in the bay redwood region and facilitating the instantiation of national culture.

In this way, newspaper culture initiated a media archive on which so much of Humboldt's future historical consciousness would ultimately come to rely. The *Times* recorded the colonial discourse of redwood settlement, preserving its rhetoric of perpetual conflict and providing future historians with classifications and discourses through which the people invading the redwoods tended to see the world and remake it. In the stories of Indian trouble, labor trouble, and trouble in the forest I tell in later chapters, the papers are a primary source, as they have been for all previous historians of the region. The dominant conflicts that rocked the region in the decades leading up to the timber wars largely work on the present through this archive's towering presence in historical consciousness. The state's self-investment in the people and markets that constituted the nation as a public performance of affective character must largely be measured in terms of this collectivizing technology. Media made national self-consciousness possible, and so interpretation of Humboldt's contemporary historical consciousness, and by extension the redwood timber wars, must begin in the voice of its public-sphere archive.

We need to treat the colonizing discourse of rights-driven markets, publics, and politics as an apparatus of power and ask how its continuous display in regional papers help set the cultural conditions of timber war. We can start by considering how the *Times* represented Anglos as citizen-subjects of what Étienne Balibar called "the nation form," by which he meant a matrix of institutions that collectively shapes modern subjectivity in the image of national ideology. Modern nation-states produce national identity with a cultural and psychological depth that Balibar calls "fictive ethnicity," which essentially means a feeling of "community instituted by the nation-state."[49] Nationality is a structure of feeling or community embodied and lived as identity produced under social conditions of state signification. It is formed within a field of power governed by state-sanctioned institutions of modern everyday life. It is crucial to note that the term *fictive* does not signify something unreal, untrue, or nonexistent but rather points to the presence of a social imaginary, in the constitutive sense that cultural theory gives this term, as I will explain in the following section. For Balibar, "Every social community reproduced by the functioning of institutions is imaginary, that is to say, it is based on the projection of individual existence into the weft of collective narrative, on the recognition of a common name and on traditions lived as the trace of an immemorial past, even when they have been fabricated and inculcated in the recent past. But this comes down to accepting that, under certain conditions, *only* imaginary communities are real."[50] That would be the very conclusion reached in Cornelius Castoriadis's *The Imaginary Institution of Society* and Benedict Anderson's celebrated *Imagined Communities*.[51]

Balibar sees the institutions of family and compulsory education as the principal engines of fictive ethnicity in twentieth-century Western nations, whereas in the nineteenth century and before, the family-church institutional dyad had done most of this work. Universal schooling under the national compulsion of enlightened social engineering produces collective linguistic identity and community that, according to Balibar, in each case "produces the feeling [in the present] that it has always existed . . . it assimilates anyone, but holds no one . . . it affects all individuals in their innermost being (in the way in which they constitute themselves as subjects), but its historical particularity is bound only to interchangeable institutions." Yet "the contemporary importance of schooling and the family unit does not derive solely from the functional place they take in the reproduction of labour power," he says, "but from the fact that they subordinate that reproduction to the constitution of a fictive ethnicity—that is, to the articulation of a linguistic community and a community of race implicit in [that nations'] population policies."

Race is essential to such language communities because they can always add strength and stability to their social project of maintaining order by positing a biologically material anchor for national identity. The geographic frontiers of a people are not in themselves necessarily enough to bind the structure of feeling for the nation across time and space. It "therefore needs an extra degree *(un supplément)* of particularity, or principle of closure, of exclusion . . . that of being part of a common race."[52] Consequently family, school, church, gender, language, and race are held to combine in the fictive ethnicity of the modern nation form. And this amalgamation is precisely what we hear in the archive of media spectacle and newspaper culture stretching back through the history of conflicts that map out the story of capital in Humboldt—a mélange of variously practical, cultural, narrative, and discursive supplements that, taken together, fairly describe an American national form of fictive ethnicity as it was differentially achieved in the redwood bay region under local conditions of Indian trouble, industrial forestry, and deforestation.

But the foregoing argument compels me to stress again the operation of mass media in the complicated engine of modern cultural colonization, for it was there that the specter of constitutional law was continuously displayed, addressing the people together, one and all, *e pluribus unum,* with an ideal image of republican virtue, calling all peoples (white and male, largely) into collective being by gathering their attentions in a public structure of feeling, situating them within a broadcast image that identified them with each other in and through that great symbolic structure—the national form of fictive ethnicity. The continuous spectacle of democratic public-sphere nationalism made locally informed participatory citizenship possible, calling to people with symbols of liberty and equality that channeled the force and fuel of their bodily labor and psychical attentions through juridical institutions that expanded the national colonial project.

Finally, the case of Humboldt teaches us to add one last cultural institution to our conceptualization of the national form of fictive identity that colonized the redwoods. In the modern American social imaginary, the concept of property forms another supplement, another extra degree of particularity or principle of closure and exclusion through which American identity knows itself and performs. The constitutional people-making machine and the media spectacle that helped establish its public and universal norms have never strayed far from this principal symbol of American virtue.

Benedict Anderson has shown how print capitalism in general and newspaper culture in particular helped make collective feelings of modern nationalism possible by establishing the experience of horizontal simultaneity—that new form of modern time consciousness in which a Humboldt pioneer, for example, who would never know and never meet more than a tiny fragment of his or her countrymen, could nevertheless develop "a complete confidence in their steady, anonymous, simultaneous activity."[53] Here again is that media link that channeled the nation and its culture of rights into redwood ecology and Wiyot territory, setting in motion the long march of capital though Indian war and labor trouble that created the conditions for timber war in the late twentieth century. It helped make this place modern by way of instating what the political philosopher Charles Taylor, among other culturally inclined theorists, would call a modern social imaginary.

The Redwood Imaginary

Cultural sociologists debate how best to interpret the meanings of social life and explain their institutionalization and reproduction, especially as they contribute to economic, gender, racial, and other pernicious forms of inequality.[54] In this sense, questions of social justice are always at the center of cultural sociology. In this book I use the culturalist concept of a social imaginary—shared symbols, values, laws, and meanings performed and embodied in the institutional repertoires of a group—to theorize the local formation of a redwood imaginary, which I define as a unique, place-based manifestation of the modern social imaginary. I strive to show how it came to embody and shape local expressions of power, domination, and resistance in redwood social history and thus how it ultimately set the conditions for timber war.

We should pause for a moment and consider the analytic content and usefulness of this term—*social imaginary*—for bringing cultural and environmental theory together in a new analytic tool for studying conflicts like the timber wars. With Taylor we can start by defining a society's or group's social imaginary as the shared knowledges, competencies, and values embodied in the various patterns of its actors—their institutions; a social imaginary, he writes, is a "common understanding that makes possible common practices and a widely shared sense of legitimacy."[55] But common understanding comes from common practice, and this circular formula constitutes the peculiar strength of the social imaginary as an analytic category: it is dialectical critical theory, a way of defining and analyzing collective cultural phenomena as complex and always emergent processes in which energetic subjects answer, carry out, and ultimately embody and reproduce the cultural structures within which they emerged and which invited them to participate in collective action and gave them so much opportunity to do so in the first place. Institutions within a social imaginary are its culture patterns—its practical, tacit knowledges performed as meaningful, signifying actions. The effects of a social imaginary on the world register in the labor these institutions direct.

A social imaginary therefore consists of *instituted* ways of acting in the same way that a language consists of instituted ways of speaking and that collective belief systems, like religions, consist of instituted ways of seeing the world. Because they are symbolic systems, people use them for signification—and because signification is material, they transform the world.

For Taylor, rational capitalism, the public sphere, democracy, and rights discourse are the vital institutions of modern social imaginaries.[56] They are what is modern in modern social imaginaries, and they show how social imaginaries are in fact moral orders, in which differing ideas and values enacted in the relatively autonomous but reciprocally constitutive spheres of everyday economic, social, and political life establish a shared way of life. They are, in brief, what we mean by modern culture.

The compound term *social imaginary* is more insightful and analytically productive than the simple term *culture* precisely because it identifies, separates, and then dialectically binds the subjective and objective elements that common usage of the term *culture* too often leaves oblique. The social is nothing if not objective and collective, so the term *social imaginary* must be read as *objective and collective imaginary*. But the term *imaginary* refers to the imagination—which is nothing if not the subject's active representation and meaning-making activity; so now it reads *objective collective representational action*. One final ingredient is necessary: in the structuralist and semiological movements indebted to the linguistics of Ferdinand Saussure, first among others, the objective collective social world is nothing if not a symbolic order; it is a meaning-making system comparable to a language system, an enabling and constraining system that individual and collective subjects put to use. Hence the term should be read *objective-collective-symbolic order of and for active representation*. There seems to be only one way to interpret this complex idea: the concept of a social imaginary defines the social as a usable system of ideal elements already up and running in institutional structures that individuals encounter as an objective moral order. Thus a social imaginary is a system of meaningful, value-laden institutions into which people are thrown and which they tend to embody and naturalize for use in making, performing, and expressing their own symbolic meanings, values, and practical lives.

Whereas Émile Durkheim called this perfomative order the world of collective representations and compelled us to treat them as social facts,[57] and Max Weber spoke of how armies and corporations originated and now characterize modern institutional spheres of psychosocial discipline,[58] the works of Karl Marx consistently turn on dialectical phrases that secure this same tenet of cultural theory: men make history, he states in the well-known phrase, but not under conditions of their own choosing. Yet seldom do citations of this powerful statement of reciprocally constitutive structure and agency go on to take note of the linguistic, cultural metaphor that he uses to explain the point: "The beginner who has learned a new language," writes Marx, "always translates it back into his mother tongue, but he assimilates the spirit of the new language and expresses himself freely in it only when he moves in it without recalling the old and when he forgets his native tongue."[59] To understand culture and its function in reproducing the imaginary institutions of society, it is necessary to know how social memory works by forgetting what has always already been there for the subject to use. The modern social imaginary is contemporary cultural theory's name for what people forget in order to live—it is nothing more or less than our everyday cultural unconscious.

Applying this dialectical model of language to all social life brings us up to date with the turn to practice in social theory. But my case study in Humboldt pushes the idea further by introducing environmental theory and history: the redwood imaginary is modernity in the redwoods, a local instantiation of the modern social imaginary in the redwood ecozone—a geographic, spatial installation of its institutional system for meaning-making lives. The idea of a modern social imaginary is more intelligible and useful if we make it a spatial, geographic, and ultimately environmental category.

Cultural Theory, Media Studies, and Environmental Sociology

The modern social imaginary—a culture system of democratic-republican polity and public-sphere media-driven capitalism—arrived on the redwood coast of northern California in 1850, setting in motion the total transformation of the region's environment and native lifeworlds into the place known as Humboldt, a built social world that colonization made significant: the place of the redwood imaginary. My central premise here is that the U.S. political culture and its textual engine in the people-making, nation-forming Constitution drove this process, making the redwood imaginary a local, place-based instantiation of its universalizing vision. I am talking not about a single cause of the timber wars, responsible for everything we will find in the conflict, but rather about a triumphant organizing address that called into being and action a constellation of institutions that, though dynamic and changing, came to dominate the redwoods.

By way of answering the questions I have asked about the timber wars—how and why did they originate and how do they work? where are they leading? what do they tell us about globalization? the nation?—my method is to combine the cultural theory of social imaginaries with elements of media studies and environmental sociology in the writing of social history. By describing the living, symbolic, and built place of the redwood imaginary as a complex and relatively autonomous cultural structure, I show how and why the places of capitalist colonization accumulate meaning-making potential over time, differentiating them as they continuously archive local history and memorialize events, especially events of unusual violence.

This program for cultural sociology expands the definition of place to include its living, symbolic, and built characteristics. By *living* I mean the institutions that express the norms and values of everyday life, as they are acted out, ritualized, performed, or practiced in the anthropological sense of that term. By *symbolic* I mean the full range of more or less codified, narrative, and written textual artifacts, for example, newspapers, journals, letters, diaries, speeches, photographs, and the history books that rest on such primary materials, as well as the stories, legends, and myths in oral circulation. And by *built* I mean the range of material artifacts, including architecture, physical geography, landscape, and the humanly modified remnants of ecology like extinct and threatened salmon runs, deforested hillsides, and silted-in bays and waterways. I treat this collective living, symbolic, and built place of the redwood imaginary as an archive of social memory that enables and constrains political action.

Turning to media studies and environmental sociology and history to develop this cultural theory of the archival redwood imaginary, I build on James O'Connor's understanding that the deepest cause of environmental history and hence of contemporary environmental movements is "a structural one: capitalist political and legal systems, capital accumulation and the commodification of social life and culture."[60] Using a word synonymous with the universal compulsion of capitalism, he writes that commodification is "the 'division of nature' into means and objects of production and consumption"; it produces "a new nature, a specifically capitalist 'second nature.'" This is nature subjected "to the discipline of the financial market," the transformation of lakes, coastlines, forests, and all biological systems into assets, the economization of all things natural, and ultimately the remaking of nature "in the image of capital, e.g., via bioengineering, factory forests and the like." And all of this was "unimaginable before social [life] and cultural life were commodified."[61]

Pushing O'Connor's language toward my idea of the modern social imaginary as a colonizing cultural system of capital, it follows that second nature is produced as the whole of global time and space fall ever more deeply into its force-laden, capillary field of scientific knowledge; it is a world resignified under science-based capitalist culture. Indeed, as O'Connor says, the ultimate effect of continuous capitalist signification is that "politics, economics, social and cultural life and environment are successively revolutionized, i.e., become more specifically capitalistic."[62] In my interpretation, second nature archives social memories of science-driven economic modernization.

Two contradictions determine how this culture system develops. The first arises from within: the well-known internal class contradiction that follows from competitive and accelerating scientific exploitation of labor and results in continuous downward pressure on prices (including wages, the price of labor), ultimately driving the system into so-called realization or demand crises. Open competition between capitals to cut the cost of production drives wages down while increasing the rate of production, leaving masses impoverished and so many products that consuming them all becomes a new central problem. This first contradiction compels the system through recurring bouts of expansion, crisis, and reorganization—a business cycle in which individual capitals are forced to continuously expand their markets and aggregate power lest they fall behind in the all-out competition. The result is a capital culture driven to continuously expand, which it does by innovative technology, speed-up, replacement of labor by machinery, the expansion of scale, vertical and horizontal merging of firms, and every other imaginable strategy to reduce the cost of production. The system survives, in other words, only by dint of the application of science and technology to everything, all the time, from here to infinity. To exist it must continuously revolutionize the means of production. Nationalism, colonialism, imperialism, urbanization, war, and the advertising system are among the developments that this analysis of capital culture's structural compulsion to expand can help us interpret.

In the New World, this first contradiction in the colonizing system of Anglo-American capitalism produced a grotesque deformation of the laboring classes under eighteenth- and nineteenth-century conditions of industrial revolution, with the labor movement of the late nineteenth century and the early twentieth emerging in response. In the twentieth century new social and cultural contradictions emerged, embodied, for instance, in the civil rights movement and the so-called new social movements of women's liberation, gay liberation, identity politics, and environmentalism. These are responses to the ongoing exploitation of communities, identities, and environments by the same systemic forces of capitalism against which early labor movements moved. They are surface signs of the deep and continuous revolution—that is, the modernization of every domain of social and ecological life—that drove mercantilism through slave accumulation and indigenous plunder into the era of unionism and eventually to that of civil rights and finally to that of today's class and race politics. The current consolidation of these cultures of resistance in the networks of anti-neoliberal and corporate globalization movements, the World Social Forum (WSF) and the global justice movement—the so-called movement of movements—are among the most recent and consequential effects of this modernization.

Together with the post–Cold War expansion of capital, the rise of this globally identified movement of movements signals the coming of an era increasingly defined by the second contradiction in capitalism, an antagonism that potentially unites every other human interest: the contradiction of global capital by global ecology. Not transcending but absorbing and extending the first contradiction, the second marks that point where capitalism begins to destroy its external conditions of possibility for production, namely, the communities of labor and environment that constitute its profit-generating capacity, including the spatial arrangements uniting these elements in built environments like cities, watersheds, states, and ecosystems. These are researchable places that embody historical change—places where the conditions of production are external in the material sense that they are not produced by the system itself but rather exist as necessary inputs of energy, including labor energy, and resources.[63]

The potential oppositional power of communities of labor and environment can therefore be described as structural; being sources of capital, they are required by capital, and so they are all potential sites of contest for control over capital. But whereas it has often been remarked that the workers have the power, raising the perennial question "Why don't they use it?" it has less often been recognized that environment is power and whoever defines it and controls the regulatory process largely determines the flow of capital accumulation. Environmental politics consequently become central first to state control over markets and ultimately to the rise of global institutions seeking to rationalize trade liberalization. In this way we see how, when modernity enters the living, symbolic, and built dimensions of place, structuring its objective potential as a meaning-making system and calling up labor, setting it to work in successive epochs of first and second contradictions to capital accumulation, it makes places like Humboldt into archival structures that structure the future of potential politics.

Guy Debord's concept of the spectacle society is useful for tying the media domain of consumer society to that of the first and second contradictions in capital culture. The second contradictions emerge by way of deferring catastrophic crises, for example, by constructing expansive credit systems, Keynesian state policies, and the advertising system. The society in which the reproduction of the conditions of production relies increasingly on expanding consumption through borrowing and aggressive marketing is the spectacle society—the consumer society. This marks a change in the mode of domination: in the words of George Ritzer, "What becomes important in spectacular society is the desirable surface of images and signs . . . the attention grabbing spectacle."[64] As Debord put it, this is society devoted to the "ceaseless manufacture of pseudo-needs." Now the market must produce consumption, and so it must situate the subject as a consumer, address it as a consumer, and elevate the value of consumption above every other value. Its horizon, again, is infinite expansion, and this sets emergent capital culture on a collision course with planetary ecology.

By describing the cultural logic of capitalist colonization in these dynamic, immanent, dialectical, juridical, media-driven, and finally environmental terms—the cultural system changes as it changes the world—environmental theory of the second contradiction offers a new beginning, not an end, for the project of cultural interpretation in places like Humboldt.

Being one set of external conditions of possibility for its value-extracting commodity circuits, the pre-Columbian geophysical environment entered deeply into the accumulating place of this redwood imaginary. As the physicality of capital's immediate environment here,

it called up unique forms of colonization, labor, consumption, and resistance and then constantly threw up new challenges for people in
each of these registers. For example, when mammoth trees invited
labor to clear-cut whole mountains, the land responded with runoff
that filled rivers and bays and prepared the future for decades of labor
in flood control, dredging, and salmon-restoration ecology. How the
spectacular rights-based culture of capitalism driving colonization of
the bay redwood region and industrialization of redwood lumber production contradicted redwood labor and ecology and set the conditions
for timber war is a tale told by murdered, removed, and concentrated
First Peoples, battered unions, deforested mountains, homeless birds,
rivers of eroded mud, extirpated mammals, extinct and endangered
salmon, and acres of silted-in bay.

Violence, Archive, and Memory

In chapters 1, 2, and 3, I pursue this local modernity in a fieldwork narrative that theorizes the public space of the timber wars. My method
hinged on living in Humboldt and immersing myself in the saturated
present tense of the timber war discourse, and in the process I came
across something quite unexpected. The physical and symbolic terrain
was laden with social memories of more distant historical events of
extraordinary violence that could not be ignored. The Wiyot massacre, the murder of redwood strikers, and the attempted assassination
of redwood forest defenders formed a historical record of capital culture in Humboldt that, together with the figures of Chain, Hill, and
Scotia, could help me to write the timber war story.

I call the events on which these stories accrue *signature events* in
the redwood imaginary. Each gathers together and records an image of
the social relations prevailing in that historical moment. There can be
no question that such images are incomplete. But the traces I uncover
and assemble allow me to construct this historical study. In chapters
4, 5, and 6, I return to the public-sphere discourse at these signature
moments of violence in the successive epochs of Indian, labor, and
environmental trouble. The social history is in this way schematically
presented through the lens of changing social and environmental conditions that seized the collective imagination in these spectacular moments. In my conclusion, I explain how the redwood imaginary, as an
archive of violence, informs social memory and structures the future
of timber wars politics.

Chain, Hill, and Scotia are highly invested symbols that serve as a screen on which individuals and groups, speaking through this archive—through this place of the redwood imaginary—project their concerns, hopes, and fears and their sometimes fanciful or practical ideas about the world. Environmentalists, sensing the ecological tragedy of capitalism, channel their life energies though such symbols and use them to produce a discourse of challenge and struggle. Loggers and subjects of timber hegemony defend what they have and what they have made by naming their world at the very same points. These are the loci of symbolic production where the cultural drama takes hold and goes public. It is here, in this public cauldron of cultural construction, that *Trouble in the Forest* sets to work encountering everything lived, symbolic, and built in the region as interested and signifying contributions to the timber war field of cultural production.

1. Power and Resistance in Redwood Country: Maxxam versus the Forest Defense

September 1998. I entered the field and engaged the redwood timber wars ethnographically at a complex moment. The campaign to save Headwaters forest had pushed the Maxxam corporation, the state of California, and the federal government into a tentative agreement to preserve the largest extant but still unprotected ancient redwood grove. Law enforcement and logger violence against forest defenders had ignited a firestorm of discourse in the mainstream and oppositional press. Nonviolent, direct-action civil disobedience against Maxxam was producing a continuous media spectacle of resistance. Alliances between local forest defense groups were strengthening, and activists were making connections between the conflict in Humboldt and what would become the globally dispersed but somehow coherently identified antiglobalization or alternative globalization movement. In this and the following two chapters, I treat the timber wars as a social space of symbolic activity—a cultural field of power and place of resistance—in which I submitted myself to the forces at work. In telling the stories of my field research, I work to reveal just how they embody and reproduce the modern social imaginary, and I explore the implications for environmental and labor movements.

I began by collecting movement literature, flyers, newspaper clippings, radio reports, and bits of conversations. I took photos, conducted interviews, and videotaped events, making a record of anything that represented an immediate voice or intervention of any kind in the discourse. Approaching this field with an open-ended question—what's going on in the timber wars? what are they and how do they work?—I let the immediate, total symbolic environment speak to me through the public-sphere discourse open to all. Part of that discourse is the physical world itself, the cultural geography and built environment that help make Humboldt a meaningful place. What is the message coming out of this struggle? I had to encounter not just what people were saying and doing—the discursive and textual production of everyone involved—but the structures of place that all their labors invoke and ultimately reproduce and transform.

Death in the Forest

Severed at its trunk by a Maxxam chain saw, the redwood fell on its historical mark, killing the Earth First! forest defender David "Gypsy" Chain instantaneously. The event of September 17, 1998, occurred on mountainous slopes near Grizzly Creek Redwoods State Park, about ten miles south and east from the hotly contested Headwaters Grove, as Chain and his comrades took the matter of forest defense into their own hands; nonviolent, direct-action civil disobedience brought the young men and women to Grizzly Mountain and so to their place in this timber war story.

Just a few acres of old growth straddling the Van Duzen River at its confluence with Grizzly Creek in central Humboldt constitute the park. California Route 36 parallels the river through agricultural bottomlands, stretching east from Highway 101 toward the death site between clear-cut patchwork mountainsides then largely owned by Maxxam. Little farming and logging towns line up along the two-lane road, Hydesville first, then Carlotta—home of the sawmill that Maxxam bought from Louisiana-Pacific and retooled for expanded redwood production after taking over Palco in 1985 and doubling the harvest rate. The logging towns boomed with new jobs and overtime pay for a workforce not unfamiliar with hard times. But by 1998 there was very little old growth left to protect on Grizzly Mountain; the boom had largely given way to layoffs and chronic litigation over questionable timber harvest planning. The tiny park is surrounded by cutover Maxxam timberlands, and the trees that Chain and his friends were defending are called residual old growth—leftovers from previous selective cuts now marked for clear-cut logging alongside second- and third-growth trees.

On the day before this signature event, activists camped at the park and hiked up the eastern slopes of Grizzly Creek to Maxxam's timber harvest plan (THP) 1-97-172. Once inside the plan, they followed the sound of heavy equipment into the cut zone to play cat and mouse with the loggers, a tactic of approaching workers as they cut, engaging them in conversation, running away when chased, and returning to reengage, thereby slowing down the operation and driving up its cost. This buys time for environmental watchdogs and citizen groups to press the courts and regulatory agencies for measures to prevent destructive and often illegal logging practices, by either coaxing the enforcement of existing laws, getting new rules passed, or advancing other preservation efforts like conservation easements.

On day two of cat and mouse, the forest defenders acted on the grounds that Maxxam was cutting endangered marbled murrelet habitat without having first completed wildlife surveys required by the California Forest Practice Rules. They asked the loggers to stop cutting until the California Department of Forestry (CDF) arrived to do its job of enforcing the murrelet rule. Earth First! videographers captured an angry logger chasing and threatening activists as they confronted workers. A logger can be heard stating that the harvest was legal and that CDF had approved Maxxam's activity, but one week after the confrontation, the facts were confirmed: CDF cited Maxxam for cutting too close to murrelet habitat at the site and illegally changing its THP without consultation.[1] In other words, before any blood was spilled, it had been a typical day of claims and counterclaims out where logging men meet forest defenders on the front lines of ancient redwood defense.

I followed these events on the Internet as if I were at the local newsstand every morning—at that time a relatively new experience in fieldwork. Corporations also reach for the power of this growing public sphere, as do corporate watchdog NGOs and social movement organizations, by generating their own Web sites and newswires. Now all parties compete with newspapers in creating and managing publics online and in print. In the timber wars we find an Internet-driven explosion of such micropublics struggling to interject. The implications of this lowering of barriers to entry into public discourse structured my experience from the start, as the total output of these voices in Humboldt is staggering. Each one puts a special claim on attentions and can function as a source for picking up cultural transmissions of established power and surging resistance.

As I tuned in to these timber war channels, first through the stories of David Chain, then in succession those of Julia Butterfly Hill and the place of Maxxam in Scotia, one scene in particular caught my attention—a logger kneeling in prayer with Earth Firsters next to Chain's fallen body. The allegorical force of that image jolted me into ethnographic action.

Reported first in the *Eureka Times-Standard,* it appeared again in the *Nation.* "The faller who'd been roaring threats," wrote Alexander Cockburn, "came up, saw what had happened and fell to his knees in prayer."[2] Then *Rolling Stone* reported the story after interviewing the logger, A. E. Ammons, fifty-two, and five other activists who were with Chain in the woods that day. Ammons revealed that he had seen his father put a gun in his mouth and blow his brains out, that he had been logging for more than thirty years, and that in 1967 he earned $64,000 but "now I gotta bust my ass to make half of that." When asked, "Is there anything you want to say to David Chain's family?" he replied, "Like what? You mean, like 'I'm sorry'? That's fucking obvious, isn't it?"[3] Later one forest defender who was at the death scene gave me a collection of video clips that included the original footage. He confirmed that the moment of prayer did occur, but it was extremely brief; Ammons fled in terror quickly after discovering the body. In interviews with loggers, logger families, and Maxxam employees who knew Ammons well, I learned that he was still working in the woods two years after the event; they all said he had suffered terribly.

The activists present with David Chain that day were Carey Jordan, twenty-six; Zoë Zalia, in her mid-twenties; Jeremy Jensen, sixteen (these went high or uphill of Ammons as the tree was cut); Michael Avcollie, twenty-eight; Erik Eisenberg, thirty; and Jason Wilson (these went low). Jennifer Walts, otherwise known as Remedy, and Chain's girlfriend at the time, was feeling sick and stayed back at the camp. The group made three cat-and-mouse approaches to Ammons as he tried to cut. When Eisenberg confronted Ammons, he charged the protesters, screaming, at which display they fled the scene. A few minutes later they approached him again; when Jordan confronted him about the cutting, he said, "Fuck it, it's our forest; we can cut whatever we want," and "I'm not going to hesitate like I did last time." The activists then left, had lunch, and came back a third time after ten, fifteen, or sixty minutes had elapsed, depending on who was making the crucial estimate. In the final and fatal confrontation, the activists split into two groups. Ammons said he thought they were gone; the forest defenders claimed he knew they were there. Either way, the tree was cut and found its mark. The day before, a group of activists had entered the site, and the same logging crew had responded to their presence by quitting operations. Instead of risking lives, they called the sheriff and had them arrested. "Yeah, Maxxam sucks," said Ammons to Carrie Jordan as the police took her away, "but I need a job."[4]

The Deal

September 1998 had already been a busy month in the timber wars. The Save Headwaters Forest campaign appeared to be concluding in the weeks before Chain was killed. A decade of escalating protest rallies, blockades, lawsuits, and political bargaining had finally produced the Headwaters Forest Preservation Act (a.k.a. the deal), through which the federal and state governments would acquire roughly 5,600 acres of Maxxam property, including 2,750 acres of ancient, undisturbed redwood forest located at the headwaters of Salmon Creek, in exchange for 7,700 acres of adjacent cutover lands and $300 million in cash, a compensation worth nearly $500 million.[5] The state would then acquire cutover lands adjacent to the old growth and establish the 7,400 acre Headwaters Forest Reserve. Earth Firster Greg King had discovered and named Headwaters in 1986, just after the Maxxam takeover; thereafter the grove functioned as the symbolic center of Humboldt's intensifying environmental conflict for the simple reason that it was the largest intact ancient grove. In contrast to the landscaped mountains of industrially managed monoculture tree farms that now surround Headwaters—on which Maxxam used clear-cutting, thinning, and pesticides to grow redwoods as an agricultural crop in efficient, even-aged stands—Headwaters Grove was a fragment of pristine forest.

Chain's death did not derail the preservation deal; only two days later, on September 19, 1998, it moved another step toward completion when California's governor Pete Wilson signed on to the act, adding $242 million in state funds to $250 million already allocated by the federal government. The deal would preserve Headwaters and several other so-called lesser cathedral groves by paying the company to abandon its constitutional Fifth Amendment takings lawsuit against the federal government, filed in May 1996.[6] But only Headwaters Grove and three hundred acres called Elk Head Spring Forest would be immediately and permanently acquired; the lesser groves would be granted temporary protection as "marbled murrelet conservation zones."

The Preservation Act was contingent on Maxxam producing an agency-approved Habitat Conservation Plan/Sustained Yield Plan (HCP/SYP) for its remaining 200,000-plus acres—a massive biological survey, timber inventory, and forestry plan produced with help from the state and federal wildlife, fisheries, and forestry agencies. The HCP/SYP amounts to a set of special new rules for timber harvest on Maxxam property. At the center of the HCP process is what the Endangered Species Act calls "an incidental take permit."[7] Using the HCP, the company ostensibly promises to protect endangered species in one area in exchange for a permit to "take" the species in other areas; the permit allows the company to destroy both habitat and individual organisms already protected by the ESA. The SYP component addresses the problem of forestry methods on designated forestry lands, presumably ensuring that they will be managed to sustain the land's ability to produce trees, profitable business, and a jobs base in perpetuity. Under the quid pro quo terms of the Headwaters deal, Maxxam would be paid to accept the HCP/SYP, and the twelve named fragments of ancient forest on its property would be off-limits to harvest for fifty years while remaining Maxxam property; together with Headwaters Grove, the fragments would constitute an archipelago of old-growth preserves in a sea of industrial deforestation on which Maxxam would henceforth be freed to harass, harm, and kill endangered species.[8]

Environmentalists saw this "mitigation" deal as noxious in principle, while business folk and property rights activists described it as burdensome government regulation. In the middle stood those of practical inclination who saw it as a compromise benefiting all. The complexity of the HCP/SYP process was multiplied by the lack of trust between forest defenders, timber workers, and the corporation.

Forest defenders felt that the deal was merely an end run around the Endangered Species Act and a big payoff to corporate criminals. Accordingly, they authored a visionary alternative to the HCP/SYP and called it the *Headwaters Forest Stewardship Plan*. They argued that because the deal would only protect isolated groves, it was inadequate as an ecological plan for the region. Their alternative proposed a 60,000-acre forest preserve that would maintain a fully functioning old-growth redwood ecosystem.[9] But from the perspective of those who made the final Headwaters deal and signed it—representatives of the Clinton administration, the U.S Department of the Interior, the California Resources Agency, the Maxxam Corporation, U.S. senator from California Diane Feinstein, and U.S. representative from District 1 Frank Riggs—everyone would win because Headwaters Grove would be preserved, the protests would end, and the community could get back to making a living.[10]

It could not have turned out any more differently. Even as the *Eureka Times-Standard* ran the headline "Headwaters in Public Hands," with politicians and mainstream environmental groups declaring that "this is the best we can get right now," Josh Brown of North Coast Earth First! told reporters that it was "a tragedy for the forest, allowing Maxxam to murder trees."[11] Forest defenders rightfully claimed that it was *already* illegal to cut the groves—they were already protected under the ESA. The Environmental Protection Information Center (EPIC) of Garberville, the county's most active local nonprofit environmental watchdog organization and the foremost litigator against industrial timber corporations in the redwood region, explained their ongoing opposition to the deal in scientific, judicial, and ethical terms. Their newsletter of September 1998 claimed that the deal would not set aside enough acreage to actually preserve the species indicated and so would not satisfy ESA prerequisites for an incidental take permit. Furthermore, the deal was unnecessary because Maxxam's reverse-condemnation suit would certainly fail to prove in court that the ESA-driven murrelet regulations had taken *all* the value of its property. The purchase price was superinflated (more than half of what Maxxam had paid for all of Pacific Lumber in the original corporate takeover, when the company's assets included 200,000 acres, the town of Scotia, a fully funded pension plan, a welding corporation and an office building in downtown San Francisco). And finally, the moral logic behind the takings suit was akin to extortion by Maxxam, which was in effect holding the ecosystem and its species for ransom.[12]

As it happened, the deal was completed in March 1999. Maxxam sold 5,625 acres to the government for $300 million, including 2,738 acres of old growth at Headwaters, approximately the same acreage of cutover lands, 307 acres of old growth at Elk Head Spring Forest (east of Headwaters), and several additional fragments of second growth. Elk River Timber Company sold 9,600 acres of cutover land adjacent to Headwaters to the state for around $80 million, of which 1,700 acres were added to newly created Headwaters Forest Reserve to serve as a buffer around Headwaters Grove. And finally Maxxam received the remaining 7,900 acres of the Elk River Timber Company's transfer.[13]

But during the media spectacle churning on David Chain's death in the autumn of 1998, this final outcome was still uncertain. Earth First! and EPIC were struggling to derail the deal, and the loggers were concerned about how their lives and communities would be affected. Would people be put out of work? Meanwhile, adding to the tension, police were repeatedly attacking forest defenders who had set up camp and occupied the death site, which they had declared a temporary autonomous zone named the "the Gypsy Mountain Free State." Law enforcement moved in several times, using pepper spray against activists peaceably sitting and lying down in an effort to block loggers from entering the site and disturbing evidence on what forest defenders described as a yet-uninvestigated crime scene.

The timber wars were still on, in other words, even though the largest extant grove of ancient trees was at least temporarily safe. The deal that ultimately created Headwaters Forest Reserve had been signed, and it did more than just protect a few endangered species—it built a psycho-geographic landmark right in the center of Humboldt's landscape of social memory. It was not the first geographic compromise formation between the hegemonic culture of timber extraction and the insurgent movement of forest defense that the timber wars produced, nor would it be the last.

Tree-Sit

As the power brokers finalized the deal, one of its most vocal detractors protested from three hundred feet up an ancient redwood tree located on Maxxam property. In early October 1998, Julia Butterfly Hill marked her tenth month of what would become an epic two-year tree-sit in the redwood that fellow Earth Firsters named Luna.

North Coast Earth First! (NCEF), northernmost Humboldt's manifestation of the loosely knit international direct-action eco-defense movement known simply as Earth First!, discovered the giant Luna in fall 1997 and immediately chose to defend it with a platform in the canopy. The idea was to have someone in the tree continuously—a tag-team effort they hoped would keep the chain saws away and produce a media spectacle publicizing their positions in the field of timber war cultural politics. Though their action was technically an illegal trespass, forest defenders felt obligated to intervene; following the logic of civil disobedience, they held that the unjust exercise of socially legitimate authority by the corporation constituted a greater threat to the common good than the small trespassing infractions they felt were necessary to make the plight of the redwoods known to the world. Obstruction of business is justified, in this perspective, when it is part of an effort to preserve irreplaceable forests of two-thousand-year-old trees and protect species from extinction—especially when the threat to the forests is the activity of corporations that continuously violate the letter and the spirit of the California Forest Practice Act, the California Environmental Protection Act, the Federal Environmental Protection Act, and the Endangered Species Act.

Julia Hill, twenty-four, joined the Luna tree-sit team with no experience and little knowledge of its deep political context. A preacher's daughter from Fayetteville, Arkansas, she came to redwood country in the summer of 1997 during her long recovery from massive injuries sustained in a 1996 car crash. After hooking into NCEF, she volunteered to take a turn in the platform, making several short stays in Luna before climbing its branches for what turned out to be the long haul on December 10, 1997. One year later—on the day I arrived in Humboldt—she was still there protecting the tree with her life and broadcasting the message of redwood defense in every way she could. By calling radio stations and news reporters and campuses around the world and inviting journalists and filmmakers to visit her in the tree, she helped Earth First! make the Luna tree-sit a global media event.[14] As one veteran activist told a reporter on the hundredth day of Butterfly's vigil, "It creates a crisis and it really focuses attention on what is happening."[15]

Daniel, Shunka, Spruce, Bonka, Fischer, Sawyer, Rising Ground, Robert, Almond, Shakespeare, Puck, Zydeco, Major Resin, Mike, Nature Boy, Seppo, and others aided the tree-sit in various ways. Like "Butterfly," these are mostly forest names by which Earth Firsters shield their clandestine activities as well as symbolically shed their mainstream cultural identities, helping them to live according to a loosely defined alternative ethic. All over the United States, in England, and throughout the world, individuals learn about the Earth First! social movement and discover that to participate, all they have to do is act; the Earth First! movement is wholly constituted by individuals acting locally, directly, and without compromise under the global principle of resistance to destruction of ecology. The actual practice of no-compromise direct action in the defense of the earth is the only criterion for participating in the Earth First! movement. Although Earth First! publishes a print and online journal, its content and management are regularly contested in its own pages as well as at the regular regional and national meetings convened by affiliated Earth First! action teams, groups, and individuals. It is a movement based on an idea, operating on a quasi cell structure and dependent on participation that begins with individuals making connections, forming local groups, and starting direct-action campaigns. There is no membership process and no toll-free number to call. Consensus decision making at the micro-level of each campaign ensures a transitory individuality of Earth First! in every locale. Such a loose form of organization ensures that the movement remains as diverse as the local peoples and local claims that attach themselves to its idea. If anything can be considered the value most cherished by participants in the Earth First! movement, it is freedom from social roles imposed by the dominant culture—namely, capitalism. The Luna tree-sit campaign was typically Earth First! in the style of its public intervention in the redwood timber wars.

Fire in the Eyes

In addition to Chain's Death, the Headwaters deal, and the Luna tree-sit, another set of powerful images percolated through the public discursive space of the timber wars as I entered its field of cultural production. Like the tree-sit, it was another Earth First! direct-action cell that addressed the timber war public in a media spectacle of resistance to Maxxam. The drama had begun more than a year before I landed in Humboldt, but its discursive reverberations through the public sphere were still strong and active when I arrived.

On the opening day of the trial in which the U.S. Office of Thrift Supervision sued Charles Hurwitz for his part in the failure of United Savings and Loan, forest defenders barged into the Scotia headquarters of Maxxam's Pacific Lumber Company and locked themselves down, intending to paralyze the company and provoke a reaction that would publicize the connection between the infamous savings and loan implosion of the late 1980s and the local scene of redwood forest destruction. That connection was Charles Hurwitz; the U.S. government had spent $1.6 billion bailing out the federally insured investors in the Hurwitz-controlled bank United Savings and Loan, and now they were about to bail him out in Humboldt with the massive payoff for Headwaters. The forest defenders' action was boldly simple. Seven activists entered Scotia in the back of a pickup truck, exited the vehicle in front of the headquarters building, then ran in the front door and sat in a circle on the office floor, inserting their arms into interlocking metal pipes. "Hurwitz out of Humboldt; Maxxam's going down, we can't have this outsider ruining our community and cutting down the trees," said a young, white, female activist named Spring into the camera of the group's videographer.

It was a stunning act of nonviolent civil disobedience, designed, as one participant said, "to create a creative tension." But the sheriff's team did not respond as expected by cutting the pipes and relocating the activists to jail. Instead a uniformed officer announced, "If you do not leave, [we will] use pepper spray or chemical mace to extricate you from your steel cases; you have five minutes."

The protesters did not flinch. After the warnings, officers seized the heads of the young women and men and one by one pulled back on the hair and forehead to gain access to the face, forced the eyes open with rubber-gloved fingers, and applied the searing chemicals directly to the orbs using cotton swabs.

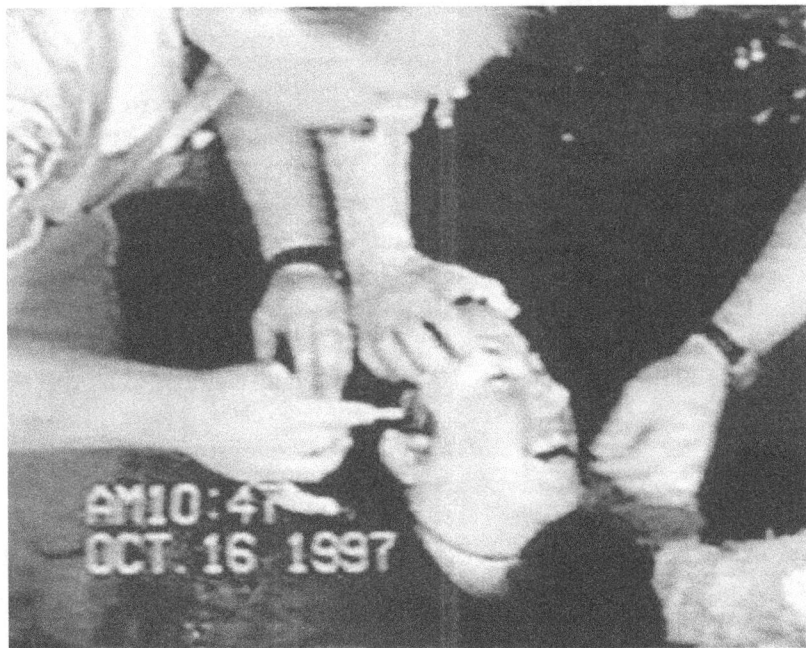

Police apply pepper spray directly to the eyes of forest defenders who are protesting the destruction of ancient redwoods by peaceably occupying the Eureka office of U.S. Representative Frank Riggs, October 16, 1997. From Humboldt County Sheriff's video; published in *Fire in the Eyes*, dir. James Ficklin, Headwaters Action Video Collective (1998, earthfilms.org). Courtesy of James Ficklin.

The effect was terrifying to watch on the police video that the Headwaters Action Video Collective (HAVC) incorporated into their 1998 documentary *Fire in the Eyes,* which publicized the sheriff's brutal new chemical "pain compliance techniques." Frames from the police video appeared repeatedly in the *Eureka Times-Standard* and circulated widely on the Internet, and *Fire in the Eyes* can now be checked out for free from the Humboldt County Library in Eureka.

Activists were appalled and united to oppose these new chemical pain compliance techniques. In the days after the lockdown at the Scotia headquarters, the same tactics were used against protesters on October 3, 1997, in the Bear Creek watershed and on October 16, 1997, at the office of U.S. First District congressman Frank Riggs. On the latter occasion, an Earth First! affinity group carrying a tree stump burst into Riggs's front office in Eureka, then sat around it in a circle and locked their arms together inside steel pipes. They begged the police to reject violence and claimed they were peaceful protesters, but by this date they knew exactly what to expect, and still they did not flinch—rather, they used the spectacle of police violence to once again further their claims. The affinity group brought its own videographer to record as much as possible before the police arrived, and police shot their own video of the scene. Lawsuits were filed, but an FBI investigation ended a year later in a ruling that the technique was legal and justified. Later, in a lawsuit against the sheriff and Maxxam charging that the two had colluded in political repression and torture in the Scotia office occupation, the plaintiffs argued that the pepper spray pain compliance technique was unnecessary and thus an illegal use of force. The forest defenders pressed their lawsuit for years, finally prevailing in April 2005, when, after two hung juries, first in 1998 and then again in 2004, an eight-member federal jury ruled that the Humboldt sheriffs had used excessive force in violation of the defendants' Fourth Amendment rights, rendering their actions unconstitutional. With each delay by the courts or by the police attorneys and each new rescheduling of the trial, the story was retold again in *Times-Standard* columns, archived on the Internet at nopepperspray.org, and disseminated through dense movement networks of communicative relations and associational life that help constitute the timber wars as a space of public discourse.[16]

The conjunction of the death of David Chain, the tree-sit at Luna, the pepper spray incidents, and the closing of the deal to save Headwaters Grove sealed my intention to enter the timber wars for an eth-nographic study of this exemplary localization of globalization. But as I have said, one image in particular galvanized my sociological imagi-nation—it assembled every element of the timber wars into the ham-mer blow of a single instant: a veteran old-growth timber faller who is cutting illegally in the twilight of Maxxam's conversion of ancient forests goes into a fit of violent rage and profanity against dialoging protesters, who are risking arrest in an act of civil disobedience and trespass on private property. He ends up killing a man trying to save an endangered ecosystem from the ravages of global corporate capital and hoping to scramble a mitigation deal that symbolizes all the ob-stacles that beset public efforts to shape local contradictions between capital and ecology. At the moment Chain dies, the perennially op-posed collective identities dissolve for a moment; loggers and forest defenders kneel together in prayer, but the gravity of the field—the cultural field of the timber wars—forces their retreat to conventional positions. A media storm erupts as voices from every position in the field address the public with opposing accounts, and everywhere this discourse turns on contestation of property rights. The instant of con-densation was irresistible.

At the first possible moment, I set out for Humboldt, carrying with me an e-mail alert put into circulation by Luna Media Services on November 24, 1998: "Please join us in celebration of Julia Butterfly's one year of continuous occupation of the ancient redwood Luna and in Protest of the Headwaters Forest Agreement and Habitat Conservation Plan." The rally was planned for Saturday morning, December 12. Ac-cording to my reasoning, it would convene every element of the forest defense—a perfect place to start my research. I arrived on Thursday, December 10, stayed in a roadside motel in Eureka, and by nightfall on Friday I had rented a studio apartment on the outskirts of Arcata for my first foray into the field of the struggle.

2. Convoking the Opposition: One Year and Counting in the Branches of Luna

California Highway 101 runs south out of Arcata along the eastern shore of Humboldt Bay, skirting the Green Diamond Company's (in 1998, the Simpson Timber Company) enormous redwood lumber mill sitting at the water's edge. The road then cuts across Eureka, the largest city in Humboldt, heading toward Rio Dell, Scotia, and Stafford, where North Coast Earth First! (NCEF) convened the rally celebrating the one-year mark of Julia "Butterfly" Hill's ongoing tree-sit in the canopy of Luna, an old-growth redwood high on a Maxxam ridge just south of Scotia.[1] On the outskirts of Eureka, I came across a hitchhiker holding a cardboard sign reading "Luna Rally." He hopped in, and I quickly learned he was a logger, a Vietnam veteran, sometimes a commercial fisherman, and a man genuinely concerned about the destruction of Humboldt's social and environmental ecology. He consented to my tape recorder and talked continuously in a serious, plaintive tone on our drive south to the rally.

"Timber workers don't think about anything but surviving," he said, explaining that he had been in Humboldt for thirty-five years doing just that. A button on his denim coat said "Veterans for Peace," but when I asked about the war, he said, "That doesn't matter; the real question is, from my point of view, I need to make a living and I work either in the timber industry or fishing." In fact, he was headed to Alaska shortly, "because the fishing industry has been wiped out here, you know—it's absolutely madness; you go out and the fish are not there." He was clearly saddened by his own complicity, stating, "I make no excuses, it's overfishing in the ocean fisheries." But it was more complicated than that, he said, "essentially the fishing has been wiped out by bad logging practices. . . . It's the road building practices in the forest." And in his estimation, "It's a crime, it's a crime to destroy the greatest migration of fisheries—it's incredible what we did, in just a few years' time."

His unbroken pathos was infectious; it overcame me with a psychical vertigo I simply describe as identification through loss. Two or more people share an experience of witnessing destruction, recognize a value in what is being lost, reveal to each other how they share that value, and in that revelation mutually recognize an image of themselves in the other, producing an identification that binds them together in an affective group of two subjects—two subjects transformed in the structure of their egos, at least to the extent that each acquires durable characteristics perceived in the other. Their collective determination of mutual loss constitutes them as a group; what is deemed lost becomes a symbolic channel through which their bond is constructed as such. Loss adds to loss as the conversation expands in recursive additions of weight to the sinking feeling the identification is producing.

In Humboldt there proceeds from this channel of communicative affection a widely shared feeling for ecological decline that pushes the moral senses toward the precipice of collective annihilation; it is an obsession that haunts the public sphere. In mainstream and left-leaning publications, including newspapers and radio, and in numberless conversations like mine with this fifty-something white logger, people convey their sense of the region's environmental and community degeneration. Oblivion is the horizon of these heartfelt encounters; species extinction, habitat destruction, unemployment, and negative social change in general become narrative symbols through which social subjects like this man and I can identify with each other as subjects who are subject to the grim and towering forces of previously obscure or misunderstood social totalities—for example, Maxxam, capitalism, or globalization, as they are actively perceived and conceived and progressively invested with psychical attentions, which can be measured in quantities of time spent thinking of these things and intensities of feeling and interest, which they construe qualitatively with values or aspersions. In the process, they identify themselves as subjects in opposition, creating the identifications that constitute psychical countercollectives of labor and environmental groups.[2]

My common ground with the logger is concern for the interlocked fate of human communities in the historical moment that we are narrating as the end of nature—it may not yet have happened, but our conversation moves us toward that idea as we confirm for each other that it is in fact happening, even as we look around and say, "Just look out there, look at what we've become, and look at the two of us," with clear-cuts dominating the mountains outside the car windows as we drive south toward Scotia, passing over the decimated Eel River, now blown out flat, hot, and wide by erosion from 150 years of logging, development, and upstream water diversions for agriculture, its young salmon deprived of cool, deep pools for safety, and its mature salmon left with few places to return to and spawn. Out here on the frontier of the redwood timber wars, the discourse is life and death, every day, all the time, with little relief on either side of the porous social-symbolic divide between the timber culture hegemony and the environmental social movement. And this concern of ours—this imaginative investiture in the narrative of decline—cannot be contained in this isolated place. It reaches upward and outward from its local ground and grasps at the world, which has long been imploding here along culture channels that saturate the timber war scene of cultural production with comparative images of faraway lands. This is the Lacandon forest; we are Chiapas villagers. Humboldt is in this way experienced as a microcosm of the world in failure.

I interrupted my passenger to note that we would soon be passing Scotia, Maxxam/Pacific Lumber's wholly owned company town, and asked him about the company's role in the ecological debacle he was describing. "As a matter of fact," he replied, "that's very interesting, because Pacific Lumber gets bad-mouthed, obviously because of Hurwitz and Maxxam and finance capital from Houston and all that, but the reality is LP, Louisiana-Pacific, was a lumber company that a lot of people worked for, you know they owned that town over in Samoa, they literally destroyed the company to get all they could immediately—they destroyed the company and then they said, hey, we'll sell it. So they sold it off to Simpson, one of the biggest timber companies; they destroyed a lifestyle; they wiped out the woods; they wiped out a good company and the workers." Indeed, I knew a bit of that story, and I remarked that his knowledge of the area seemed extensive—but he refused my gesture. "No," he said, "not a lot of knowledge, a lot of bitterness."

Why? "Because this country, this area, is the finest temperate rain forest on the face of the earth. There's nothing better for timber growing on the whole planet. It should have gone on forever as timber country. People could have lived for the foreseeable future cutting timber; but that's not happening because, well, finance capital, Harry Merlo, what happened to LP—and you know what people think about Pacific Lumber. We're going to this rally now because of Pacific Lumber, but essentially the same thing Pacific Lumber is doing now already happened with LP."

In 1998 the Louisiana-Pacific Corporation still owned the last operational pulp mill in Humboldt, located on the Samoa Peninsula adjacent to Eureka, as well as the flake board plant on the outskirts of Arcata, which has the distinction being Humboldt's most highly polluted industrial site. By the summer of 2002 both these facilities had been sold off. Louisiana-Pacific had completely cut its redwood timberlands by the 1980s and mostly sold out by the 1990s, both in Humboldt and in Mendocino County, where it once held more extensive acreage.

By point of comparison, Harry Merlo of LP had been the first Charles Hurwitz of Mendocino; he was the out-of-towner CEO whose bottom line called for crushing local unions and leveling the forest, generating high social costs from environmental externalities and finally deindustrialization, even exporting an entire factory to Mexico. He sold out in Mendocino largely to the Fisher family of the Gap clothing fame, the second Hurwitz of Mendocino. The Fisher family's simultaneous ownership of the Mendocino Redwood Company, which it formed from the remnants of Louisiana-Pacific, and the Gap and Banana Republic clothing chains, which had long been targeted by labor activists for using sweatshop labor in places like Saipan, united Mendocino's antilogging resistance with antisweatshop labor activists in the burgeoning antiglobalization movement in the same way that Hurwitz's acquisition of Pacific Lumber and then Kaiser Aluminum established a bond between redwood forest defenders and United Steelworkers.[3]

At the Seattle antiglobalization protests of late 1999, Gap stores in the city were targeted because they symbolized what both unionists and forest defenders see as a signal problem with globalization: companies made rich by exploitation of U.S. labor and environmental resources take their wealth overseas as a form of power to exercise over foreign communities of labor and environment, cycling the profits back home to be exercised in the United States as additional leverage over remnant communities of labor and environment. A global cycle of accumulation is created that avoids accountability by moving between and above national regulatory systems. Communities of labor and environment at home and abroad are left degraded. In 2008 the cycle turned again when the Fisher family's Mendocino Redwood Company (MRC) picked up the pieces of Pacific Lumber Company after Maxxam put the company into bankruptcy. But by this time things had changed. MRC had successfully changed its image by favoring selective logging practices over clear-cutting, preserving old growth, restoring streams, and gaining sustainability certification from the respected Forest Stewardship Council. One reason that the bankruptcy judge agreed to MRC's acquisition of Pacific Lumber was its promise to extend these sustainable forestry practices to the new company it would form in Humboldt from Pacific Lumber—the Humboldt Redwood Company.

But you're a timber worker yourself, I said to my passenger. Are you working for one of these companies now, or are you laid off and out of work like so many others? "Yeah, there's a lot of people out of work," he said ambiguously, "and I got a choice whether or not I want to, as a displaced timber worker, get a job doing some kind of rehabilitation work, but more likely I'm going to go to Alaska to do some fishing." Do you think that most other timber workers are like you, concerned and engaged in these matters? "No," he answered directly, "most timber workers are ignorant and scared because their whole life is cutting in the woods or working in the mills or whatever, and they can't . . . it's hard to understand that you're screwed, that you're just a victim of finance capital." Then he described the scheme under which it makes economic sense to borrow money cheaply and buy any timber company with land that has been sustainably logged and its resources conserved. A company that has saved its old growth will be trading under value on the stock market and will thus appear as a prime target for acquisition by corporations that seek out and liquidate underperforming assets, even if that means razing a forest and dismembering a community of labor, converting its potential to perpetually nourish local culture into payments on newly created debt. Labor and environment are pushed ever further toward the status of mere means to an end from which both are alienated by design. "You know," he said, becoming the first among many who would tell me the same story about the mythical company that finance capital destroyed, "Pacific Lumber had a reputation as the only company that did it right," but "that was all destroyed by Maxxam—Maxxam was no different from LP, no different than the timber industry."

Concerning the acquisition by Maxxam, I mentioned the charges that Pacific Lumber's fully funded pension fund had been raided by the new managers. How could the employees let that happen? "They're very dominant over the workers," he said. "You do what the company tells you and don't think about nothing." You would think they would form a union, I said. "There was never really a possibility of there being a union there," he replied, "because there was some kind of a family, everybody thinks, 'Wow, we're all together in this town.'" He went on to describe with obvious respect the intimate bond these people shared. "There is a whole culture of people," he said, who "worked for generations in Scotia; that was their family, their life, Pacific Lumber was all they ever knew, or wanted or cared about, because it provided everything—and when Maxxam came in, they tried to say, 'We are simply continuing as it was.'"

We were passing Scotia. From the highway heading south, you look down to your right over the roofs of two giant lumber mills, the factory, the power plant, the drying yard. Around the little town center is a tightly held enclave of small yellow houses extending for about a mile along the banks of the Eel River, about 270 in all. "Still a nice place to live," he said with a nod. "Nice houses." But Scotia is a "paternal" company town, he said, the last one that he knew of in the area, probably anywhere, and sadly "there is no way of knowing what the future of Scotia is," because "Maxxam is interested in short-term capital gain; they're going to get everything they can, as quick as they can, then they're going to sell it off." When we pulled off Highway 101 at the Stafford exit a few minutes later, he noted that the crowd was already in the hundreds. We got out of the car, and he melted right in. I never saw him again.

For this embittered, out-of-work veteran logger, Scotia is a symbol of a golden past—once admired but now being lost, a whole way of life on the chopping block of global finance. In the months that followed, I learned the depth of this narrative and its constitutive role in forming the opposition to Maxxam. Humboldt's resistance to globalization, not limited to forest defenders, is grounded in the experience of corporate invasion. People are obsessed with Charles Hurwitz, Maxxam, the power of corporate accumulation to disassemble communities of labor and the environment with seeming impunity, and the bottom-line motive of profit that appears to be driving the conflict. Hurwitz puts a face on the region's collective sense of anxiety about the future—giving the redwood imaginary both a villain and a possible hero—"if only he could be made to understand." He is a symbol of both fear and hope for environmentalists and loggers alike.

I had no chance to discuss the hitchhiking logger's concern for the figure of Julia Butterfly Hill. In our twenty-minute drive, he never brought the subject up. Nor did he mention the killing of David Chain. But he did place Scotia at the center of a struggle that cannot be understood outside the context of deep labor history. In future months of field research—in interviews and conversations with loggers, their families, and timber industry supporters—I learned that most everyone carries a charged mental image of Hurwitz and his company. I learned that Scotia inhabits the center of the hegemonic logging culture's symbolic universe. And I learned that by explicitly linking these struggles together, Julia Butterfly's media-savvy tree-sit posed a major public challenge to the man, the company, and the moral legitimacy of Humboldt's domination by corporate timber.

Earth First!

When I entered the crowd, the forest defense rally had already begun. I walked to the front and set up my video camera about twenty feet from the makeshift stage. My camera was one of at least thirty in the crowd. There were local small-channel corporate news cameras, independent filmmakers, and at least one lecturer from Humboldt State University using a handheld camera. Others had 36 mm cameras or audio devices to make their own record. Having a camera did not single me out. It felt like history in the making, and everyone wanted to spread the word and the image.

Darryl Cherney was on the makeshift stage singing "Maxxam on the Horizon," the lyrics of which tell the story of how Charles Hurwitz purchased Pacific Lumber, the litany of legal charges and lawsuits brought against his business dealings, and his goal of harvesting every ancient tree he owned in Humboldt. Cherney is a long-time Earth First! organizer and victim alongside Judi Bari in the infamous car bombing of 1990. After holding up a pair of handcuffs and a miniature model of a colonial-era stockade while leading the crowd in a chant—"Stocks and bonds for Hurwitz, Stocks and bonds for Hurwitz"—Cherney introduced the song as "the life story of Charles Hurwitz in two and a half minutes."

In the town of Kilgore, Texas
Was born a tailor's son
From the killing of the Indians
He learned how the West was won

His name was Charlie Hurwitz
And he terrorized the land
His killing field was Wall Street
And his gang was called Maxxam

Maxxam's on the horizon
And there's dollars seen in every redwood tree
Maxxam's on the horizon
Ain't no one safe with Hurwitz runnin' free.[4]

It was funny, hard-hitting musical social criticism, and the crowd loved it. Cherney's musical antics embody what forest defenders in particular, but the direct-action movement in general, including anti-globalization groups like Art and Revolution, whose name says a lot, think about the role of music in social movements. In the words of the deceased Earth First! organizer Judi Bari, "Music is a really good organizing tool. . . . I think a movement that's held together by music is way stronger."[5]

At this rally and others, Cherney practiced what might be called the biopolitics of resistance; his musical persona doubles as a personal public address system and adds the special effect that music has on human emotions to the symbolic resources of the movement—it binds people together. Music is used as a collective, ritual address that reaffirms emotional bonds of those converted and offers a mode of conversion for newcomers like myself.

In the terms I develop in, through, and for the case of Humboldt, music is understood as a rhetorical style of identificatory resource mobilization; it channels the psychosocial economic attentions—that is, the interested, attentive, affective labor—of subjects of resistance into a world historical project of reversing the symbolic order of established reality. It aims to rename the world. The patriotic country music and nationalist anthems that reign on Big Red 92.3—the only FM radio station my car could tune in while in Scotia—perform the same type of identificatory service for the town's prevailing culture of blue-collar conservatism and the region's logging culture as protest music does for the movement. Whereas the music of the movement is negative and critical, expressing all the complaints that constitute the forest defense idiom, the symbol-laden country music and patriotic songs of nationalism reaffirm timber culture in a positive register, celebrating working-class masculinity and militant nationalism.

Scotians and timber supporters in general often respond directly to the challenge of environmentalist rhetoric with the music and rhetoric of patriotism, salutes to the national flag, and appeals to symbols of freedom that never range far from the civil rights of property. On the other hand, forest defenders have been seen dragging the flag on the ground and hanging it upside down, an expression of willful dis-identification with the nation that must be read through the immediate context: "If Maxxam is what the nation means, I'm against it." They want to change the nation. By reversing the symbol that established power has claimed, they simply and publicly imagine reversing the power. Publicity, again, is the crucial dimension of this tactic for the social movement of forest defense.

Also present and recording at the rally was KMUD, the local Redway-based community radio station that has long supported the environmental movement. "K-mnd" broadcasts *The Environment Show* and other programs sympathetic to forest defense, unionism, anti-globalization, and the gamut of global peace and justice movements active in the area. And Headwaters Action Video Collective (HAVC) was there, at one point taking the stage with camera in hand to let the crowd know they were making a movie. Their previous production, *Luna, the Stafford Giant,* documented the origins of the tree-sit and doubled as a promotional video for the effort.[6] Some days after the rally, Robert Parker, director of Luna Media Services, gave me a copy of *Luna* and a press kit titled "Headwaters Forest Information Packet"—the last of dozens he said he had passed out to journalists. Inside were feature articles on Butterfly from *Time, People,* the *San Francisco Examiner, Newsweek, Good Housekeeping,* Italian *Elle,* the *Los Angeles Times,* the *Times* (London), *Le Monde,* and *Jane* magazine, as well as numerous articles in the *New York Times,* the *San Diego Union Tribune,* the *Eureka Times-Standard,* and other papers.[7] Lunatree.org was the Internet address given for hooking into the campaign and movement.

The redwood forest defense, it turned out, was on the cutting edge of the digital revolution in social movements in the years leading up to Seattle 1999 and the Internet-driven explosion of antiglobalization activism that captured the world's attention in the early years of the twenty-first century, before 9/11, the so-called global war on terror, and the invasion of Iraq drew much of the movement's energy into antiwar activism, at least temporarily. We should echo Marshall McLuhan in saying that the medium of turn-of-the-century video activism is integral to its message.[8] The message coming out of Humboldt was not just "Deforestation! We must resist!" but "The whole world is watching this specific forest being cut down right now, almost in real time—look what we can do with this technology!" A whole new business environment was created as the movement went digital, and before long Maxxam would have its corporate Web site up and running and trying to compete in this new public sphere. Police also got into the act and started showing up at rallies and civil disobedience armed with their own video cameras. Before I completed my fieldwork, the network movement born in Seattle had spawned the Independent Media Center (indymedia.org), the media movement most emblematic of the Internet's capacity to transform globalization by arming resistance with the organizational wizardry of instantaneous global image messaging not beholden to corporate interests.[9]

The omnipresence and convergence of these technologies in the forest defense campaign are one manifestation of its collective will to transform every level of individual, local, state, federal, and international collective consciousness. The great hope is that showing people what is happening will help create an ecological understanding of the interconnections between ancient forest destruction, species extinction, industrial forestry, the power of multinational corporations, and the unfolding consequences of so-called free-trade globalization, in which *free* largely means free from accountability for externalized costs of production—free, in other words, from labor organization and environmental regulations that restructure unhindered price competition in capital culture. Public intervention in the same global semiosphere through which and with which free-trade globalization propagates itself as the desired ideal is the movement's prevailing tactic. Simply put, the movement works to discover the facts and package them for visual and narrative projection into public-sphere discourse.

This means broadcasting, publishing, and uploading photographs and video not just of clear-cut ancient forests and salmon streams filled in by landslides and other scenes of destruction associated with industrial logging, but also of testimonials by committed agents of resistance and the courageous, solitary acts of defiance and movement-building solidarity by local citizens who dare to reach out across social divisions of race, class, gender, and profession. Making connections across all social divisions—especially the gulf between workers and environmentalists—is a dedicated practice. Forest defenders know how corporations and the state have long employed strategies of divide and conquer for stifling challenges made from below.

But such images are narrated as well, and the story told can be summarized as follows: the destruction of local public-trust values by agents of globally vested private property rights. Environment must be defended locally, but global support and consciousness are needed—an emergent public is every campaign's most sought-after resource. It is commonly held among forest defenders that if people on the outside could only *see* the destruction of these forests and hear the evidence against Maxxam, they would not be fooled by the company's ideological screen, its greenwashing commercials, its bought-and-paid-for timber science (comparable to "tobacco science," as several of my contacts described it), and its all-around claims to good environmental stewardship. Environmental writers including David Helvarg and Sharon Beder have cited Maxxam's hiring of the global public relations giant Hill and Knowlton as an early example of how corporations were learning to invest in greening their public image, a phenomenon that shows how the forest defense movement, like every social movement, must be understood as a struggle for power over the mental environment.[10]

Although the rally's symbolic discursive atmosphere—its narrative environment—was saturated with images and concepts of extinction, destruction, deadly force, corporate malfeasance, and bloody martyrdom, the mood was celebratory, something I came to understand was characteristic of Earth First! protest events. The rally is one of the campaign's strongest rhetorical practices, consisting of a variety of performances. Musical acts that convey multiple positions and complex political consciousness are interspersed with speeches by activists from every conceivable position in the field. The timber watchdog nonprofit Environmental Protection Information Center (EPIC) of Garberville in southern Humboldt usually has representatives on hand to report on legal developments in its preservation efforts and lawsuits. Other nonprofits and ad hoc organizations such as affinity groups engaged in particular microstruggles often take the stage to plead for resources, money, supplies, attention, and people power. But most of the cries are for network formation. And there is never too much of this serious business before the rallies return to revolutionary folk music. These social movers know their strongest asset is an invitation to pleasurable identification through playful group experience that counters or mitigates the overwhelming tendency for identification through loss and death that might otherwise cripple the movement's intelligence with pessimism.

Earth First! rallies are what Mikhail Bakhtin and his followers in cultural studies might call a Rabelaisian politics of the carnivalesque, where individuals live for a moment outside the established culture system's imposed hierarchy of social roles, participating in a festival of renewal that hopes to regenerate solidarity through laughter, music, and ironic criticism—they mock official culture while remaining dead serious. Of course, the change they envision is permanent and reaches the core of American lifestyle, namely, its excess. The message rarefied to a slogan might read: "Act locally by living more simply while thinking more globally and having more fun while you're at it." This is revolutionary pleasure as a new ethic of community song, dance, art, and resistance.[11]

At the Luna rally a woman could be seen carrying a not entirely whimsical sign over her head reading "Free Scotia": combined with the marketplace good humor, the folk-musical atmospherics, and the eco-populist rhetoric of resistance to global corporate power banging in the air, it was hard not to feel the crowd collectively acknowledging the brief genius of this catchphrase. It sums up the complex understanding that logging culture in Humboldt amounts to a corporate timber monopoly on established reality.

Drawing its meaning from the immediate semiological environment, the sign could be read "Free the Scotian people from the ideology which forms their unconscious consent to Maxxam's social and ecological malfeasance." But most forest defenders understand what the hitchhiking logger told me—that woodsmen and their families have little power over their paternalistic employers. Seen as a critique of class power, "Free Scotia" means "We'll be on your side when the time comes for you to stand up and claim your power as the collective structure that Maxxam relies on for the labor energy it uses to profit by blasting value from the forest into redwood commodities." Many forest defenders might not articulate such theoretical language, but most I encountered claim solidarity with any workers who stand up for labor rights *and* sustainable forestry. As for the workers and residents I interviewed, they know that change is imminent and that the company's bottom line works against their own interest. "I'm just

Sign held by forest defender at the one-year Luna tree-sit rally, December 12, 1998.

getting mine before the trees are gone and they shut it down," one Maxxam mill worker told me; at the time he was thankful that instead of letting him go in the last round of layoffs, the company had kept him on at the desk in the Scotia museum. "Free Scotia" is not an attack against workers or an arrogant presumption; it means that a worker and environmentalist alliance starts on the ground in the real places where labor is done in the world.

By reducing an entire sociological analysis of corporate social control and timber hegemony to a two-word phrase, the sign is a perfect example of the culture jammers' first principle of communication. The Ruckus Society's "Media Manual" put it like this: an effective direct action "is like a freeway billboard, designed to hammer home one—and almost always one—message"; it is a message that "with one glance . . . is (or should be) unmistakable." Each action and phrase should take the hegemonic charge of established processes and terms and transform them into movement resources. We can feel the term "free" in "Free Scotia" taking this turn. We should not be surprised to learn that the Ruckus Society was cofounded by Mike Roselle, a founding member of Earth First!, and it was not incidental that I found copies of this "Media Manual" in circulation among Humboldt's forest defenders.

When Cherney ended his song and left the stage, Robert Parker of Luna Media Services stepped up and gave a rousing account of Julia Butterfly's mission. Luna, he said, "is a symbol of our resistance to the Pacific Lumber Company and the Maxxam corporation and their rapacious forces. Luna is a beacon of hope for our future that one day we can have a sustainable forest and a sustainable economy." The crowd broke in with howls of approval. "It's only one tree, but look at what it means to us, look at what it symbolizes," he said, "it reminds everyone of what else is out there." There are many thousands more redwoods slated to be "sacrificed under the Headwaters Forest Agreement and Habitat Conservation Plan," he said, and Luna symbolizes the forest defenders' continued opposition to that plan. Activists oppose the deal precisely because it *sacrifices* thousands of acres, trading off the best of the best (Headwaters) for just about everything else. "Luna is symbolic of hundreds and thousands of other steep and unstable slopes that are not even being addressed by this Habitat Conservation Plan and are going to be destroyed and in turn destroy more salmon and more towns," he continued, referring to the hamlet of Stafford, California, which at one time would have been visible from the site of the rally. A mudslide originating in a Maxxam clear-cut on December 31, 1996, destroyed seven of its houses.[12]

The destruction of Stafford symbolizes how poor logging practices hurt people directly, but most of Maxxam's land was uninhabited, mountainous, forested watersheds. The real power of the Stafford symbol in Parker's address was the equivalence it proclaimed between the homes of people and the homes of salmon. Bad logging destroys both human and animal habitation. Repetition of the Stafford narrative by forest defenders formed a community interest between people and salmon—we are all in this together, we are all under assault by the same corporation.

Parker then gave the crowd some recent history of Maxxam in Humboldt: "A lot of people, including company official president CEO of Pacific Lumber John Campbell, have said that all of the attention we're getting and this very action [the Luna tree-sit] itself is bizarre," said Parker in a rising voice. "What I think is bizarre is the fact that we're cutting down two-thousand-year-old trees. What I think is bizarre is that there are not seven homes standing over here in this clearing anymore. . . . What I think is bizarre is the fact that this company was allowed to operate for the past three years committing three hundred violations of the California Forest Practice Act and only recently had its license suspended."

Then he spoke the words that have resonated with me throughout my investigations, helping me recognize how the concrete articulation of property law in Humboldt is a symbolic expression of modern individualism and a cultural constant in the region's libidinal economic body politic—a specific formation in its legitimate authority and the very anchor of governmentality in the redwood social imaginary. "Julia has been called a criminal," he said, "as have many of us, for trespassing on Pacific Lumber land. But let's look at the record here, let's look at Julia Butterfly's criminal record—she doesn't have a criminal record. Let's look at Pacific Lumber's in contrast; nine misdemeanor-level criminal convictions within the past three years for violations of the California Forest Practice Act. *Who's the real criminal here? Who's trespassing on who?*"

The emphasis was his, and it was well received by the hollering crowd, because it identified a basic fact of hegemony in Humboldt. On one side you have fines, jail terms, and chemical pain-compliance measures for individuals trespassing on private property in explicitly political direct-action civil disobedience that potentially raises the cost of production and embarrasses the company as it possibly mobilizes public support for the movement. On the other side, regulatory agencies and law enforcement look the other way and allow repeated corporate violations of the California Forest Practice Act, the Endangered Species Act, and the California Environmental Protection Act, the cumulative effects of which include the killing of endangered species, elimination of their habitat, sedimentation of rivers and Humboldt Bay, and the loss of salmon spawning grounds—in short, the widespread destruction of manifold public-trust values, or what is the same thing in this case, their conversion into profits for private accumulation.

Parker's juxtaposition of these criminal records exposed the logic of domination at work. The institutions of property, represented by capitalists—in this case Maxxam—cannot stop themselves. They profit by feeding on public-trust values while relying on social movements to regulate their appetite and so ensure their future.

Direct-action environmental media campaigns like the Luna tree-sit work by way of their worldwide transmission of such juxtapositions; they articulate the movement's collective desire for ecological challenge to capitalist accumulation and realize their effects in the redistribution of consciousness. In this way, they function as an information feedback mechanism for the dominant culture system, without which the ecological conditions of its reproduction would likely be consumed more quickly. The system would, in other words, destroy itself sooner if not for the forest defense. Direct-action campaigns attenuate the so-called second contradiction.

In terms that Alberto Melucci has given to social movement objectives, Parker's juxtaposition—"who's trespassing on who?"—reverses the symbolic order and creates a site for resistance to grow.[13] When such reversals are broadcast, consumed, filtered, transformed, and absorbed in the repertoire of rhetorics and practices constituting a movement's symbolic, expressive, and therefore material idiom, they aid the formation of distinct and effective counterpublics.

In the trenchant terms of Michael Warner, a counterpublic forms itself by engaging in public address and always maintains, "at some level, conscious or not, an awareness of its subordinate status. The cultural horizon against which it marks itself off is not just a general or wider public, but a dominant one."[14] Every new counterpublic exists only to the extent that people withdraw some portion of their attention from previous concerns and invest it anew in the new collective object or project, the new counterpublic.

Describing avant-garde art movements and reiterating Warner's notion of self-generating publics, Sven Lütticken says simply that each public sphere is "a fiction that creates its own reality."[15] This is true for discrete social movers, local campaigns that feed into larger movements, and the big movements themselves, like the environmental movement. Lütticken speaks of nested publics of relative autonomy in terms appropriate to the complex layering of movement publics that accrue in the timber wars: "While each of the specialized domains [i.e., Weberian spheres of art, science, and law] has its own, internal semi-public space—constituted by art journals, or scientific conferences, for instance—they also require a general public sphere to mediate between themselves and the rest of society. It is through this *Öffentlichkeit* [public] that the smaller spheres have effects on the 'outside world.'" Movement counterpublics use the space of writing, speaking, rallies, and actions, hopefully reported as news and disseminated across other media, to address, create, and transform their own publics, colonize more general or wider publics, and so influence society and the world at large by feeding back grassroots sentiment into the political body—the so-called general public.

We must remember, in taking this approach, that individuals, as addressees, are not and cannot be merely passive receptors of such an address or indeed of any discourse or system of laws whose force they encounter. On the contrary, cultural consumption is always cultural production, in the words of Michel de Certeau, and for Jacques Derrida reading is always (re)writing.

Such an understanding is crucial for the linguistic and communicative metaphors that social theory uses for culture and society: the objective order of institutions, encountered by every individual as a structure of meaning-making systems that mediate human engagement with the world, functions not just as the repressive and organizing, governing force of law but also as the very condition of possibility for participation in social exchanges that constitute cultural practice. Just as a system of language is transformed and reiterated by every speech act it makes possible, so are social systems of institutional structures transformed and reiterated in every act of social exchange, each of which necessarily fails to fully meet the institution's implicit demand to repeat its ideal form, but which nevertheless maintains its form, at least to some extent.

By viewing tree-sitting as a collective form of public address, we can relate it to the cultural turn in social theory and social movement studies. As a tactic, any material effect it achieves will register in the forest, where trees might be saved and watersheds restored, or in the consciousness of those whom it manages to reach. Alberto Melucci has described such efforts to change consciousness as the primary goal of new social movements in the age of information. They strive, he wrote, to "reappropriate the capacity to name through the elaboration of codes and languages designed to define reality, in the twofold sense of constituting it symbolically and of regaining it, thereby escaping from the predominant forms of representation."[16] Movements operate at this level because naming is central to the social construction of reality. Representation controlled by dominant models, he says, governs people's relations to the world in ways that perpetuate their domination. Representation and reality tend to lose their distinction, and domination sinks below the threshold of consciousness. In this sense, domination proceeds by colonizing the subject—by moving inside it, inhabiting its consent to the dominant culture's naturalizing, unconscious patterning of the desires and ideas that channel the energies of its labor on the land.

All social movers must contend with this power of internal social compulsion. They themselves have been named and grew up in a world that had already named everything and was already operating as a self-reproducing system of naming and representation. But their actions reaffirm what history has proved—that critical self-awareness and sensitivity to emergent conditions, combined with the shock of contingent events, can transform the tendency of individuals to become passive ideological machines of social reproduction into a tendency to become subjects of world historical social action. As emergent subjects, they can develop self-consciousness and begin to intervene, for example, by asserting themselves symbolically in the politicized timber war field of cultural production. They can start the hard work of renaming their world and providing alternatives that can change or even break the cycle of representation that creates and maintains concrete social relations.

The Luna tree-sit, for instance, considered as a collective gesture by the forest defense, is a complex operation in counternaming. It inscribes the tree with a personal name and thus symbolically extracts it from the categorical domain of private property. The tree becomes a place—a center around which to build a community of resistance, an immediate locale of critical practice with a global imaginary. By personifying the tree, befriending it, and risking their lives for it, defenders challenge the dominant cultural code of capitalist enterprise, which signifies the tree as a commodity ready for legitimate conversion to profit. By adopting the tree and making it a home, they make it a sacred place where the tree-sit collective can live and perform its rejection of the dominant worldview and set up what amounts to a mini-commune behind enemy (property) lines. This place can then function as a broadcast system for publicly pronouncing the alternative ethos or model for life that projects an alternative future for everyone.

Further, each redwood tree-sit amounts to a public invitation to join the effort, to cast away allegiance to dominant structures, even if for only a moment, and to join the community of resistance. Together with Earth First!'s other rhetorical forms—protest rally, street carnival, cat and mouse, lockdown, banner hang, and road blockade, to name a few—tree-sitting constitutes a unique language or idiom in the manifold discourse of forest defense. All these forms are signifying performances, meaningful productions of cultural material that actively redistribute the power to name within the redwood imaginary.[17]

Luna Media Services and the whole Luna tree-sit crew produced and maintained a distinct movement counterpublic for more than two years, transforming the horizon of Humboldt's forest defense. The movement imaginary and thus the wider redwood imaginary in which it is nested had once again been permanently marked—and not only because the tree, Luna, was eventually purchased by Julia Butterfly Hill for $50,000, therefore entering the landscape of social memory as a new psycho-geographic landmark in the built environment. The movement subculture of tree-sitting gained immeasurably by this action; its success was the movement's as well as Butterfly's, and its story will be told as long as tree-sitting survives as a cultural practice of resistance.

Butterfly

Even within Earth First! unique voices emerge, singular idioms that can tell us much. When Robert Parker finished speaking, he called Julia on his cell phone and held the device to the microphone. "Hello, Julia Butterfly," came the electronic response from the forested ridge overlooking the rally, and the crowd howled the signature Earth First! wolf cry loud enough for her to hear over the telephone. "Love, it's all about love," she said, before breaking into a song about love, the lyrics of which mostly were lost in static as the phone began to malfunction.

When the signal cleared, she was describing the symbol of Luna, perhaps unconscious of the fact that before her vigil was over, she herself would become the campaign's most powerful symbol, at least for a while. "Luna standing for over a year since she was longed to be destroyed by Pacific Lumber is the most amazing symbol of love that there possibly is," she said, effortlessly suturing the rhetoric of love to that of death and destruction in a provocative syntax that bore the impression of religious understanding untroubled by contradiction. "The fact that this earth gives us life every day," she said, "even though we do everything in our power to destroy that life, is love in its most amazing form." In a voice both plaintive and preaching, pleading and authoritative, earnest and enthusiastic, Julia Butterfly Hill captured the hearts of this crowd as she had many others and thousands, perhaps even hundreds of thousands, of individuals during two years

Julia Butterfly on the cover of an undated, photocopied pamphlet in circulation in Humboldt, December 1998.

of tireless radio, television, and rally appearances by cell phone, video link, and print media quotation.

Her words to the rally interpellated everyone. They called everyone into identification by articulating what Ernesto Laclau and Chantal Mouffe call "chains of discursive equivalence"—in this case equating every human being on the planet by using the most sweeping possible language of inclusion. This is the rhetorical style of environmentalism, the knowledge production of which is by definition totalizing and inclined toward the planetary scale.[18] Its central concept of ecology is to the discourse of natural science what globalization is to the discourse of political science and economics—a signifying technology that leaps over disciplinary boundaries to embrace the totality. Each becomes a totalizing historical discourse of community that reaches out for the other. Today global studies struggles to integrate environmental logic into its political economic language, whereas ecologists have long sought to extend their reasoning to political economics. Leaning against the stage at the Luna tree-sit rally, just beneath the microphone that delivered Butterfly's message to the rally and thus to its public extended by news cameras, filmmakers, and radio reporters, a hand-painted sign spoke out in big black block characters: "Extinction Is Forever," it said, in terms exacting another message of global subjective equivalence.

The main stage at the Stafford rally that celebrated Julia Butterfly's one-year anniversary of living in the branches of Luna, an ancient redwood on Maxxam/Pacific Lumber property, December 12, 1998.

Julia's voice continued, rising and falling in the practiced cadence of the pulpit as she used it to make the audience her own. "The amount of commitment from people hiking up and down this mountain to people donating food and supplies and money to help keep the phone going and doing outreach and the groups working to save all of the forest and protecting the earth and our lives: it's all happening because of love," she said, "because we love the earth and we love life and when you love it you protect it and you take care of it and you cherish it; it is one thing that all of us share in common no matter who we are, what size we are, what our gender is, our religion, anything and that is that we all share the same environment—whether we're here in the redwoods in California, the forest in Mexico, or in the forest in Australia, we're all sharing the same environment and therefore when we affect one part of the environment we're affecting the rest of life and we're affecting everyone else and what Pacific Lumber/ Maxxam corporation is doing here is not just affecting their private property, what they're doing is affecting ancient forests that have been here long before white people ever stepped foot on this country, they're affecting the air that we breathe the water that we drink and the very health of our lives and private property and our government has absolutely no *right* to destroy the environment that each and every one of us share." Her speech was breathless—animated, driving, gripping, allowing precious little reflection from moment to moment—but when she finally paused, the crowd again raised the Earth First! wild animal's howl.

In Julia's address to the crowd, private property, an institution that the nation's founders used to protect the individual from the intrusive powers of government and thus helped make into a singular symbol of American liberty, is resignified as the symbol of something else: accumulated power—power accumulated to the degree that it now governs governments, to an intensity at which it now forms a certain governmentality of its own. As articulated by Hill, it is an institution embedded in race and gender relations and shot through with the power inherited by aggregated redwood capital from Humboldt's colonial period of primitive accumulation through Indian genocide.

What right does private property have? she asked. Why does that question make sense to everyone here? How could property *have* a right?—because in Hill's discourse, which at least in this moment well represented the forest defense more generally, "private property" is the name of corporate power. Technically, the rhetorical move is metonymy: the term "property" is used to stand for Maxxam, which it-self is meant to be understood as just a particular manifestation of corporate capital. In this view, property culture has become a subject in itself; it has agency to the extent that it takes on a life of its own in the system of accumulation that feeds it and drives it forward. Hill sees private property exercising control over people, so she effortlessly substitutes the right that empowers the agent for the agent itself, a metonymic shift that expresses a widely if often unconsciously shared analysis of the movement: namely, that the consensus legitimating the right of private property is exactly what determines the power of Maxxam to make history in Humboldt. Maxxam is the face of prop-erty, power, and hegemony in the redwoods.

From this first moment of my immersion in the timber wars as an immediate field of cultural power, the concept of property imposed it-self with this kind of lucidity and plainness of expression. Property is, very simply, the contested symbol at the center of the timber wars— just as it had been, as we will see, in the redwood labor wars of the early twentieth century and in the Indian wars of the nineteenth, al-beit in different ways in each period.

"I thank you all for coming out today in the celebration of love and the amount of commitment that has helped save this over one thousand year old absolutely beautiful magical ancient redwood tree," said Julia Butterfly, interrupting the revelry inspired by her last rous-ing flourish, her voice now escalating even further, raising the stakes yet again, "but you know what, there's more, because Luna is just a symbol of all that we're fighting for." This is the conscious language of symbolism that both simplifies and complicates interpretation of the forest defense, forcing us to recognize how individual subjects trans-mit collective voice and desire without therefore losing their individu-ality. Against the hegemonic signification of Luna as property, Earth First! had given it a name, elevated it to the status of a living being with inherent rights of its own, and publicized its plight; they made it into a countersymbol through which a counterhegemonic subject could possibly and in fact did actually form.

Now she was directing that symbol back on the public in a bid to multiply its effect. "We have to come together louder than ever before," she continued, and now is the time, "because consciousness is shifting people; let's take this momentum and let's run with it, let's go to the people who supposedly have control over our lives and demand that they respect our life, respect our environment, and therefore respect our lives; we need to take legislation and make our own legislation, if they're not going to do it right, well, let's demand it, let's get in their face, let's not back down, let's make our demands on the quality of our life and the demands are for a healthy environment—that is where all ancient and old-growth forests are protected forever as the beautiful legacy and historical and international treasures and monuments that they are; that clear-cutting must be replaced with sustainable forestry; what we do today affects others, it affects their futures and it affects the quality of our lives; asking for sustainability is not asking for too much; we must ask that steep slopes be permanently protected, seven families' homes in Stafford are no longer there because of the disrespect of cutting on steep slopes." Then, without pause, the forest defenders' emergent symbol of love made the connection to its prevailing symbol of sacrificial death, harnessing the symbolic force of martyrs in the making to the identificatory project of building the movement's new public: "Does somebody else have to die before they start doing things right?" It was a reference to David Gypsy Chain—killed in the act of redwood forest defense.

In explaining what she meant by "doing things right," Julia Butterfly continued her work of rhetorically building a chain of equivalence between just about everyone in Humboldt. Her prescription articulated every position into a totality deformed by Maxxam. Here is how she put it: "No more dumping of herbicides, it's poisoning the land and the water and the people; all of these things are affecting our lives, we have to take a stand, we need to take a unified stand in love, raise our voices, raise our bodies in love and demand that the health of our environment and the quality of our lives be respected and protected because private property and the government have no right to destroy our lives—if we come together in love, then just as surely as that has helped Luna stay standing for over a year since she was marked to be destroyed, if we come together in that same love we will see real and permanent protection for our forests, but we have to do it now, we have to do it together, and we have to do it loud: LOVE!"

That final shout produced another, definitive moment of collective identification in the crowd. It was contagious. The acceleration of the energy was palpable—a real Durkheimian moment of effervescence in which the crowd's ritualized copresence with its sacred totem accelerated the pace of social interaction toward a psychosocial frenzy, reaffirming the movement as a community of obligation to the forest and, beyond that, to what Luna was said to symbolize—the global environment. Identifying with each other through the symbol of Luna under the influence of the heroine tree-sitter's public address, the crowd was viscerally interpellated as a collective: we became the endangered, yes, but the defiant and refusing as well. In the image she projected, none of us can or ever will be safe on our own. We need each other. We are one with the town of Stafford, the salmon, the murrelet, and the redwoods themselves—a community degraded by corporate capital but hopeful and rising in dialectical measure.

A Man of the Crowd

When Butterfly hung up the phone, Parker introduced several members of the Luna tree-sit ground support team, the individuals who shouldered the daily responsibility of making the protest action happen. They carried supplies up the mountain, ran supplies up the rope to the platform, maintained the two platforms on which Hill lived, ushered journalists and media camera personnel up and down the mountain, and managed media outreach and communications.

As the team ascended the stage, the mood was ecstatic in recognizing their work for the Luna tree-sit, Butterfly, and the movement, but it quickly and dramatically plunged as one of them stepped up to the microphone, nervously pulling his hat down over his brow, so that his eyes barely peered out. Parker announced the man as Dan, a young, tall, thin, somewhat pale and bearded white man who had been the first defender to climb Luna; but when he spoke he did not thank the crowd for its recognition. Instead he said glumly, but with mocking jocularity, "Um, Gypsy mountain is being raided by the police again, right now; let's play a little game—Simon says, 'Gypsy mountain is being raided,' Simon says, 'Everybody working on Gypsy mountain go to the stop sign.'" He pointed toward the traffic sign not far behind the makeshift stage. Another member of the Luna support team then stepped into the gaping hole that this man had just ripped in the feel-good collective assembled by Julia's words. "Let's hear it for everyone who has done anything for Luna," he said. "Everybody has done *something*."

The crowd tried to roar back to life on this intervention, but it was a weak effort. Another young man from the ground support team stepped up to the microphone and addressed the crowd in a somber voice. "I wanted to say a quick thing about what's going on at Gypsy mountain," he said. "[It] pertains to what's going on here because the first time I went up to Luna and the first time any of us ever went up there you'd step off the road next to the mudslide and you'd just step into a totally different world: that forest doesn't exist anymore, and it's been a really difficult thing for me to come to grips with, that this place is gone." When the speaker paused, the collective weight of six hundred people in a dead silence imposed itself. An inattentive child could be heard in the background talking freely against the open expanse of funereal quietude. In the space opened up by this rhetorical turn, I recalled the words spoken by David Brower, a founding member of the U.S. environmental movement and a hero to many, who famously said that whenever environmentalists win a battle, it is only temporary—but when they lose, it is permanent. When an ancient grove is cut in Humboldt, the loss is permanent.

Reminding everyone that the forest immediately around Luna had been cut while Julia was in the tree, the speaker continued: "There's a lot of attention and energy directed at Luna, but for some reason it was a half step behind the destruction, and right now at Gypsy mountain—where our brother Gypsy was killed—there's a beautiful grove of trees right behind the stump of the tree that was killed and that in fact killed Gypsy and I feel that right now is a good time where people can come together and understand what it means to have a forest still standing when you walk away from it, because Luna is a beautiful tree and it's incredible Luna is there and Julia's efforts, but there was a whole forest there and it breaks my heart to think that it is gone." As a man in the crowd, I became a part of its family funeral. Luna's hold against the chain saws was a symbolic channel of identification through loss and a last stand for justice—and the tree-sit as spectacle was a wide public call to its specific and transient but powerful community.

For many forest defenders, working to save a forest means growing to love it and then watching it be destroyed. Such pain was to be a common refrain among the activists I met over the coming months. But here, at the first rally I attended, I was new to the collective grief that individuals publicly share over the destruction of forests. Understanding this identification—this living bond with the ecosystem—is imperative to understanding the entire rhetorical practice of Humboldt's forest defense.

The reverence with which ecology is recognized can serve the same function for imagination as do gods in another form of collective identification, in other communities—it stands before the subject like a mysterious, all-powerful, and disseminated force, conjuring through community a response of respect for familial authority and collective obligation. Like the symbolic function of the American flag in the center of Scotia, which calls on people to identify with the hegemonic national subject and reflect the core values instituted by the juridical state, the forest is an emblem underneath whose gaze the defenders' disparate attentions are gathered up into new collective subjectivities and made available for conscious, collective agency to develop and use. There is always shared grief organizing these subjects, and the forest rallies are always part funeral, but they never fail to end in a call to action intended to empower individual participants.

Next onstage was an older white man who introduced himself as a representative of Veterans for Peace, a group based in Garberville, a small town in southern Humboldt. "It amazes me to think that, at a time when people are just beginning to contemplate the meaning of entering a new millennium, Julia Butterfly Hill has just spent the last year sitting in a tree that has been standing since the *last* millennium," he said, "and I think the world is now starting to go, 'Wow, that's deep.'"

Everyone laughed, but in fact the world *was* taking serious notice of Butterfly's heroic stand against the outlaw corporation. And the veteran's main point was that Julia's message—and the larger message of Humboldt's forest defense—was in fact reaching the world. "I bring you greetings from thirty-five hundred miles away, from the small villages of Chiapas," he said, from "where I have just returned. The villagers there, the indigenous people, were involved in a very similar struggle to save their forest and their way of life, and I was surprised that wherever I went, they knew about the Headwaters forest, they knew about David Chain, and they knew about Julia Butterfly, so I bring greetings of solidarity." This was most likely no exaggeration; there are many examples of local expression and collective resistance, but the case of Humboldt has been particularly well publicized—the timber wars have been a global spectacle for years.

Continuing his message of global solidarity, this bearded, fifty-something veteran continued in a stoic, direct, and convincing tone: "I want you all to know that this particular part of the struggle here in Humboldt is not unique—you are not alone—you are still part of the worldwide struggle to save human rights and to save the planet that we live on." Over time I would learn that his use of the term "we" and his inclusion of the local forest campaign in a unified vision of global struggle is nearly universal among Humboldt's various forest defenders. They see themselves as, and publicly claim to be, agents in a globalizing collectivity of insurgent resistance against an already globalized collectivity of politico-industrial and capitalistic forces of social and ecological malfeasance.

Such a totalizing vision is one of the forest defense's great strengths, not one of its weaknesses, as mainstream media commentators have suggested when they deride local campaigns and especially the anti-globalization movement for being "against everything." But the message is hardly muddled just because it matches the forces of domination one for one in complexity—a specific point of resistance for each point of domination. White, middle-class, educated forest defenders, landless Mexican peasants, and *ejido* farmers are in fact equivalent, at least to some extent, before the globalization of capital, not just because the resisters draw the chain of equivalence into consciousness and pronounce it to be a public project, but also because capital permeates these places of resistance with equivalent force. Property instituted in the redwood forest is in fact equivalent to that instituted in the Mexican rain forest under the rules of the North American Free Trade Agreement. Globalization, in other words, is the great motor of equivalence, drawing the circle of possible counteralliances wider and wider with every new round of privatization.

Tuning in to the spectrum of rhetorical frequencies generated by local campaigns, from Humboldt to Chiapas and around the world, one begins to understand the clear vision of environmental, economic, and social democracy being articulated by the new social movers: every individual element in the resistance originates in a concrete local refusal of the system, which is usually seen as a distant, overarching decision-making entity that is extracting values from local communities of labor and environment. The absent but overpowering system of capital operates by establishing a flow of value out of the local and into the global, without an equivalent return. The operation of Maxxam in the redwoods fits this modus operandi exactly.

"This is just a couple of days after the fiftieth anniversary of the Declaration of Human Rights," continued the globe-trotting veteran with increasing excitement. "Our country seems to be very strong in supporting human rights; however, there seems to be a problem in that we don't really concern ourselves so much with the human rights of other Third World countries and indigenous peoples right here in our own country—so I wanted you all to remember that we are all part of the same struggle, it goes much farther beyond all of us here, and as veterans, those of us that have served our country, we know full well the sacrifices that are required, there are people being killed, there are people being wounded, and the timber wars are a war just as much as any war our country has fought; this is an internal war, it's us against them, we have to preserve these forests and I'm sorry that some people's lives have been endangered but we all have to remember the struggle and stay focused on what we're doing and remember that the whole world is with us, the whole world is watching, in countries all over the world they know about this issue, so don't think that no one cares, always keep your chin up, keep your fist in the air, we're going to win this battle." His speech was effective—the crowd clearly agreed with his call to arms.

Using the term "timber wars" directly, in the vernacular of the struggle, the veteran seemed to know at that moment what it would take me some time to fully understand: that this local struggle really was worldwide, that the local campaigners really do identify their struggle against Hurwitz with Zapatista villagers and Nigerian peasants huddled outside the armed gates of U.S. oil institutions, and that they are busy actively and consciously creating a community of resistance under the collectivizing symbols of the redwood forest, Julia Butterfly, David Chain, Judy Bari, the endangered salmon, the spotted owl, the marbled murrelet, the shrinking logging communities, Maxxam, Scotia, and the devastated indigenous nations whose claim on this place goes back furthest of all—all the remnant communities that capitalist colonization of the redwoods has produced.

The dead, the dying, the threatened, and the helpless remnant communities of people, animals, and ecology of Humboldt form points of identification through which the struggle is made in psychosocial and economic terms. By sharing the primary human bond of love with the forest, equating its destruction with that of other peoples and places under assault by the same or similar forces, and focusing their attention through these symbols, individuals make the movement exist as a collective force; its form, in concrete terms, is the culturally material entity of shared meaning embodied in specific points by the logic of identification. This is where and how the group exists, in these shared moments of subjective equivalency channeled through the collectively identified symbols.

Chain, Hill, and Scotia inhabit such imaginary points, but it is the full histories of the takeover, the Headwaters campaign, the spotted owl, the marbled murrelet, the car bombing, and the deal that fill in the horizon of their timber war performance—they provide the context. They are the events, individuals, conflicts, and processes that together form the practically lived, imaginatively narrated, politically built, and otherwise symbolically performed material of resistance through which are enacted the individual identities and group formations that, as the movement, confront the symbolic order of established reality with an alternative, challenging code.

The Alliance for Sustainable Jobs and the Environment

At the Luna tree-sit base camp after the rally, people were crowded into a small shack that sheltered a communal kitchen. Warmth, music, and food made the atmosphere festive. There was drumming in one corner and conversation everywhere. My companions quickly abandoned me, and because I knew no one, I took a position and merely waited for something to happen.

By chance I fell into conversation with Darryl Cherney and John Goodman. Cherney is perhaps the best known of the forest defenders. He was in the car with Judi Bari when the bomb exploded. His musical talent and penchant for public speaking at forest rallies and public hearings give him wide name recognition both in the movement and in the dominant logging community. With Greg King, he cofounded Humboldt's earliest Earth First! group.

Goodman, on the other hand, was rather a newcomer. At over six and a half feet tall and wearing jeans, cowboy boots, a wide-brimmed cowboy hat, and a red, white, and blue United Steelworkers T-shirt, he stood out dramatically from the other forest defenders. As we talked about the growing alliance of environmentalists and labor, it became clear that historical circumstances were pushing these two men together. They were living links in the emergent global chain of equivalence between environment and labor that the corporation created when it seized control first of the Pacific Lumber Company in 1985 and then Kaiser Aluminum in 1988. First Maxxam stirred up Humboldt with its drive to liquidate the county's largest privately held stands of ancient forest. Then it stirred up the steel industry when it locked the Steelworkers out instead of negotiating in good faith with the union. When the Luna rally occurred, the strike against Kaiser had been on for eleven weeks—thirty thousand workers were out at the company's Washington, Ohio, and Louisiana plants. Kaiser put in one thousand replacement workers and kept the plants running,

United Steelworkers at the one-year Luna tree-sit rally, December 12, 1998.

using laid-off Palco workers and retirees as scabs on the West Coast. The Steelworkers wanted wage and benefit increases, job security, and limits to contracting outside the union, while the company wanted to eliminate as many as nine hundred jobs.

Maxxam's combined operations in lumber and metal were creating the social conditions for an alliance between two militant publics— labor unionism and environmental activism. Just a few months after the Luna rally, Cherney and Goodman confronted Hurwitz at the annual Maxxam stockholders' meeting in Houston, on May 19, 1999. They presented him with a document titled *The Houston Principles* and demanded that he take responsibility for the working peoples, communities, and environments that are jointly impacted by his corporation. The Alliance for Sustainable Jobs and the Environment had been born. I reprint the document that the group presented to Hurwitz here in full because it says so much about the forest defense and the channels through which it combined with the labor movement and went global during my field research. Read carefully, it reveals the extraordinary vision and theoretical alacrity of the forest defense and its new labor allies. Their understanding of capitalism—the system of culture driving commodification and occupying the redwoods—is here manifest and seen to be working through the tactics I have described as the forest defenders' idiom of public address.

The Houston Principles

Whereas:

The spectacular accumulation of wealth by corporations and America's most affluent during the past two decades has come with a huge price tag.

Corporations have become more powerful than the government entities designed to regulate them.

The goal of a giant, global corporation is to maximize wealth and to wield political power on its own behalf. Too often, corporate leaders regard working people, communities and the natural world as resources to be used and thrown away.

Recognizing the tremendous stakes, labor unions and environmental advocates are beginning to recognize our common ground. Together we can challenge illegitimate corporate authority over our country's and communities' governing decisions.

While we may not agree on everything, we are determined to accelerate our efforts to make alliances as often as possible.

We believe that:

A healthy future for the economy and the environment requires a dynamic alliance between labor, management and environmental advocates.

The same forces that threaten economic and biological sustainability undermine the democratic process.

The drive for short-term profits without regard for long-term sustainability hurts working people, communities and the earth.

Labor, environmental and community groups need to take action to organize as a counter-balance to abusive corporate power.

The environmental and labor advocates who have signed these principles resolve to work together to:

Remind the public that the original purpose behind the creation of corporations was to serve the *public interest*—namely working people, communities and the earth.

Seek stricter enforcement of labor laws and advocate for new *laws* to guarantee working people their right to form unions and their right to bargain collectively.

Make workplaces, communities and the planet safer by reducing waste and greenhouse gas emissions.

Demand that global trade agreements include enforceable labor and environmental standards.

Promote forward-thinking business models that allow for sustainability over the long term while protecting working people, communities and the environment.[19]

My first day of field research ended with that conversation between labor and environment. Although I did not know it at the time, my path was being set for the duration of the study. The voices of loggers and forest defenders were pointing me back to the history of capital in Humboldt. They remember the indigenous peoples who first claimed this place and the structural power of people who labor to change what it means in the present and secure its linked social and ecological future.

In later research, their words were my guide as I struggled to place the Indian wars and the labor trouble within the same frame as the ecological conflict. That frame, it turns out, could be constructed directly from the theory embodied in the practice of forest defense. Its global consciousness and media-savvy activism see the institution of property as embedded in a parasitical system feeding on local values and driving the convergence of environmental and social (labor) interests toward a totalizing vision of opposition to globalizing corporate capital.

They know that the power of corporations leans back on constitutional authority of free property right, and they know how that power is active in organized modes of production and behavior—namely, in the commodity circuit of redwood production that accumulates as power not just over labor but over the media spectacle that can prepare mass attitudes, at least to some extent. Thus their own strategies intend to meet the corporation on the same grounds. By addressing the public through spectacular media with direct calls to action that reverse the symbolic order of established timber culture, social movers united in forest defense make the singular engine of that culture their own—they reverse the power of spectacular media and use the constitutionally programmed expansion of rights discourse to display their own vision that another world is possible.

The prevalence of Chain, Hill, and Scotia in my story follows from my focus on public discourse and the public life of symbols for reasons I hope are abundantly clear: their extravagant tactics and media consciousness are successful in gaining attention and getting word out. But numerous other collective actors helped form the intricate mosaic of relational forces that make up the field. There is a deep organizational environment that structures Earth First! in Humboldt and conditions its possibilities for action.

At the local level, these other collectivities include but are not limited to EPIC, the Headwaters Action Video Collective (HAVC), the Humboldt Watershed Alliance, the Northcoast Environmental Center (NEC), the Trees Foundation, Sanctuary Forest, the Mattole Restoration Council (MRC), the Salmon Group, the Mattole Forest Defense, Salmon Forever, Friends of the Eel River, Friends of the Van Duzen River, and Friends of the Mad River. Other organizations with consequential roles but not based in Humboldt include the Mendocino Environmental Center (MEC-Ukiah), the Redwood Summer Justice Project (Santa Rosa), and the Bay Area Coalition for Headwaters (BACH). Though each is independent, together they form a loose network that collectively constitutes the resistance movement, often combining their efforts and offering support to each other's campaigns to transform forestry methods, halt streambed destruction, stem water pollution, and otherwise intervene everywhere that corporate land users externalize costs on the public at the expense of public-trust values. No single element has any meaning in itself outside this deep social context. Neither do any two of these organizations have the same internal structure, function, or mission. And they feed into the national and global dimensions of environmentalism and antiglobalization in unique ways.

A dense network of complex communication flows through this super-rich associational environment, complicating my approach to the study. How could I possibly represent a field of production so manifestly overdetermined by the interests of this many agencies, actors, interests, and passions? It was not an option to describe each organization in detail. Rather, I seized on the concept of events. The stories I could tell of events in the making can be made to convey the logic of the field and give the reader a feel for the situation.

I began from the stories of Chain, Hill, and Scotia because their images leaped out of the local media context into circulation within the global semiosphere and captured my attention, calling me into the field. I then learned of these other groups and their roles in the wider conflict. What made the storied events of Chain and Hill stand out was simply their successful publication as stories, as events, as something spectacular over and against the opaque (ideological) background of naturalized authority that had been achieved by timber culture in Humboldt, especially as that authority was embodied in the Pacific Lumber Company and Scotia. Maxxam's purchase of Pacific Lumber cracked the surface of that background authority and set the conditions for massive resistance. It is there, in the shattered peace of this timber war field of power and place of resistance, that we can now take stock of the hegemonic forces in operation.

3. Everybody Needs a Home: Speaking Up for Workers, Owners, and the Company Town of Scotia

After the forest defense rally celebrating Julia Butterfly Hill's completion of her first year in the canopy tree-sit platform, I surveyed the industrial landscape of Humboldt, traveling the roadways and photographing the built environment, talking to people, reading the papers, gathering literature from every position in the redwood timber wars, and planning for long-term participant observation, interviews, and field research. Nine months later I returned and spent two years living on the scene, immersing myself in the deep social context of redwood timber culture, the forest defense movement, and the archives of Humboldt's perpetual public-sphere conflict, which includes the landscape of social memory that all these forces had built over 150 years of capitalist colonization.

Eventually I seized on the idea (really, a method) of treating these three domains as elements of a single cultural structure: the dominant culture, the culture of resistance, and commingled textual, material, oral, built, and geographic records (the archives) together form a structure that structures the timber wars. How could I best convey my experience living within and learning to use this cultural structure? Cultural theory provided the foothold I needed: this cultural structure operates as a language, enabling and constraining my research experience just as it does the experience of big timber operators, loggers, and forest defenders in their respective pursuits. All of us must learn to put the same stories and landscapes to use in our meaning-making lives. Whereas for me Chain, Hill, and Scotia are theoretical ciphers of globalization and resistance, poised on the cusp of twenty-first-century ecological meltdown and providing organic access to one specific scene of revolt, a symbolic locale that embodies and exemplifies the rise of reciprocally determining labor and ecological antagonisms across the horizon of world-system capitalist expansion, for the forest defense they are symbols around which to rally to their local cause and raise public consciousness. But Chain, Hill, and Scotia are also symbolic sites of elevated concern and psychological investment for the industry folk and the loggers. The men and women whose labor energy animates the redwood commodity circuit, as well as the managers, executive officers, and owners who struggle to shape the aspirations of workers and extract their labor values, all enter into the public arena of the redwood timber wars through these same symbolic points of psychical fixation.

Whereas the forest defense is a positively literary movement, an ecstatic eruption of images, narratives, poetry, films, science, and political supplications, and the company had its Web site, paid spokespersons, and public image handlers, the loggers have fewer resources and less access to and power over mass media. In the 1990s, forest defenders built a strong public challenge to timber hegemony using every tool in their rhetorical and practical toolkit, compelling workers and company men to respond in defense of the dominant culture. Forest defenders dragged them as well as the citizens of Humboldt into a raucous public-sphere discourse, waking them from the more or less quiet sleep of their natural attitude, often seemingly in anger at having been disturbed. By tuning in to the channels of their public reaction and listening to the agents of the company and some sympathetic workers and citizens defending the Headwaters deal, the company, and the dominant culture of industrial redwood production against the environmental challengers, we can achieve a fuller understanding of the redwood imaginary—the living, symbolic, and built structure that structures the cultural performance of the redwood timber wars.

The Voice of Authority

Around the time of the Headwaters deal, the company came out strongly with editorial comments in local and regional newspapers. It used its commercial Web site to host a news service reporting on the struggle with activists and environmental regulations. It deployed official spokespersons to ensure that the company was quoted with authority in the press, contacting or responding to the *Eureka Times-*

Standard and other media outlets every time an issue came up. It hired scientists, lawyers, public relations people, and corporate officers to appear in force at the public hearings convened by the various regulatory agencies involved in the THP approval process. It produced a continuous stream of newspaper, radio, and television advertisements that did not merely promote the company's products but tried to craft a public image of overall good corporate citizenship based on responsible labor, community, and environmental stewardship.

Being the most widely circulated local paper, the *Times-Standard* is the closest thing there is to a collective voice for this dominant culture. By striving for a journalistic ideal of presenting unbiased coverage of events, its pages function as an index of established authority and reality for two reasons. First, in the broadest sense, the local news reports are slanted by their unspoken commitment to "official" news sources. With the advent of each new protest or bit of important company news like another round of layoffs or a court challenge to environmental regulations, its reporters routinely called Maxxam's official spokesperson Mary Bullwinkel, who kept her office in the Scotia headquarters. She worked closely with management to shape the public perception of the company, and that meant making herself available to the press for continuous comment on events in the timber wars. Her voice is omnipresent in the redwood archive of public culture that accumulated throughout the 1990s. She was continually cited as an authority on events. Second, unlike the various forest defense and watershed protection groups, the timber workers did not have their own publications, and neither did Maxxam put out its own serial publication with input from the public or its workers. The company did have its Web site, but again timber workers had no forum there—it was merely the voice of management, an advertising spectacle aimed not just at selling product but at countering the environmental challenge at every opportunity.

During the Luna tree-sit, for example, Pacific Lumber used TV commercials and print advertisements to respond to its environmental challengers. One exemplary ad ran in both the *Times-Standard* and the timber trade journal *California Forests*. Its text is superimposed over a background photo of an iconic, modest, wood-framed family home, seen from the front with an inviting porch and a closed front door; to the reader's right side, a vertical row of three inset photos shows an owl, a raccoon, and three salmon—positioned according to the natural order of things, which is clearly one message the ad

Pacific Lumber advertisement in *California Forests* magazine, official publication of the California Forestry Association, October–December 1999. This advertisement also ran on the editorial page of the *Eureka Times-Standard*, October 16, 1999.

is designed to convey. The bird lives above the forest mammal, beneath which the fish swim. Their natural place is the forest, but here their background is a quaint wooden house, by which substitution the dominant culture of lumber production, the company, and the industry are themselves naturalized. "How to choose a home? . . . very thoughtfully," begins the text. "We all need shelter. . . . It's true for people—and it's true for owls, murrelets and salmon, too." The raccoon, owl, and fish look directly into the reader's eyes, putting human and animal on a common visual axis that reinforces the rhetorical chain of equivalence established by the textual "we." "Birds, fish and other animals don't have nearly as many choices," the copy continues. "That's why those of us at The Pacific Lumber Company go out of our way to protect the forests and streams on our property. We work hard to make sure wildlife can find a home in our forests and streams." The reader is identified with all the other humans and animals of the forest, all of which rely on this benevolent master, the corporation, for the primal need of shelter. We are all endangered species; like the owls, murrelets, and salmon, we readers need protection. In this way the company verifies its status as master signifier of the law—paternal possessor, protector, and ultimately benefactor with an implicit power to destroy what it wants of the animals, the people, and all our homes. Its authority as decision maker is the silent assumption on which the force of the rhetoric builds. The Palco logo occupies the lower right corner under the inset photos, reinforcing the friendly pole of the company's message, for who cares more for you than your *pal*? And again by extension, the industry is a pal that invites us inside of its commodity circuit: "Join us in celebrating National Forest Products Week, October 17–23, 1999," writes Palco in the lower left quadrant of the ad.[1]

But why advertise like this? And why at this time? Perhaps because forest defenders were attacking the company's legitimacy by publicizing the Stafford landslide, which had recently cascaded down a mountain south of Scotia from a Maxxam logging road through a clearcut, sending a wave of mud downhill and destroying several houses. People had to run for their lives. The resistance was using a principal American symbol of sovereign property—the sacred private home—to portray the company as a threat. Maxxam must work to maintain its grip on this symbolic ground. By creating the chain of equivalence—a false, or rather *fictive,* unity between owl, raccoon, salmon, people, home—and substituting this happy continuum for the real scene of conflict, the ad helps the company to paper over a hard political truth: its new green corporate culture of sustainable forestry is a compromise with environmental and labor constituencies presently identifying their interests and combining to resist corporate deforestation. That is the story never told: every regulatory limit to infinite conversion of ecological values into capital for export to large outside ownership is a concession hard won by forest defenders—the company's conservation programs are all concessions of political power to the forest defense, concessions that rechannel the flows of accumulating value. The greening of Maxxam expresses the movement's success.

The forest defense compelled Maxxam to enter the public sphere in this way—to advertise justifications for what had long been taken for granted. The image it broadcast into this already saturated field of image production mobilized narrative and iconic elements that situated its public and its workers, putting them back in their place: the place they had always occupied in the mythic stories of redwood country, the place of passivity and dependence on great paternal lumber companies and their timber baron forefathers who always took care of everything just fine, according to their own rules. In this way the company sought to bring the reader into the Palco sphere of affective influence, about which we learned something from the hitchhiking logger, quoted in the previous chapter, according to whom the company's legendary care for labor and land had helped make Humboldt great.

A Voice for Timber Workers

The *Eureka Times-Standard* is one of the few places where workers who so desired could speak and be heard, and during the lead-up to the Headwaters deal in 1999, they almost invariably defended the company. For them logging is an endangered way of life, and the company's haggard image, under siege by environmentalists, required defense. They also publicized their support for the timber industry by answering reporters' questions about the ongoing controversy, although a leading environmental reporter in the area told me of his constant difficulty in getting Scotia residents and Maxxam loyalists to talk. In this way loggers did manage to maintain a symbolic presence in the public sphere. They and their peers in the timber-dependent economic community regularly contributed to the opinion columns and editorial pages and showed up for rallies to support the industry and at logging conferences, logging-related sports and community events, public hearings on forestry rules, and county supervisor meetings. For example, in the spring of 2002, Maxxam/Pacific Lumber's new CEO appeared before the board of supervisors and publicly called on it to requisition the federal government for Homeland Security Funds for use in support of local law enforcement in its war against what he called the "ecoterrorism" of the forest defenders. Forest defenders flooded into the next board meeting and demanded an apology for calling forest defenders terrorists, after which the drama carried on for several weeks in the editorial pages, in letters to the editors of county newspapers, and in conversations among activists.

Dropping in on the continuous newspaper forum of claim and counterclaim in the days just after the one-year Luna rally, we find supporters of the dominant logging community speaking out—keep in mind that this is the rational public sphere at work, distilling the values, shaping the opinions, and garnering the psychical attentions that all parties are eager to bind to their cause and make part of their respective historical collective subject. The *Times-Standard* of December 22, 1998, carried a letter titled "Comparison Draws Fierce Response," in which an angry Jim Hornback rebutted a previous letter sympathetic to the forest defense. Hornback excoriates Scott, the letter writer, for misleadingly and shamefully comparing "the actions of David Chain and his so-called willingness to risk injury and death to those who fought for our nation's independence, the abolition of slavery and the elimination of segregation." Scott's "outrageous statement stinks of tripe," wrote Hornback, before turning to attack the figure of David Chain himself, dead three months at the time of the letter's publication. Dr. Martin Luther King, George Washington, and Abraham Lincoln "were probably not smoking dope," and so "the attempt to compare the unlawful, disrespectful, eco-groovy, unemployed, affluent, mostly white, non-tax paying activists" with "our forefathers (men and women) . . . degrades the memory of the above mentioned." The attempt to associate the actions of forest defense with patriotic American icons and events often evokes such contempt among Humboldt's timber-friendly citizens.

Hornback's portrayal of Chain represents what I came to know as a standard negative view that many timber supporters have of forest defenders in general. A narrow concept of the law tends to suture their criticism. Pot smokers and trespassers cannot challenge the legitimacy of property-owning and taxpaying Americans, for this is a nation of law. Hornback's appeal to law masks the truth behind Scott's favorable comparison of contemporary civil disobedience to the founding acts of American nationalism. Symbolically targeting property and power created and maintained by unjust laws that serve class privilege is arguably the oldest and most respected American political tradition.

The nation did in fact emerge from the throes of nonviolent direct-action civil disobedience—from events symbolically powerful enough to bring masses over to the revolutionary cause. We cannot forget the Boston Tea Party or the way that Thomas Paine's revolutionary pamphlet "Common Sense" called for Americans to reject British rule because it embodied the ancient tyrannies of monarchy and aristocracy, two foundations of power that have no ground in the people and thus no place in a free nation's constitution, the very basis of which is the aggregation and consummation of the people's will in a form of free association that voluntarily gives up some dimension of private civil right, including property right, to secure a democratically determined public right and collective interest. Forest defenders are correct in comparing their civil disobedience to the dissenting tradition from which the founders emerged. Both the Boston Tea Party and Paine's revolutionary pamphlet were spectacular displays aimed at building a counterpublic using media channels of collective attention, and the founding representatives to the constitutional convention institutionalized such direct-action media displays as constituent functions of the public sphere when they wrote the Bill of Rights to be a juridical engine of free speech, press, and assembly. With that founding text, they created juridical engines of a free public sphere that has guaranteed to this day that public calls for resistance and spectacular nonviolent collective actions of civil disobedience with a message would always be protected technologies of display and address that the people could use to self-organize, take history into their own hands, and thereby participate in the democratic mandate of self-governance.

From Humboldt to Seattle 1999 to the ongoing antiglobalization movement, this cultural heritage of protest is actively mobilized in the public-sphere spectacle of resistance. Forest defenders draw consciously on statements of the strategic philosophy: "Unjust laws exist; shall we be content to obey them or shall we endeavor to amend them?" wrote Henry David Thoreau in 1848, explaining that paying taxes makes one complicit with a state engaged in the mass hypocrisy of enslaving one-sixth of its population while simultaneously claiming the banner of freedom and democracy for some of the others.[2] The answer, of course, in the American tradition of voluntary association, is a matter of conscience, and that is what Mr. Hornback had encountered in the redwood timber wars. The forest defense is a movement of conscience, the objective of which is the refusal of unjust forestry laws and unaccountable corporate power, which activists correctly see as allowing the commons to be destroyed for the profit of a distant absentee owner. Nothing angers the men and women who speak from the dominant position in the timber war field of cultural production more than hearing environmentalists described as patriotic Americans engaged in a freedom struggle against illegitimate, unaccountable power. It is no wonder, then, that they are continually angered.

The Chief Executive Officer

Hornback's colorful attack on David Chain's character came at the very moment in which another exemplary public debate in Humboldt's public sphere touched the constitutional foundations of American liberty—the struggle over the Headwaters Grove preservation deal. On December 19, 1998, John Campbell—a lifetime employee, elevated to president of Pacific Lumber in 1989 and CEO in 1993, who stewarded the company through Redwood Summer and the Headwaters deal— published a "Guest Opinion" column in the *Times-Standard* (also printed in the *Sacramento Bee*). Under the headline "If Headwaters Plan Gets Rejected, Everybody Loses," Campbell reviewed the Habitat Conservation Plan and framed it as a constitutional question. He used the article to take a threatening stance toward the local communities of labor and forest defense. Remember that Campbell published these comments during the endgame; the forest under debate was at that time the largest intact grove of the final 1 percent of unprotected ancient redwoods—96 percent had already been cut down, and 3 percent were protected in parks.

"The last best chance to preserve the Headwaters Forest and protect thousands of additional acres of wildlife habitat is still being debated," he began, "when it should be moving into the final phases of implementation." Then, in an ominous tone, he appealed to the deep-seated rural fear of manipulation by mysterious urban agents: "Powerful forces are arrayed to block the plan the Clinton and Wilson Administrations have developed for protecting these old-growth, no matter what the cost would be to the environment." Here he represents the corporation as a defender of the forest, its protector against the danger of environmentalists who hope to block the plan. The "defeat of the plan would leave Pacific Lumber with little alternative except to proceed with a constitutional takings lawsuit against the state and the federal government." The deal is necessary, in other words, to save the forest from the company, and the forest defenders are blocking it. The circularity of these arguments, while not lost on the forest defenders, often passes for common sense among supporters of the industry. Campbell obfuscates one major fact: it is already illegal for Maxxam to cut the grove—that is what drives a takings lawsuit, a regulation that a property owner contends has taken all the value of said property.

The Habitat Conservation Plan (HCP) at the center of the deal had been written by the company and for the company with the help of regulatory agencies. It was, essentially, an out-of-court settlement for the Fifth Amendment inverse-condemnation or "takings" lawsuit brought by Hurwitz. If the deal was completed, the HCP would eviscerate the forest defenders' legal strategy of using the courts to force state and federal forestry and wildlife agencies to do their legislatively mandated jobs of enforcing the Endangered Species Act and conducting an environmental review of the cumulative impacts of government actions, as required by the National Environmental Protection Act, the California Environmental Protection Act, and the California Forest Practice Act. It would replace these established tools of regulatory control with money-driven deals between businessmen, politicians, and regulatory agencies.

In an acknowledgment that the forest defense tactics of spectacular display had reached a national audience and transformed public consciousness, Campbell attempted to blunt their effect with his own public spectacle. "This is the moment for California and the rest of the nation to look past the banner waving and shouting," he continued. "The plan will put an end to more than a decade of confrontation and turmoil on the North Coast." Sixteen hundred Maxxam workers "and their families," he continued, need this deal for their future, and so does the county. Activists have fought for years to preserve Headwaters, he complained, but now they are fighting against its last and best chance. "If the opposition succeeds," he reasoned, the "constitutional takings lawsuit against the state and the federal government . . . could wind up costing taxpayers hundreds of millions of dollars" more than the plan itself. Of course, in that event the biggest loser would be Maxxam, who would not receive the $480 million compensation package and would have to fight an unprecedented Fifth Amendment court battle in which victory was hardly assured.

Other citizens also injected opinions into the debate. Art Wilson, for example, wrote to the *Times-Standard* editors in October 1998: "In the struggle over how our timberlands are managed, traditional industrial forestry advocates often say that their private property rights are being violated by those advocating taking better care of our forests. Shouldn't we also be asking whether there are responsibilities that go along with those rights?" His words show how the redwood timber wars are a public struggle that always comes squarely to rest on two essential juridical conditions of liberal society—free speech and property rights. Art Wilson and John Campbell, citizen and CEO, are both appealing to the same discursive foundation of legal and constitutional rights. But whereas Campbell claims the Constitution as a bulwark protecting private liberty with property rights, local folk like Art Wilson cite its democratic provisions for political equality in the legal determination of public goods, in this case environmental quality.

It is not hard to see how capital accumulation structures the function of these conflicting objectives of the Constitution. While the takings lawsuit was written and filed by corporate lawyers, working with essentially unlimited budgets that represent Maxxam's historical accumulation of labor and environmental values, local folk like Art Wilson are left writing to the local papers and mobilizing congressional write-in campaigns. As citizens and activists, they took time from their working lives to engage the law. In addition to speaking out in the public debate, they also challenged individual timber harvest plans, attended organizational meetings, put out environmental publications, and dreamed up clandestine, direct-action civil disobedience like logging blockades, marches, and banner hangs—hoping through all of this to slow the machine, save the ancient forest, transform consciousness, raise awareness, and spread the word about forest and species destruction. They worked to build a counterpublic against the Headwaters deal.

Two days before the Luna rally, the *Times-Standard* front-page headline reflected the success of the movement in creating counterpublic momentum and stiffening the challenge for timber supporters: "'Butterfly' Still Out on a Limb." A large photo, taken from the air, showed Julia standing with her arms stretched out to the sky at the very tip of the tree; beneath her, hung against the blue tarps that provided a makeshift shelter for her sleeping platform, a banner proclaimed, in the command conjugation characteristic of advertising, "Respect Your Elders." "I take it day by day and prayer by prayer," she was quoted as saying. "I still feel like there's more for me to do."[3]

Summing up her message for the public, Hill told a *Times-Standard*

Times-Standard

The North Coast's daily newspaper since 1854

146th Year, No. 344 — THURSDAY, Dec. 10, 1998 — 50 cents

Wal-Mart explains drilling

Company unaware it needed permit

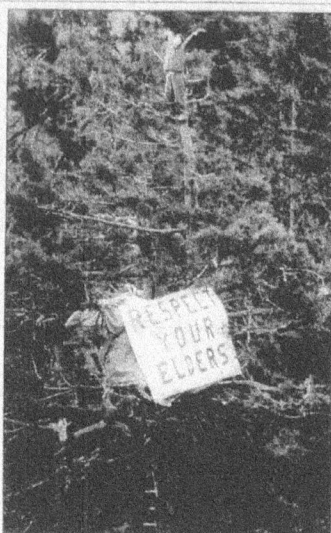

'Butterfly' still out on a limb

Julia "Butterfly" Hill, atop a 1,000-year-old, 180-foot-high redwood tree near Stafford, welcomed aerial visitors touring the Pacific Lumber Co. territory. The tree-sitter took to her redwood home or redwoods and plywood one year ago today.

Panel moves closer to impeachment vote

Guilt conceded in killing of boy, 11

Thomas Fox is wheeled into court on Wednesday where he pleaded guilty to killing Danny Williams, 11, in 1997.

No suspect in fish kill, agency says

The Times-Standard

WEATHER
Partly sunny with increasing clouds

SPORTS

Personal Communications Services for the North Coast

CAL NORTH CELLULAR

#1 in Coverage
One Number Express
Flexible Rate Plans
Cellular · Paging
Voicemail

3012 Sectral Street, Eureka · 444-2253

WHAT'S INSIDE

www.times-standard.com

On December 10, 1998, the *Eureka Times-Standard* reports on Julia "Butterfly" Hill's year residing in the upper branches of Luna. She would remain for one more year and descend on December 18, 1999. Courtesy of the *Eureka Times-Standard*.

reporter that the redwood timber wars cannot be seen as a struggle of timber jobs versus the environment, as it had been continually framed by timber companies and their supporters and sometimes the media. Rather, she said, it is about unaccountable corporate power being exercised over labor, environment, endangered species, and everything else. Hill said she had spoken to loggers and believes they feel powerless against the overwhelming authority of Maxxam—a common belief in Humboldt, in fact, where timber workers' unionization is almost nonexistent.[4] If things continue as they are, she said, with the corporation clear-cutting the ancient forest as fast as it can, the loggers "and I will be standing side by side in a clear-cut and we'll both be crying," because there will be no more jobs and no more forest either.

But according to John Campbell, the report continued, Butterfly and the other protesters are trespassing on private property, interfering with company operations, and endangering themselves, all the while deceiving their own constituency. "I think she comes and goes now," Campbell was quoted as saying, the first of many times I would be told by timber supporters that Hill was not really what she claimed to be, she comes down all the time, she meets men in the woods for sex, she goes out for dinner, she was seen in Eureka, she had been seen dining at the Scotia Inn, and so on; the timber community was truly invested in refuting the myth she was starting to become. Several days later, buried in the *Times-Standard* coverage of the big rally for Luna that ran under a headline "Community Disapproves, Stays Silent," Mark Koble, a recently laid-off Maxxam employee, told a *Times-Standard* reporter, "I think she (Hill) should be jerked from that tree." The people in charge have been "too lax," he continued. "I oppose this method, when they break the law—part of the solution is for the environmentalists to understand that there are sets of rules and we have to abide by them."[5] Koble thinks of himself as an environmentalist, according to the report, and he thinks the company should comply more strictly with logging rules—a point underscored by the fact that, when the interview took place, the company's logging license had been suspended for the rest of the year as punishment for repeated violations of the Forest Practice Act, leaving its mills dependent on logs bought from independent contractors.

Other residents of Scotia conveyed similar sentiments to the reporter. "I don't think they've gotten any point across," said one, while another just gave a thumbs-down sign and then rotated his finger around his temple, smiling and shaking his head. But the paper's main article on the rally reported that approximately six hundred people had listened to Butterfly's cell phone call. "We need to take a unified stand in love," she was quoted as saying, and Robert Parker, her media coordinator, added that Julia had no plans to come down.[6] But Maxxam spokeswoman Mary Bullwinkel reiterated that the tree-sitters are trespassing. "It's a little like being held hostage on your own property," she said. "They have no respect for the law." She said that the situation "compared to having a visitor sitting on someone's porch and refusing to let the homeowner in through the front door" and telling the homeowner he or she should not be living there anymore. "We're frustrated that these types of activities are continuing," she was further quoted as saying. "It's a form of terrorism."

It occurred to me then that the label of ecoterrorist continued to exercise local currency, even after thirteen years of hard social struggle had generated not a single case of violence against timber workers and almost a decade after North Coast Earth First! explicitly renounced violence in general, as well as the monkey-wrenching method of tree spiking practiced by other ancient-forest defenders in other regions. This question of violence is essential to the timber war field of cultural production and has been at least since the Federal Bureau of Investigation began its secret and ultimately illegal campaign against Earth First! At what point does civil disobedience become sabotage and sabotage become violence? Unfortunately this study will not be conclusive, but belief in the existence of violence seems to be distributed according to social position.

During my extensive fieldwork in the months to come, I became increasingly aware of the continuous accusations of terror made by employees of the company and some timber workers and began trying to track their function in the timber wars. Industry people describe trespassing, logging blockades, cat and mouse, trespassing, tree-sits, and property destruction as violence. But forest defenders call these tactics peaceful, nonviolent direct-action civil disobedience; the explicit question of what constitutes property destruction and whether or not such destruction is violence is contested from both sides. Within wider Earth First! circles, a continuous debate takes place around the wisdom and potential of using property destruction as a tactic for so-called radical environmentalism—but NCEF!, under the guidance of Judi Bari in the late 1980s, explicitly committed itself to nonviolence and publicly renounced property destruction as a method. No Earth Firsters in Humboldt have been convicted of violent crimes or even significant property destruction beyond petty vandalism, yet still the idea is propounded that the group is violent and actively carries out sabotage on the timber industry. While it is true that holes have been dug in roads and barricades built, these are not tactics of terror comparable to those of recent infamy in Iraq, London, and New York—events that lend the term "terrorism" its contemporary potency.

At least part of the explanation resides in the organizational forms and practices undertaken by Earth First! It operates on a quasi-cell structure, using affinity groups operating independently, without hierarchical leadership structures, and thus shows its debt to historical anarchist movements and tactics. Because its members and various constituent groups often work at the margins of a law they deem illegitimate or failing or inferior to a more compelling law, they are vulnerable to being labeled antilaw and so treated as lawbreakers, destroyers, and so ultimately terrorists. And because much of their work is clandestine, they are always at risk of disruption by clandestine methods. Much of their art is to play this middle zone to effect positive change—that is, change they approve of—without triggering a public reaction that could set their movement back. Whereas saboteurs themselves become vulnerable to sabotage, NCEF! has tried to step back from that precipice and retain their moral legitimacy by insulating themselves from accusations of violence.

Talking with Maxxam spokeswoman Bullwinkel in her Scotia office, I noticed a book on her shelf titled *Ecoterror: The Violent Agenda to Save Nature*. Researching the title, I found it was published by Ron Arnold, a leading voice for the so-called wise-use movement and the director of the Center for the Defense of Free Enterprise (CDFE), which describes itself as "an educational foundation for individual liberty, free markets, property rights and limited government" and dutifully maintains a Web site that archives news from around the nation and the world on the challenge of radical environmentalism.[7] And as it turns out, around here charges of ecoterrorism are almost always embedded in the economic arguments and property claims propounded by groups like this. In Humboldt the symbolic center of this conservative, patriotic property culture is the company town of Scotia.

Scotia, the Town of Concern

We should not be surprised that Scotia became a central scene in the timber war drama. As a place in redwood country, it stands apart, the last of its kind—an authentic company town. For a century before Maxxam swept in on the wave of globalization, this hamlet on the banks of the Eel River in central Humboldt had been a machine of industrial extraction, a fortress of capital at the edge of its American empire. If culture truly does enter nature through labor, as the environmental social theorist James O'Connor has written, then capital culture entered Humboldt through Scotia.[8]

What accumulated here was more than just capital—moral authority accrued within the redwood culture that helped fuel the conflict that erupted when Maxxam's new management abandoned the company's culture of good corporate citizenship. Over time, for the timber folk of the bay redwood region, the company had become a symbol of everything good about capitalism. By 1985 Scotia was nothing less than a self-representation of redwood capital culture, an institutional, economic, and architectural structure of self-identification by which and through which the dominant culture could know and understand itself. Then Maxxam's arrival sparked an identity crisis. With this in mind, we should pause and consider how Pacific Lumber built Scotia and managed to operate from colonial times to the present, building good feelings for nation and property and coming to occupy the sacred center of Humboldt's timber culture.

With the help of the famed San Francisco financier William C. Ralston, who maintained a controlling financial interest while the others ran the company, Pacific Lumber was incorporated with about ten thousand acres in 1869, most of it acquired for $1.25 per acre under the Morrill Land Act of 1862. Though lumbering operations did not begin until 1882, by 1886 the company employed 150 men and by 1888 was Humboldt's largest producer, shipping twenty million board feet and employing 300 men.[9] At that time the township of Forestville occupied the present site of Scotia, and the company rented the land, but in 1888 it bought the location and changed its name to Scotia. By that year it was already the third most populous town in the county, and by providing better-than-average bunks, cookhouses, and plumbing at this early stage, the company was already becoming a bastion of moral authority in the redwood region.[10] Over the coming decades, as most other companies on the redwood coast deforested their land of ancient trees, degraded their watersheds, extirpated their salmon runs, alienated their workers, and struggled against unions, Pacific Lumber built a reputation for stewardship of its forestlands and its workforce. It sold major tracts for park conservation, reseeded with nursery-bred seedlings, used selective-cut methods to preserve ancient trees even on cutover lands, kept large tracts of pristine forest in reserve for the future, and kept the unions out with a company town strategy of providing the best wages and conditions in the industry.

One family can largely be credited for presiding over the making of Scotia. After a period of complex reorganization, Simon Jones Murphy captured a controlling interest in the company and consolidated it with other holdings to create the Pacific Lumber Company of Maine in 1905.[11] With directors hailing from Saginaw, Michigan, and Chicago—midwesterners who headed west with their knowledge and capital after running down their own local supply of timber— the new company reported $10 million in capital stock and combined output through Scotia's two mills at 500,000 feet per day.[12] Foreseeing the impending completion of the Northwestern Pacific Railroad in 1912, Pacific Lumber announced an investment program in Scotia that would pour up to $500,000 into the town over the next several years.[13] Two years later the company had amassed a total 65,000 acres of redwood stumpage, including 48,099 acres of virgin timber, and with Scotia's two mills producing 110,000 board feet of lumber per day, it was already Humboldt's largest operator.[14] Its company town strategy was off to a flying start. By building and maintaining a town for the benefit of its employees, the company benefited itself, first of all by making it possible to maintain a large workforce far from any urban center while keeping all the adjacent land and resources under its own control. Simon Murphy's sons would run the company and build up the town throughout the twentieth century until their deaths, after which its surviving directors took the company public on the New York Stock Exchange in 1975, exposing it for the first time to the explosive powers of debt-wielding global financiers like Charles Hurwitz, who ten years later combined his knowledge of newly deregulated bond markets with his Wall Street connections and pried Pacific Lumber loose from the remnants of family control. When nostalgic company men lamented the loss of Pacific Lumber to Maxxam in the post-takeover days of turmoil, it was the Murphy family's stewardship that they missed.

But only by understanding how the Murphys' and the company's efforts in well serving its land and its employees were profit-driven projects can we begin to fathom the company's symbolic status in the days of the redwood timber wars. As we go through the litany of company-town-building efforts and employee benefits that built this status and identified so many people in the region with the company's well-being, we must keep in mind that the capitalist imperatives of keeping costs down were deep-seated motives—especially labor costs. Continuous efforts in dividing up labor, increasing scale, and mechanizing production kept the company growing and profiting. The company's program for good labor and environmental stewardship was at the same time a model of good business. At Pacific Lumber good corporate governance and community building were one and the same.

When the company built a schoolhouse in 1915, it served its employees well—and kept them nearby where their labor was needed. In 1918, during an upsurge in International Workers of the World unionism across the lumbering communities of the Pacific Northwest, it instituted its first employee benefits system (the Continuous Service Compensation Plan, offering 2 percent yearly bonuses after one year of service, increasing to 7 percent over the next four years, and remaining at that rate for the employee's life of service). This program did much to secure the loyalty of labor, helping to fend off strikes and to keep the mills running.[15] When the strikers did come to Scotia in 1919, they were led out of town by a delegation of loyal war veterans now working for the company.[16] In that year the county's lumbering companies returned to the nine-hour day with little resistance from unions, and this from a workforce that had brought it down by hard struggle from twelve to ten and finally to eight in the previous year.[17]

Likewise, when the company's First National Bank of Scotia was completed in 1920, it kept workers' funds in the company coffers, providing security for both at the same time.[18] By 1923 it had built the Scotia Inn and was bringing the public in on tourist trains to admire the operation. When it erected a hospital on the site, it eliminated the need of transporting injured workers twenty-five miles up to Eureka, improving workers' lives dramatically and making the company even more attractive to workers in what was and still is one of the most dangerous jobs around. In these same years, the company opened a saloon as well, to the obvious benefit of the rough-living lumbermen who came to spend their wages at the company bar. The company store worked the same way: to the benefit of both parties, it sold all kinds of household goods, furniture, and groceries, even offering fresh beef from the company's own thousand head of cattle, fattened on its own rangelands.[19] At this time the company paid its men in paper and metal scrip usable only in these company businesses. In addition, by offering rents of just one dollar per year to both a Protestant and a Catholic church, it spoke to the workers' spiritual needs while keeping them close to home.

By way of securing its status as environmental stewards while advancing profits, in 1931 the company sold close to 20,000 acres of land for preservation in parks, a fact that it continuously recalled in the years of conflict over Headwaters when its ongoing care of the land was being publicly debated. Less well known are the details of the long fiscal negotiations it took to cement a profitable transfer, which began at least as early as 1925. Company president John H. Emmert finally agreed in principle in 1928 after long worrying that the Save the Redwoods League would initiate condemnation proceedings if some deal were not reached.[20]

The company's public standing received another boost in 1935 when employees formed a band, whose thirty members received company money to order green uniforms in 1938. They continued playing patriotic songs and big band favorites up through the 1985 transition, marching in parades throughout the county and entertaining the company at its legendarily massive, company-catered Fourth of July picnic, convened every year on the banks of the Eel River.[21] Those were the days in which the company announced its new methods of power-felling the giant trees with pride—the gasoline-powered drag saws and diesel-powered Caterpillar tractors having entered the woods around 1928, dramatically increasing the productivity of individual workers through mechanization, expanding the scale of production, and creating wealth-driven civic excitement all though the 1930s.[22] In 1940, when the company purchased large holdings in the Yager and Lawrence Creek basins behind the town of Carlotta, its benevolent holdings stretched twenty-one miles long and seven miles wide, including 22,000 acres of pristine old growth, on its own account a twenty-two-year supply of wood.[23]

There is more. The company entered almost every aspect of its employees' existence, inserting itself into the traditions and leisure that compensate for the sacrifices of a difficult working life. For example, Scotia's six-hundred-person Winema Theatre, completed in 1920, was used to host an annual children's Christmas party since it opened in 1925, with company presents for every child.[24] The company created a retirement plan in 1941, of which it paid three-quarters of the cost. In 1950 it instituted a life insurance plan, by which time the company owned 131,000 acres and employed 950 men and rented them rooms at seven dollars per month. In 1956 Pacific Lumber closed down the hospital, but a doctor and a dentist stayed on the site, while more serious cases were sent to the new hospital in Fortuna, a few minutes to the north. In 1959 it opened a new recreation center with an indoor pool for residents, with an indoor basketball court, weight room, and game room as well—this in addition to the men's club in town, with its card room, pool tables, and locker rooms.[25] By 1957 the company owned 172,000 acres and advertised that its sustained-yield logging policy harvested only mature trees and only in areas in which 50 percent or more are so defined. Company foresters marked which trees were to be left in every harvested area, and trees were felled uphill in such a way as to avoid damage to the young trees and slopes as much as possible.[26]

In 1961, when the company instituted a college scholarship program that awarded four years of college support to one employee family member per year, based on scores of the National Merit Scholarship Test, it had nearly completed its nest of corporate care, offering core support in all the institutions of modern everyday life: in government, religion, economy, family, and leisure, the company mediated its workers' lives.[27] Perhaps the apex was reached in 1965 when a long-term disability insurance plan was added to the list of employee benefits at no cost—the end of life itself was now brought within the company's reach.

By 1961 this reach was wide indeed. With Stanwood A. Murphy as president and E. M. Carpenter as vice president, the company was operating in Scotia, Eureka, and Elk River with sawmills, planing equipment, chippers, dry kilns, and plywood-manufacturing equipment keeping 1,425 men hard at work and enjoying the best wages and working conditions in the region—and not incidentally putting out 395 million board feet of redwood and 180 million board feet of fir.[28] Its combination of advanced forest stewardship and good benefits and living conditions for workers built unparalleled feelings of respect for Pacific Lumber—by now widely and affectionately referred to as Palco.

Until 1950 the company owned all the businesses in Scotia, at which time it built a strip mall shopping center and began leasing the stores to private businesses, maintaining the control of ownership but relinquishing the task of managing the establishments, including a laundry, a butcher, and service station. The company store was finally out of business, but corporate control of the town and workers was still online.[29]

According to Vice President Carpenter, the decision to get out of the business of running these establishments was made in part because "we were still living with the song lyrics of you owe your soul to the company store. As a result of this stigma, we felt it would be in the best interest of the company to lease out the stores to outside businessmen in order to prevent this thought from continuing on. We were deriving little profit from the stores and it just seemed like the most sensible step to take at the time."[30] As these words make clear, Scotia had always been an irreducible architectural and cultural defense against "this kind of thought"—the town itself had been built as a form of corporate control over labor.

By 1971 the company owned 285 houses in Scotia, maintained them with a fifty-man crew, and rented them to 1,450 workingmen at substantially undermarket rates.[31] At one hundred dollars for a well-kept four-bedroom, two-bath home, it was no wonder that one writer called Scotia "Paradise with a Waiting List" in an article for the *Saturday Evening Post* in 1951.[32]

But change was on the horizon. When A. Stanwood Murphy died in 1963, his controlling interest in the stock of Palco passed to his son Stanwood A., who was appointed chairman in 1971, only to die two years later of a heart attack.[33] During his tenure, he created the A. Stanwood Murphy Memorial Scholarship program, under which any employee's child who completed high school and won acceptance to a four-year college was given five hundred dollars per year. This particular element of the Scotia strategy continued to resonate in the politics of the redwood timber wars throughout the 1980s and 1990s—whenever the company was discussed and people went over what had been lost, they tended to bring up the fact that Pacific Lumber even provided college scholarships. Though the program only came into being in the late 1960s, it seemed to shape contemporary feelings about what the company had always meant to the community. But without the firm hand of the family patriarchy for guidance, and with all those forest resources in reserve, the company went public in 1975, opening the possibility of corporate takeover in the coming era of fast-money acquisitions in the loosened credit markets of the post–oil shock, post-Carter Republican administrations.

In 1971, on the eve of these great changes, Vice President Carpenter reflected on the Scotia strategy for social control and image management in the age of growing environmental consciousness:

When we first started, the town was of a necessity due to our location. Through the efforts of the Eddy family and the Murphy family
we continued to improve on the town to the extent that we were so
far in we decided to keep right on running the town. It really is
owned more as a part of the Company operation than anything
else. We merely look at the town as just one more of the benefits
we offer our employees. . . . We are in a constant search for good
potential workers, particularly in the area of management. And
there is a real demand among various companies for all of the good
workers available, so we have to compete for them. We feel that the
opportunity to get good housing at a low cost will help us to get top
people over other companies. Particularly those individuals just out
of college who are looking to the possibility of raising a family. . . .
One of the biggest factors for us to continue operating the town
is the great amount of tourist attraction we draw. When tourists
come into town they cannot help but to see the town and we feel
that they will judge the mills and our entire operation by what they
see around them. As a consequence we feel it is much easier for us
to keep the houses up and present a good outward image of the
whole operation by continuing to run the town.[34]

In Scotia, providing for labor and providing for business are one
and the same—the dynamism of capital culture in the redwood region climaxes here, in the architectural fusion of life and labor with
the necessities imposed by physical nature. By 1981 there were 1,173
people living in Scotia, with employees who wanted to get in having
to spend two years on a waiting list. In that year company spokesman
Stanley Parker told local reporters that "Scotia is the Pacific Lumber
Company's pride and joy."[35] With Palco cutting 120 million board feet
per year, about as much redwood as annually grows on the property,
he suggested that there was at the moment no incentive to sell the
town. If Parker had any inkling of the trouble that was brewing, he
showed no signs of it here.

According to the local historian Ray Raphael, when Maxxam
reached a merger agreement with the Pacific Lumber Board of Directors on October 22, 1985, taking a controlling interest at $40 per
share and acquiring the company at probably 50 percent of its value,
the community was apprehensive.[36] By way of showing how strongly
people felt, Raphael published an interview with local citizen and
timber supporter Bill Bertain, whose father had lived and worked in

Scotia since 1920, operating the laundry until 1988, when he turned it over to Bertain's brother. Bertain explained how the atmosphere in Scotia got more and more poisonous after the takeover. When Hurwitz started dismantling the company, for example, selling the welding company for $325 million and the San Francisco office building for $30 million and tripling the cut rate while ramping up working hours and making the men come to work earlier, he sparked a resistance movement that reached beyond the environmental community. Workers and citizens of Scotia and Humboldt also tried to fight back. Bertain, a lawyer, tried to get to the bottom of the takeover and expose illegal actions taken by Maxxam in acquiring control, going so far as to help organize local folks into an ultimately unsuccessful effort to buy back the company in the interest of the workers and the people. In response, Hurwitz forced his brother out of Scotia. "Most people in town are still pretty intimidated," Bertain told Raphael.

> There's a lot of fear. You can lose your job if you complain too much. It's a real sad situation. The people know that the faster they work, the sooner they'll be out of a job. They used to feel that they could work there the rest of their working lives and now that security is gone. They realize exactly what is going on. Some people, on the other hand, have developed some degree of nastiness towards the environmentalists. That's understandable, but lots of their anger is directed in the wrong direction. If Hurwitz hadn't accelerated the cut so blatantly, few people would have ever known about the "Headwaters Forest." The people who feel that Campbell and Hurwitz are the good guys argue the property rights line. I believe in property rights too, but I believe you have to acquire property by fair means, not by fraudulent means. And I think you have a duty as a steward of the land to do it right and think about future generations. To be a free man, you're going to have to fight.

When I interviewed Bertain at his Eureka office in 2001, he was still inflamed about the outcome. His principal motive was social justice, he said, as he passed along documents and clippings alleging Maxxam malfeasance and the efforts of the so-called PL Rescue Fund, which had briefly challenged the company's rule in the late 1980s and early 1990s with an employee stock ownership plan (ESOP) that attracted four hundred employees and family members to one organizational meeting in September 1988. According to the plan's cochairman, Pete Kayes, though "the vast majority of the employees support the concept of the ESOP," it will be difficult to succeed because "there hasn't been an employee organization at P.L. for forty years."[37]

In the same year, Maxxam published an advertisement in the *Times-Standard* that appeared in the form of an open letter to the community. Whereas in the past "we felt no one needed to be reminded of the obvious ongoing care and concern the Pacific Lumber Company has for the community which has, for 120 years, graciously hosted our operation," this year "we have concerns about certain trends. . . . There are griping pockets of dissent which seek to overthrow or dismantle this company and its tradition of more than a century of community care. . . . Just think about the sharing and giving we have accomplished together in the past 120 years . . . before you sign any petitions, just think, that is all we ask of you this yuletide season."[38] The Scotia strategy had clearly broken down; the workers were stirring, and the company was compelled to enter the public-sphere struggle and shore up its programmatic identification of their lives and well-being with the company line.

In the midst of this controversy, an earthquake struck Scotia in 1992, causing a fire that burned the company's central shopping district to the ground. The disaster initiated a two-year discussion and building project that resulted in a new town center—an architectural rebus that reflects in its surface the hostile semiotic environment from which it emerged as another attempt at making the town a tourist attraction and tool of image management. With raw redwood pillars and rough-cut siding and green trim done with impressive workmanship, it resembles very closely a cabin built of Lincoln Logs. Given the necessity of shoring up its crumbling corporate image among both environmental and employee challenges, it is hard to imagine how any other color scheme could have been chosen. The new public face of the town was built to publicize its new green corporate self-image.

The cumulative effects of Maxxam's takeover of Pacific Lumber demoted the operation in the eyes of many; it fell in rank from great corporate citizen to a pernicious outside corporate raider. Lifetime employees were now setting aside the famous orange Palco jacket, once a symbol of pride among workers, in shame. One longtime Maxxam/Pacific Lumber worker, whose father had worked his whole life at Pacific Lumber and still worked there, told me that "it has got to the point where you don't want to be seen in a Palco hat anymore, you don't want to be seen wearing a Palco jacket anymore, because you just don't want to have to deal with it." Whereas "it used to be a

Hobys Market and Deli, Scotia's town center, 2001.

sign of status, because if you worked for the Pacific Lumber Company everybody knew you had a good job, you had a pension, you had benefits and your kids were taken care of and you had a job for a lifetime," now "people are bitter about that . . . people who have lost their jobs years ago and hate that company, because they had a bad experience there and because of the position that's been out there in the media."[39]

A few conclusions can be drawn from this short, schematic, and impressionistic history. The town was necessary at first because of the location; nature ruled, and it was physically and geographically essential to have the town on-site. Later company management conceived of it and ran it as a program for labor control. Finally it was used as a public statement of the company's ecological integrity and land stewardship as well. If space warranted, it would be relevant in this regard to describe in full the company's fun public salmon restoration display in Scotia, its demonstration forest, and the self-produced video history sold at its private museum of its own corporate history. But instead it seems appropriate for a laid-off timber worker living in the shadow of Scotia to have the last word.

I met him in 2001, and we sat together in a café as he talked on a day that the company had announced another unexpected round of summer layoffs. That morning radio news reports were quoting Pacific Lumber's president John Campbell blaming the layoffs on "forest defenders who cut off the log supply with frivolous litigation." Cynthia Elkins, spokeswoman for the Environmental Protection Information Center, countered that the big job-producing logs were all gone—Maxxam had cut itself out of business. With this public-sphere spectacle of charge and countercharge as a backdrop, I asked the forty-something white man about his life as a mill worker and how he related to the new layoffs. "The timber industry around here is totally downhill from what it used to be," he said. "I worked in a sawmill for fourteen years, and that's the reason I'm not working in a sawmill anymore, it's for the same reason." There are just not enough jobs? "That's right, there are just not enough jobs. The whole environmental thing came up about the spotted owl, that's when my company shut down, I was consequently laid off."

But the spotted owl thing was ten years ago, I said. "Right, it's kind of a sore spot for me, because I had a really good job. The only reason that this town exists is because of Pacific Lumber. It's like a bedroom community. It's the only reason it's here. At one time there was 156 sawmills in this county, now there's only four." What is causing the decline? "The environmental issues are basically what's going on. The Earth Firsters and all that stuff . . . has really taken its toll." But the environmentalists say there are no more logs out there, I interjected. "The logs are there, but nobody is allowed to cut them anymore. . . . Do you know what happened to Headwaters forest? . . . PL was basically just forced to give that land up. And it was a major part of their harvest . . . so that's kind of a sore spot with me. Because when all of that was going on, the mill that I worked at was basically supported by PL. Our whole existence would ride on their logging. . . . I have to say, once Maxxam took over, that was their downfall right there."

Were you ever in a union, I asked him. "Never," he replied. "As a matter of fact, the mill that I worked at said that if the unions ever came in there, they would shut down. That's just the way it was. Nobody ever said anything about it or anything because that's just the way it was." When Maxxam took over, he said, "Nobody around here wanted that to happen," but "Hurwitz had the money, so there's nothing anybody can do about it." You can "talk to anybody that lives around here and works at PL—most of those people have been there for thirty years or more—and they don't want to give up their jobs," he said. "Most of them are in pretty cushy positions, they've been there so long that, that they're just instituted into it, you know what I mean?" Yes, I said, identifying with his plaintive tone, feeling for myself how the dilapidated state of Rio Dell mimicked the effect of his story of decline. "Like I say," he went on, getting into it now, channeling his anger and wounded pride with intelligent reserve, "it's a sore spot with me, because I put fourteen years into this mill that I worked at and I had employee stock in the company, I just lost it all. . . . The reason I'm so down on all of this is because now I'm stuck in a dead-end job. All the time I put into the mill there, that was like my career—now I'm stuck working at a goddamned ranch milking cows." By way of comparison, "PL is like the sacred god that you can have—you know I would go to work for them in a second, if I had a chance to get on there I would do it, because you're set for life, you go to work for that plant and you're set for life, as long as you do your job and do it right."

Workers' housing in Scotia, 2001.

But when Scotia's big mills finally closed in 2001 and its famous old-growth milling machines were auctioned off in 2002 because, as the company sees it, it had run out of big trees, employees were not feeling set for life anymore. Said spokesman Robert Manne to the *Times-Standard*, "This company is bleeding"—having cut 250 employees since the 1999 Headwaters agreement.[40] By 2007 the Carlotta and Fortuna mills were also shuttered, and the company declared bankruptcy on January 18.[41] Over the next year and a half, all parties trained their attention on the court struggle over competing reorganization plans. The company wanted to sell off 22,000 acres for elite 160-acre mini-estates, to be marketed as an ecological alternative to massive subdivision by keeping the population density low and preserving big trees. A second plan was put forward by debtors who were still owed more than $800 million, an obligation largely secured by the company's timber inventory, the value of which depends on the very environmental issues and regulations at the heart of the controversy. The debtors wanted to sell the land and inventory to the highest bidder. A third and victorious plan was offered by the Fisher family's Mendocino Redwood Company (MRC) of Mendocino County. The judge ruled on July 8, 2008, that its proposal to acquire Pacific Lumber, retool the Scotia mills, privatize the company town, and cut costs by eliminating managerial and labor redundancy was in the region's best interest. Ironically, the judge cited MRC's proven track record of sustainable forestry under the auspices of the Forest Stewardship Council's Certified Sustainable Forestry program in making his decision, though Mendocino had been almost completely deforested of ancient trees by the land's previous owner, before the new company gained certification.[42]

The bankruptcy outcome cemented Maxxam's fate. Hurwitz lost control and would soon be out of Humboldt—a major victory for the forest defense. He had issued junk bond debt to buy the company in 1986, ramped up the cut rate, stripped the land to pay the interest, and sold Headwaters Grove for $480 million in a deal that put restrictions on logging the land. For twenty-two years, he did not pay down the debt but just bled the region with interest payments.

Still at the center of the ongoing struggle in Humboldt sits the company town of Scotia—a living relic—still working, but not that hard, and though its museum stayed open through the summer of 2008, the mill tours had been canceled and the salmon hatchery display was left unattended while the new owners planned their ascent.

Timber Is King

Just how deeply does the dominant culture of modern capital run beneath these discursive and architectural surfaces? We can look at one of the landscape's most prominent features and use it to think about the modern social imaginary in Humboldt.

Eureka is home to the county courthouse. Above the main entry stands a tile mosaic, thirty-six feet tall and twenty feet wide, depicting a logger with ax in hand, his foot on a log, and his settler family behind and close by to one side: father, mother, and child together form a master signifier of the crucial libidinal engine of modern economics—the nuclear family. Several symbols of technological progress skirt the edges—a steamship, a factory, and row houses. The overall effect is to stamp the county's official bastion of legal authority, its architectural monument, with the legitimate face of timber culture. But perhaps the deepest, furthest-reaching message is not the picture itself but its location. It stands above the main entrance. This is the face of the law in Humboldt: timber is king.

The county jail stands behind the courthouse in these pictures, a visual symbol of the violent force backing the dominant order. Upon entering the glass doors under the logger's feet, the visitor is greeted by several big, darkly lit, but sumptuous paintings of redwood groves lining the main corridor. The building's dedication plaque is straight ahead, honoring the "early pioneers of Humboldt County, their descendants and those who came after," not those who came before—not those who lost everything, but those whose "pride and vision carved out of the forest a community of substance ever building for the future." Yet because it is "further dedicated to public service in the name of democracy to promote and maintain liberty and justice under law based on universal equality as the foundation of human dignity," it constitutes a promise to everyone, not just the triumphant, that a place will be held in which to stand before the law and make a claim on the national identity. It is not without symbolic effect that, were the courthouse to be demolished, Indian Island would be visible from the spot where these photos were taken, and that the forest defenders are regularly held in the jail and tried for trespassing while protesters gather on the steps below the mosaic narrative of big timber culture. It might also be pointed out that the logger stands with his back to Indian Island, looking east toward Humboldt's national origins, not west toward its colonial horizon. In Humboldt's saturated space of

The exterior of the Humboldt County Courthouse features a tile mosaic commemorating logging in the area.

historical surfaces, even this architectural accident can take on profound significance.

Not far south along Fifth Street lies a small logging museum. Numerous outdoor technology exhibits tracing the mechanization of
lumber production surround the building and dwarf its presence,
dominating the visual landscape of open space here at Eureka's Fort
Humboldt State Historic Park. Established in 1853 and abandoned
in 1870, the fort's role in removing and concentrating Indians into
reservation camps is recounted in a museum that has been built into
the site's old hospital building. The stories recounted here complicate
those told by the numerous popular histories of logging that are sold
in the nearby gift shop. But the park's public literature, introductory
window display, and Web site still say that the fort's mission was
"to assist in conflict resolution between Native Americans and gold-
seekers and settlers who had begun flooding into the area after the
discovery of gold in the northern mines." And the sign inviting driv-
ers on Highway 101 to pull into the park ironically reads "Free Logging Museum," without mentioning the larger, newer, critical Native
American exhibition.

These surfaces embody and so now illuminate the deep permeation
of physical Humboldt by industrial redwood production—they are
elements in a symbolic order that draws on every aspect of the built
environment, including the deforested landscape, the silted-in rivers
and bay, the acres of second and third growth, the town of Scotia, the
factories and mills, the ghost towns, the mothballed railroad system,
the road system, the park system—everything produced by the culture
system of capitalism as it colonized the redwoods.

As Michael Warner has written, social movements can be thought
of as counterpublics at work in the context of dominant publics. From
this perspective, the public appeals made by the forest defenders, the
company, the workers, and their community of support can be treated
as instances of public political discourse; we must now extend this
idea, conceptualizing the range of objective symbolic, geographic, and
physical architectures we have been discussing here as forms of public
address. As Warner might have put it, each element in this narrative
landscape of historical memory and perpetual social conflict specifies
"in advance, in countless highly condensed ways, the lifeworld of its
circulation." When a voice emanates from one of these positions, intended for each and all to hear and to see and to visit—and it is precisely this nondirectedness that constitutes it as a public address—it

not only says, "'Let a public exist,' but, 'Let it have this character, speak this way, see the world in this way.'" And the speaker of that discourse then "goes out in search of confirmation that such a public exists, with greater or lesser success—success being further attempts to cite, circulate and realize the world-understanding it articulates." The continuous symbolic mobilization and countermobilization of the timber war stories can be read in this way—they all come with lifeworld specifications explicitly and implicitly inscribed. They all, as Warner put it, run their ideas of the world "up the flagpole and see who salutes"; they all "put on a show and see who shows up."[43]

Cultural geography and the built environment operate in the same way. Open a sawmill, build a company town, and see who applies for a job. Organize a rally and see who attends. Occupy a tree and see who comes to help. Conquer the land, build a courthouse, dedicate it to the law and to universal human rights, install public art like the logger mosaic, and see who shows up to answer its call to step up and fight to define the terms of person, property, and production for life—definitions that will determine the flow of accumulating values originating in land, labor, and physical nature. Social movements can be measured by the publics they are able to call up and maintain—and so can objective symbolic orders of naturalized performance and the built environment like the one we have here in redwood country.

The Struggle Is a Public Space of Discourse

Just as numerous watershed protection groups and environmental defense nonprofits and watchdogs work within the forest defense, each developing a unique idiom of intervention in the timber war field of cultural production, so do elements of the dominant culture such as Maxxam, the other large industrial timber corporations, and various individuals and groups who support the timber industry and thus occupy the center and right of that political spectrum. Each intervenes in the discursive field with symbolic acts and articulations that function to maintain the symbolic order. This culture too, like the resistance, is a hodgepodge of idioms—it can only be branded hegemony with the caveat that it be understood as a plurality of voices united in concern for redwood production, property culture, and nation, just as the forest defense is a plurality of voices united in concern for sustainable forestry, global ecology, and social justice.

Each struggle is a space of discourse, defined by the circulation of claim and counterclaim, address and rebuttal, public and counter-public. Just as the Internet should be understood as a new public space for political discourse, so should we understand new struggles as openings of new spaces for new political performances and new movements as new idioms of resistance. Collectively, movements send messages that get answered back, creating new fronts or theaters of struggle. Again, Michael Warner captures this idea precisely: "Public discourse, in other words, is poetic," he writes, meaning "not just that a public is self-organizing, a kind of entity created by its own discourse, or even that this space of circulation is taken to be a social entity," but also that "all discourse or performance addressed to a public must characterize the world in which it attempts to circulate, projecting for that world a concrete and livable shape and attempting to realize that world through address."[44] Then, in what I believe should be an important contribution to understanding the spectacular function of mass media in constituting the power of collectives to be and thus act collectively, he says that the success of a public address "depends on the recognition of participants and their further circulatory activity." But the problem is that "people do not commonly recognize themselves as virtual projections. They recognize themselves only as already being the persons they are addressed as being and as already belonging to the world that is condensed in their discourse."[45] Today our public spheres are dominated by the address of corporate-dominated consumer spectacle, which colonizes the cultural stuff and the spaces with which and in which citizens, workers, activists, and CEOs work themselves up and make values worth performing. That is why the great struggles like the timber wars have turned into public-sphere contests to capture attention and identify actors with causes and projects. Even the nation, the big project, the big modern movement, needs to be seen in these terms of discursive projection and media technologies of spectacular, psychical identification.

From this perspective, the timber wars can be seen as a public space where the performative cultural politics of forest defense square off against those of established corporate power and property rights. The fight revolves around redwoods because that resource ecology dominates the region and feeds corporate capital's limitless compulsion to bind communities of labor in commodity circuits, but it does not stop there. It draws in other related concerns. In an expanding circle of consciousness of relationships that is perhaps the most essential characteristic of ecology movements and environmentalism in general, and which increasingly dominates the concept of globalization, forest defenders tend to expand the domain of their struggle to all of Humboldt's rivers, its species, the quality of its air, and every aspect of its communities of labor. They make the struggle into a spatial technology for making public-interest claims all the way up the scale from redwood environment and labor to the economy, politics, race, class, gender, property, law, the state, globalization, and planetary ecology.

We have noted the entry of David Chain's controversial death, the Luna tree-sit, Julia Butterfly, the role of Earth First!, Scotia, Maxxam, the cultural form of the corporation, the institution of private property, and the system of capitalism itself into this new space of struggle; these were the storied images and events that dominated this field of cultural production in the fall of 1998. The narrated images of David Chain's killing, corporate domination, species extinction, ancient forest destruction, political corruption, Butterfly's feminist ecoheroism, civil disobedience, watershed contamination, and the chemical torture of nonviolent activists formed a reservoir of cultural material and a structure of collectivized psychical attentions (objects) out of which and into which both individual identities and group formations could be and were being channeled and built in that moment. Each of these illustrated story lines adds content to the overarching story that people are learning to tell: corporate timber culture dominates the field of power, while communities of resistance are hard-pressed to stand firm in the transition to globally financed capital culture, and timber folk get stuck in the middle.

By way of illustrating how passionate this public discourse became in the meaning-making lives of this uncomfortable middle, we should briefly pause to follow up on the stories of David Chain and Julia Hill. Both activists left behind a material legacy that was subsequently attacked in a vicious and desperate symbolic fashion. Over Thanksgiving weekend in November 2000, the tree that Butterfly occupied from December 1997 to December 1999, and which she bought from Maxxam for $50,000 to ensure that it would be conserved forever, was deeply cut around the base of the trunk with a chain saw. When the images hit the public sphere, no one seriously doubted that the gesture originated in the timber community. The press release carried by Hill's Circle of Life Foundation described the attack as deep, precise, potentially fatal, and most likely carried out by an experienced tree faller.

Emergency crews rushed to the scene, and professional arborists debated how best to suture the wound and help the victim survive. Great steel brackets were ultimately bolted around the base of the tree, stitching up the lesion.[46] Front-page photos in the *Times-Standard* showed Butterfly kneeling in the redwood duff with her tearful face turned up to the canopy and her hands on the gash.[47] Timber supporters also chimed in. Wrote Matt Morehouse of the attack on Luna, Hill's "two year criminal trespass has encouraged this brave act of civil disobedience. . . . Unfortunately that protest will be treated by the media and the law much more harshly than those who engendered it. Miss Hill and her socialist cohorts are bent on stripping all good citizens of their rights to own and use roads, trucks, cars, guns, fireplaces and most other accoutrements of civilized life."[48]

In the case of Chain, killed by a redwood tree cut by a Maxxam logger, his mother brought a wrongful-death lawsuit against the company, the settlement of which included an undisclosed sum of money, permanent preservation of the tree that killed him with a hundred-foot buffer zone, a roundtable working group charged with discussing the timber wars in a constructive fashion, and the creation of a monument to Chain on a portion of Pacific Lumber property in Humboldt. When I visited the monument near Grizzly Creek Redwoods State Park in September 2007, a park ranger guided me to the remote plot. The monument consisted of a large piece of granite, leaning against the base of a redwood, with a bronze plaque bearing Chain's bust in relief against redwood scenery and various forest defense scenarios. Inscribed on the plaque were the words "May this memorial stand in honor of a forest defender and his commitment to non-violence." It had recently been vandalized with a power drill through the forehead and doused with red paint, a symbolic remurdering of the slain forest defender. As the ranger on duty explained, no one doubts the origin of this gesture either, but neither does anyone question the legitimate pains of the workingmen and -women whose livelihoods had always hung in the balance.

The culture of rights that American constitutionalism articulates in these active and laboring voices of Humboldt illuminates how the local, place-based redwood imaginary, in its complex and unique instantiation of the modern social imaginary, embodies the dual code of modern Western liberal nationalism, in which institutionalized private property stands over and against—as a specific defense against— the public interest. That simple formula, inscribed by the founders in the durable texts of enumerated rights, formed a juridical engine that resonates today in the immediate idioms of environmental conflict. Our understanding of the timber wars must begin from this point: that the colonizing culture of capitalism sailed into Humboldt Bay on this discourse and developed its tendencies, occupying the land and putting labor to work. In the chapters that follow, I revisit the incipient redwood imaginary at three signature moments of extraordinary social violence, tracing the emergence of contemporary historical consciousness and Humboldt's landscape of social memory through the Indian wars and the labor trouble that set the symbolic and material conditions of the redwood timber wars.

4. Indian Trouble:
The Colonizing Culture of Capitalism

Force is the midwife of every old society pregnant
with a new one. It is itself an economic power.

—*Karl Marx, 1867*

Somehow, pioneers, in their history as in their lives,
have always had a pretty hard time of it and been
exposed to the charge of doing some very hard and
wicked things.

—*E. H. Howard, Humboldt pioneer, 1881*

In the early darkness of Sunday morning, February 26, 1860, a group of men
whom history has left unnamed quietly slid their boats beneath a
strong north wind and landed on the shores of Indian Island—a wind-
swept, marshy patch of land on which Humboldt Bay's ancient indigenous peoples held their yearly world renewal ceremonies. The Wiyot
Indians hosted the ritual throughout the long decade of gold rush,
timber boom, and so-called Indian trouble that came to Humboldt in
1850, when the Laura Virginia land company rediscovered the narrow passage into the bay on April 14 of that year—five months before
California was annexed to the United States of America.[1]

What happened on the island that night still haunts the redwood
imaginary. Conversations chill when the scene is described. Books
recount the horror. Yearly vigils are held near the spot by remnants
of the Wiyot people, whose ancestors were decimated that night.

Recently the tribe has purchased a portion of the sacred site and plans to resume its ceremonial renewals, but today there lie only ruins of the sawmills and shipyards and settler farms that moved in after the slaughter. An interpretive center under tribal control is planned by surviving Wiyot now living on a small reservation overlooking the bay from its high southern bluffs.

Not far from the island, Eureka's free Historic Logging Museum sits on the grounds of Fort Humboldt State Historic Park. The fort was established at Eureka in 1853, built high on a bluff overlooking the bay from the east, as part of a federal effort to manage the Indian trouble. The old army hospital at the site was recently renovated to house a museum that tells the official story of cultural invasion, broken treaties, lethal repression by settlers, and ultimately the massacre on Indian Island and the war of extermination whose endgame it triggered. State-ordered removal of all the redwood tribes to reservations ended the all-out war that raged between the massacre in 1860 and 1865.

Today Indian reservations and casinos stand at the margins of the county, driven from the center in a socio-psycho-geographic expression of how the colonizing culture imagined the place of Native Americans in its new order of things. While some Indians of various descent stood with forest defenders and United Steelworkers at local antilogging protests, California Board of Forestry meetings, and anti-globalization rallies, others have kept their distance, and the tribes in general exploit their resources as does everyone else, even doing some commercial logging, though Native landholdings are largely inland, beyond the narrow strip of redwood ecology that hugs the coast.[2]

These complex relations form the backdrop of Humboldt's contemporary timber wars; they suggest that a return trip is necessary through the region's nineteenth-century theater of primitive accumulation—accumulation of capital by force, as Marx described it, prior to and constitutive of capital as a legitimate cultural power over labor in the so-called free wage-labor markets that came to dominate production for life in the redwoods.[3] After taking the land, production of lumber for exchange created surplus value, otherwise known as profit, or simply more capital, which timber operators used for more control over labor and for pressing land claims. They resisted the unions, mechanized, built company towns, merged and grew through land acquisition, bought media space for advertising and for the mass preparation of attitudes favorable to industrial timber interests, and in the end they sold out to the global corporations.

Removal

Native Americans resisted leaving their homeland for reservations where they would have to live on someone else's ancestral land, thrown together with people of differing languages and customs. By the 1860s, they had no choice. Large numbers of Indians were rounded up and held at Fort Humboldt before being sent away to reservations.

The Corral

Hundreds of Indians were held prisoner at Fort Humboldt, awaiting removal to reservations. Guarding them became a problem as more were brought in. Responding to Indian complaints about white brutality, and to prevent the prisoners from running away, the soldiers built a corral to confine the Native Americans.

The old military hospital at Fort Humboldt in Eureka now houses a historical museum of Indian trouble, including this installation depicting the white community's attempt at a "final solution" to the Indian wars in 1862. Remnants of the bay redwood region tribes were concentrated at the fort, held under lethal conditions of deprivation and abuse in a corral built for the effort, and then removed to reservations.

Primitive accumulation got this dynamic system off the ground or, rather, sank it in the ground. From the perspective of each new place that it colonizes, accumulation by force installs the system, a crucial first step in making capitalism what it is—a self-sustaining, inexorable place-making force of cultural occupation and expansion. The trouble for pioneer colonists and early timber operators in Humboldt, of course, was that Indians already occupied this place. The shores of Humboldt Bay on which the Laura Virginia Company landed were Wiyot territory; the surrounding river valleys and mountains were the place of Yurok, Karuk, Hupa, Chilula, Whilkut, Nongatl, Mattole, Sinkyone, Lassik, Wintun, Chimariko, New River, Konomihu, and Tolowa peoples.

Listening in on this colonizing culture, I draw lines of association out from the media representations of the Wiyot massacre to the colonial voices, government documents, military records, early histories, maps, and other cultural documents that elevate the event and ensure its place in the production of contemporary historical consciousness. These sources do not transparently reveal the concerns of Humboldt's colonizing population, but they do form an archive of utterances from which we can begin to interpret how its subjects knew their world at this historical moment. The cares, worries, practices, and psychical investments embodied in the archive created by the signature event of the Wiyot massacre show how pioneer knowledge was power that organized labor and remade the redwoods in the national image. As we will see, it bears the mark of the unconscious culture engine of rights discourse that drove the invasion and established a modern social imaginary in the redwoods.

Sign of the Times: Primitive Accumulations of Capital Culture

The first public-sphere media report of the event appeared in the pages of the short-lived competitor to the *Humboldt Times,* the *Northern Californian,* published weekly on the north shore of the bay at Union Town, now Arcata. S. G. Whipple, the editor and proprietor, just happened to be in Eureka on Sunday morning after the massacre, en route to San Francisco on the steamer *Columbia.* He left his assistant editor Bret Harte in charge of the paper. Whipple sent a statement back to Harte in Arcata, who published it in the paper's Wednesday edition. It is the closest we have to a firsthand account of the scene (italics in this and the following quotations are mine).

Indiscriminate Massacre of Indians:
Women and Children Butchered

. . . On Monday, we received a statement from our Senior, at Eureka en route for San Francisco. He says: "About 9 o'clock, I visited the Island and there a horrible scene was presented. The bodies of 36 women and children, recently killed, laid in and near the several ranches—they were of all ages, from the child of but two or three years to the old skeleton *squaw*. From appearances the most of them must have been killed with axes or hatchets—as the heads and bodies of many were gashed, as with such an instrument. It was a sickening and pitiful sight. Some 5 or 6 were still alive and one old woman was able to talk, though dreadfully wounded. Dr. Lee, who visited them and dressed the wounds of those alive, says that some will recover if properly cared for. It is not generally known that more than three *bucks* were killed—though it is supposed there must have been 15 or 20. It is thought that the bodies of the men were taken away by Indians this morning as four canoes were seen to leave the Island.

On the beach south of the entrance it is reported that from thirty to fifty were killed. It is also reported, that at Bucksport, all were killed that were there. I passed sight of them about 11 o'clock and saw the ranches on fire. It is also said that the same has been done at the several ranches on Eel river. No one seems to know who was engaged in this slaughter, but it is supposed to have been men who have suffered from *depredations* so long on Eel river and vicinity." It is said that some jerked beef, about 100 lbs., was found in one of the Indian ranches on Indian Island and on south beach.[4]

Although it was later suggested that the beef in question was actually seal (the bay Indians did not eat beef), the mention of "depredations" and cattle production gets right to the heart of the matter.[5] Because cattle depredation was the public rationale for the slaughter, it will be instructive to consider the productive relations in which these capital crimes were embedded.

After just three years of white colonization, wild game was already becoming scarce and cattle populous on what had been the indigenous commons. By 1853 systematic destruction of the region's elk, deer, and bear populations by settlers forced the state to recognize that if it wanted to keep the peace, it would have to supply the natives with meat. That year California Superintendent of Indian Affairs Edward

Beale reported to the U.S secretary of the interior that white encroachment on Indian-occupied lands and corruption of "the system of beef delivery to Indians" had led to starvation.[6] Beale's "garrison plan" for "the colonization of California Indians on reservations," purportedly for their own protection, shows what was at stake in the struggle to occupy redwood land and industrialize its timber production and agriculture: white appropriation by preemption, school land sale, and homestead took Indian territory and left tribal peoples hungry.

While Indians driven into the hills and onto corrupt reservations starved, Humboldt produced 80,000 pounds of butter, 2,000 pounds of cheese, and 3,604 cattle in 1856, an increase from 1,812 head in 1854. That number rose steadily to 19,205 head in the pivotal year of 1860.[7] Salmon were harvested in masses by equally entrepreneurial whites at the Eel River mouth. By 1857 they had achieved industrial production of 50,000 pounds of smoked salmon for local consumption and 2,000 barrels of cured salmon for shipment to world-system markets that stretched to Australia, China, the Sandwich Islands, and New York.[8] The first agricultural statistics compiled by the new county's assessor office in 1854 reported cultivation of 2,500 acres, including grains, potatoes, corn, and orchards (the orchards having been planted but not yet bearing).[9] The situation of shark fishing in the bay yields the same impression: on a single day in June 1856, the crew of the *Sam Slick* caught enough to yield 300 gallons of shark oil for machinery and lighting. The next year a fleet of ships eagerly greeted the sharks as they entered the bay.[10] By the 1860s no appreciable shark fishery could be sustained.

During this period, industrial timber extraction became Humboldt's number one commodity for export. In 1854 native forest yielded 27 million board feet of lumber to the hand saw and oxen, with 187 boatloads shipped into the world system through the mouth of the bay, so much, in fact, that a market glut ensued that, when combined with a general recession, shut down many of the area's mills—an economic shock from which the industry did not recover for years.[11] Here we see how redwood capital was in chronic crisis from its inaugural run, regularly expanding under competitive pressure and then suffering contraction when overproduction drove prices down until markets collapsed.

As white pioneers struggled to ramp up their achievements, fattening cattle on native lands, Indians living in the mountains took what they could from the branded, private-property herds. In years before the Indian Island massacre, the natives occasionally killed white men as well as their cattle, usually in one-to-one reprisals for the killing of their own, for the violation of native women, or for encroachment on their territory.[12] Meanwhile Indians lived in constant fear of collective punishment for untried crimes. Wrote the *Humboldt Times* on the morning before the massacre: "Indians are still killing the stock of the settlers in the back country and will continue to do so until they are driven from that section, or exterminated." Two head of stock had been killed just that Wednesday, with twelve others driven off but later recovered. "We learn that the settlers on the Van Duzen are preparing to abandon that section," the report continues, "unless some protection of their property is afforded and speedily." According to the editors of the *Humboldt Times,* the Humboldt Dragoons, a company of volunteers commanded by Captain Seaman Wright, had been unable to provide the needed security.

In a letter of May 11, 1860, to the *San Francisco Bulletin,* the pseudonymous "Exodus" of Humboldt claimed it had been one year since any white person had been killed by Indians. The man killed was Ellison, and his story, embedded in a flurry of letters concerning the Wiyot massacre, says much about social conditions that shaped the signature crime and how it incited the population into an all-out war of extermination that ended in the forced removal of the natives into concentrated camps—the reservations.[13] The letters were published in the San Francisco papers, not in Humboldt's, for reasons that readers will better understand as they grow more familiar with Humboldt's public culture in the 1850s and 1860s.

On Tuesday after the event, the *Bulletin* reported the arrival of the steamer *Columbia,* sailing down the north coast with newspapers from Umpqua, Crescent City, and Humboldt: "She brought information of an indiscriminate and horrible massacre of Indians by a gang of white ruffians." Reported J. A. Lord, passenger on the *Columbia* and an express messenger of Wells, Fargo and Company: "Not less than 200 Indians were killed—men, women and children—on this Sabbath morning," in simultaneous massacres by "farmers and graziers of the Eel river country" who have "suffered Indian depredations." Although it was said that peaceful Wiyot Indians living on the bay "furnished arms and ammunition to those in the mountains" and sheltered them in their war with the volunteer dragoons, this accusation was never taken seriously, for the Wiyot lived close by the colonizers, working with them and for them on the shores of the bay. As evidence of these friendly relations, we have the May 11, 1860, letter of Exodus to the *San Francisco Bulletin:* "In many cases these Indians were useful. They were divers hands at the fisheries; they were harvesters, aiding whites in getting their grain and bringing them berries, fish and clams; they were packers and guides to the mountain trains; while their wives were of much service to the ladies of Eureka on their wash days and in other household duties."[14]

The good ship *Columbia* also carried Humboldt's Sheriff Van Ness to San Francisco. He gave his account to J.R.D., contributor to the *Bulletin:* the massacre had been committed by "a body of men, some 40 in number, from Eel River. They rode through the South end of the bay in the night, hitched their horses, took Capt. Buhne's boat, crossed to the south shore and killed all the Indians they found. They then proceeded to Indian Island and commenced an indiscriminate slaughter." The reason, said Sheriff Van Ness, was that Indians in the mountains, in league with Bay Indians, "are constantly killing cattle. . . . This is but the commencement of an Indian war in that section of the country."

Another letter, signed "Anti-Thug," appeared in the *Bulletin* on March 13, bearing accusations that the federal government was to blame: "All of this was done within a stone-throw of the United States barracks; but let me assure you that the United States barracks and United States Troops are as useless here, as they are any where. Notwithstanding, there is a petition signed by all the families on Eel River, praying for Major Rains to protect friendly Indians, yet he lets things rip." Major Rains was then U.S. commander at Fort Humboldt and a known Indian sympathizer; he stood accused by the settlers of not protecting their property—namely, their cattle free-ranging on what had been native commons. Anti-Thug had visited the massacre on the morning after: "History fails to record anything like it," he wrote, "and I blush when I know that these things have been done by beings calling themselves Americans." Then he named the leader of the gang directly—but the *Bulletin* censored the accusation, awaiting fuller proof. In his letter of February 26, published in the *Bulletin* on March 3, Charles Rossiter wrote that "the massacre was headed (as reported by an Indian and believed by a majority of people) by a white man named Brown." But no one was ever brought to trial, and at any rate, accusations by Indians were not admissible in Californian courts at that time.

Another letter on March 13 from Humboldt to the *Bulletin,* this time signed "Eye-Witness," made a direct, verifiable accusation: The Indians had been murdered "by lawless white men belonging to Christian community, without cause or provocation, calling themselves volunteers of Capt. Wright's company. . . . Our Sheriff says, 'served them right!' and the tone of the newspaper called Humboldt *Times,* advocates such principles. United States soldiers had been sent out into the field to protect citizens and property, but were withdrawn from cooperating with men murdering women and children."

Public criticism of the U.S. military for allowing the Indian trouble to spin out of control grew quickly in these weeks, compelling Major Rains to address the massacre's growing public—by which I mean everyone tuning in to the media's accumulating correspondence. Rains defended himself in a letter to the *Bulletin* on May 24, blaming the violence on Captain Wright's company of vigilante dragoons.[15] As evidence of his positive efforts to protect the Indians, Rains quotes a letter of command he had written to his field officer Lieutenant J. W. Cleary (Sixth Infantry U.S.A.), who was out and about trying to limit violence by the cattlemen's militia, one of whom was Ellison, the last white man killed by Indians in Humboldt before the genocidal massacre. According to Rains, Ellison was a dragoon who had, in his quasi-official capacity as a recognized volunteer, distinguished himself among the Indians, who knew him as one of a party of four whites who had fired on a group of Indians who were carrying off some poached beef, killing two of them. Ellison, in other words, was an Indian killer who was himself killed while out killing Indians.

But "the plan of the volunteers [dragoons] killing all the Indians to check cattle-stealing is evidently perfectly absurd," wrote Major Rains to Lieutenant Cleary, "as I have been assured again and again by different persons that there are 3,000 of these upper Eel Indians alone and perhaps 10,000 in the country and its vicinity, all told. . . . Now, if thirty-five men in three months kill three Indians, it requires just 250 years, at that rate to kill them all on Eel River and 700 to rid the county." The military, it turns out, was trying to rationalize control of the Indians: militia leader Captain Wright and his volunteer Indian hunters were just making things harder.

Rains ordered Lieutenant Cleary to persuade the Indians to root out the cattle poachers, turn them over to the militia, and promise that they would stop killing cattle; he would then take that promise to the white volunteers and their cattlemen financiers and beseech them stop their bloody attacks on the Indian men and their women, children, and elderly. "The hostility of these Indians is questionable," wrote

Rains. "They came to kill cattle not from malice, but because they find it difficult to subsist." In fact, the Indians were starving while, according to the letter of Eye-Witness, "in a circle of about 25 miles there are ten or twelve [white] persons living and about 2,000 head of cattle that are not in any enclosure."[16]

These public letters reveal the growing desperation of natives in the years leading up to the winter of 1860. In the midst of food shortages, volunteer militias were hunting them down—economic relations of primitive accumulation that can help us interpret the colonial discourse occasioned by the massacre as we now return to pages of the *Northern Californian* and the *Humboldt Times* and listen more closely to their immediate report of the event.

Of those killed, wrote guest editor Bret Harte in the *Northern Californian* in his first news report of the event, "many of them are familiar to our citizens. 'Bill,' of Mad river, a well known and rather intelligent fellow, has proven a faithful ally to the white men on several occasions and—has had his wife, mother, sister, two brothers and two little children, cruelly butchered by men of that *race* whom he had learned to respect and esteem."[17] Harte took it upon himself to publish a scathing editorial alongside this news report in a separate column (italics mine).

> Our Indian troubles have reached a crisis. Today we record acts of Indian aggression and white retaliation. It is a humiliating fact that the parties who may be supposed to represent *white civilization* have committed the greater barbarity. But before we review the causes that have led to this crowning act of reckless desperation, let us remind the public at a distance from this savage-ridden district, that the secrecy of this indiscriminate massacre is an evidence of its disavowal and detestation by the community. The perpetrators are as yet unknown. . . .
>
> The people of this county have been long suffering and patient. They have had homes plundered, *property* destroyed and lives of friends sacrificed. The protection of a Federal force had been found inadequate and when volunteers have been raised and the captured savages placed on reservations, by some defective screw in the Federal machinery they have escaped.[18]

On the following Saturday, the *Humboldt Times* of Eureka—the region's first media organ, established in 1854—published its report under the headline "Indian Massacre." The paper had dominated the county's public-sphere media until Whipple's *Northern Californian* mounted a brief competition in 1859–60, after which the two papers merged into the principal pioneer chronicle of the period and the primary source for all historians of colonial Humboldt. Its pages performed a continuous media spectacle of address to the new Humboldt public, calling it into being and action and archiving its trace in the redwood imaginary:

> It may well be imagined that this unexpected attack on the *diggers* so near town, accompanied by such a terrible and indiscriminate slaughter, produced considerable excitement here on Sunday morning. . . . As there were only *squaws* and children at these places, except two old *bucks,* it would appear that the design at first was to kill only the bucks. . . . There are men in this county, as there may be elsewhere, where the Government allows these degraded diggers to roam at large and plunder and murder without restraint, who have become perfectly desperate and we have here some of the fruits of that desperation. They have had friends or relatives cruelly and savagely butchered, their homes made desolate and their *hard earned property* destroyed by these sneaking, cowardly wretches. . . . Their brethren in other parts of the State, many of whom approve of hanging up white men without "due process of law" for much less crimes than these diggers have committed, heap ridicule upon them and shed crocodile tears over the "poor Indians." . . . If *in defense of your property* and your all, it becomes necessary to break up these hiding places of your mountain enemies, so be it; but for heaven's sake, in doing this, *do not forget to what race you belong.*[19]

In this report, the *Times* editor Austin Wiley used the dominant language of his era and his community, while dissenting voices were forced to address the public in distant San Francisco. And the story of how the community reacted to Bret Harte's more sympathetic editorial can help us understand why: angry citizens ran him out of town. Under fear of his life for having publicly defended the Indians, Harte departed for San Francisco aboard the *Columbia* on March 28.[20]

A crude system of classification is discernible in Harte's and Wiley's public discourse. Racial signification of the Indians as inhuman (bucks are male ruminant animals) and primitive (diggers are technologically backward root gatherers) is linked to an appeal to state and federal governments for assistance in neutralizing the threat. Women and children are awarded a unique status among the victims in a gendered morality that consistently refers to itself in racial terms. Their deaths are somehow more atrocious. Wiley himself was an advocate of extermination and a veteran associate of Indian killers by the time of the massacre, and here he exhorts his racial public directly: "In defense of your property . . . do not forget to what race you belong."[21]

Both poles of the region's editorial spectrum—the sympathetic and liberal Harte and the murderous conservative Wiley—used common terms of nation, race, god, gender, and property to represent the native as other and obstacle to the total white project. What matters is how the *Humboldt Times*—the voice of the people, the dominant technology of collective identification in the moment when the structured and structuring media archive on which the region's historical accounts would rely was being built—refers to itself and its project of establishing the nation in redwood soil as a masculine, Christian, and racial effort to shore up the institution of property. The power-laden language of this public-sphere spectacle of violence put Humboldt on display to itself and the world as a collective—in this case a morally outraged frontier society in which public culture builds a sense of unity by dint of the cultural practice of racial signification. Articulating concepts of religion and gender through legal claims of property right, colonial mass media gave tangible flesh-and-bone content to the local image of state authority by appeal to the highest law of the land—its authoritative, people-making discourse helped give the people of Humboldt what Étienne Balibar calls a fictive ethnicity.[22]

The editorial positions of Wiley and Harte mirrored the nation's ambivalent constitutional program of guaranteed private rights and state obligations to guarantee public interests. Both demonstrate publicly their republican virtue by professing deep humanity and sympathy for the victims, but neither could question the ultimate propriety of their collective occupation of Indian lands and the consequent destruction of Indian life. Their impulse to universalize—embedded in the philosophy of the law to which they appeal—was brutishly offset by racial particularism. The only question they could imagine was how best to accomplish the work of subjecting that "savage-ridden district" to the civilizing law of their own sacred nation. That was the law to which conservative, landowning, cattle-running Anglos appealed when they made public claims of "Indian depredation"; that was the law to which outraged, humanistic settlers appealed for state intervention in the Indian killing fields; and that was the law that both federal armies and volunteer militias, in their own way, mustered to protect in the place-making, people-making racial property war that established rational, accountable property as the basis of capitalist colonization in Humboldt. Humboldt's future is white capitalism, not tribal communalism. And this signature event of the Wiyot massacre proved over time to be a powerful attractor of historical consciousness, engendering an archive that has grown over decades to stand not quite alone but definitively above that of any other event in colonial times.

American Humboldt: Market Revolution, Media Spectacle, and the National Imaginary in the Redwoods

These traces of the massacre remind us that the year is 1860.[23] The United States is a racial police state on the verge of civil war, caught in contradiction between its voracious economic appetite for the cheap slave labor fueling its economic transformation to commodity production and the ideals of its revolutionary freedom struggle, which had motivated its founding fathers and animated the rising tide of moral opposition to the slave market's institutionalized servitude, torture, and murder.

Humboldt was not unaware of the profitable trade being made of human life, as revealed in the *Humboldt Times* by the regular reportage of slave accumulation. Its issue of October 28, 1854, conveyed some "peculiarly interesting" statistics:

> Cotton Slave Statistics: The Southwestern *News* makes up from the Census Reports some very important statistics, peculiarly interesting to the Cotton growing and Slave States, South Carolina, Georgia, Florida, Alabama, Mississippi, Texas, Louisiana and Arkansas. The whole area is 6,662,185 square miles, of which 21,675,682 acres are improved land. The whole number of slaves is 1,797,768 and whose average rate of increase for the last ten years is 54.46 per cent. The number of bales of cotton made is 2,204,521, averaging 1,197 bales of cotton per thousand slaves. Average number of acres of improved land per bale is 10–12.[24]

These mechanical figures stun the senses, which identify more readily with individuals than populations. They tell a harsh story. The New World was vast, but its riches lay dormant; from the perspective of the capitalist system, it had plenty of land and resources but lacked the labor power necessary to realize their value. Enslaving life energies provided the system with that elementary input. American Manifest

Destiny reaching through the continent to Humboldt was a knowledge engine running largely on this slave labor fuel, the value of which transformed the land into commodities for exchange: the expansive energy of U.S. nationalism, expressed in the success of its westward push to the continental shore, was largely and literally objectified slave body power. And if market revolution as the motor of westward nation building must be read in this context of slave accumulation, so must the production of the nation's local variation that is our object of investigation—the redwood imaginary. It would be peculiarly interesting to know if Fort Humboldt soldiers doing "conflict resolution" during cold Humboldt winters wore slave cotton socks while they put lethal force behind the laws of the colonizing culture.

Concurring with numerous American historians, Charles Sellers describes the market revolution that swept westward in the early 1800s as an era of rapid institutional transformation for the whole of American life. At breakneck speed in the Jacksonian period, institutions of state, religion, family, gender, race, media, and property amplified the power of markets to deliver economic goods.[25] The allure of riches, made real by slave energy, seduced subsistence peasantry into savvy business production of cash crops for exchange in competitive markets, valorizing a new acquisitive culture of accumulation that rocked the family scene, which had for decades been the basic unit of household production and a tireless forge of yeoman subjectivity. As men were increasingly drawn out of household production into the market, women were increasingly relegated to the growing isolation of the private family interior. This was a deep change in the social machinery constitutive of collective psychical character in the rising new world.

The ideological coordinates of this changing family structure included the well-known "cult of domesticity," which promulgated by way of explosive print capitalism and pulp-fiction-reading publics the concept of family as a warm and peaceful sanctuary from cold and cruel public markets. Family was increasingly seen and experienced as a special place for the special roles of woman as childbearer, nurturer, sensual lover, and psychological specialist for the patriarchal authority, economic specialist, and public, political man that the market was making of the husband.[26]

In the new American family, this ideological division of labor forged a new subjectivity—one divided at the core like the institution that formed it and therefore well suited for a modern existence of labor alienated from life, work alienated from pleasure, public alienated from private, and man alienated from woman. The offspring it produced were increasingly virtuous, hardworking, disciplined, and patriotic—in other words, more specifically capitalistic. They were subjects of a system of efficient production that channeled ever more psychosocial bodily energies into labor for market production.[27]

Of course, this degree of institutional change took time to achieve. Fragmentation of the extended family and its reformation into the dominant modern nuclear form proceeded over decades and centuries of industrialization and urbanization, processes that the changing family form both made possible and reflected.[28] Seventeenth- and eighteenth-century sons had generally stayed closer to the family scene, moving out after marriage onto the father's land in a patrimonial system that preserved patriarchal authority by granting sons ownership of land while denying it to women.

Early American subsistence economies inherited this sexual culture of psycho-economic authority from European household production, but the new world conditions ramped up its force. "Cheap land," wrote Sellers, "held absolutely under the seaboard market's capitalist conception of property, swelled patriarchal honor to heroic dimensions in rural America."[29] Describing the father's role and investiture with authority, he continued, "Fee simple land, the augmenting theater of the patriarchal persona, sustained his honor and untrammeled will." For all its cultural change, America's market revolution preserved this patriarchal character in a complex system of land tenure whose federal laws unambiguously promoted white male privatization of land. Land system, personality, and social authority merged in a patriarchal property culture wave that swept across the vast interior and sailed around Cape Horn before passing through San Francisco on its westward path into Humboldt.

We have already seen that the market revolution was in fact both a racial and gender formation, in the special sense that Michael Omi and Howard Winant gave that term when they wrote that racial formation "is the sociohistorical process by which racial categories are created, inhabited, transformed and destroyed. . . . Racial formation is a process of historically situated projects in which human bodies and social structures are represented and organized."[30] But is this not precisely the work that the nation's founders entrusted to the institution of property, which they designed and enumerated constitutionally to foment market revolution? Can we see market revolution as a project representing and organizing bodies and social structures for specific kinds of work, organizing them by knowledge—that is, representations—and representing them by means of social institutions? The institution of legitimate property ownership, for example, the paragon of freedom in the pantheon of American symbols, does just that. It was closed to blacks, Indians, and women in varying degrees and ways at this time; it was a complex social representation of them and thus a cultural formation of them, an image that placed them outside the system of authority. The state made land system laws that instituted symbolic authority in the living categories of race, sex, and gender inhabited by the white male.

The Protestant ethic was another energetic psychical supplement to the racial state land system's gendered culture drive toward market revolution. It "revitalized the market's initial surge," wrote Sellers, with "Calvinism's thrilling promise of divine encounter." Here Sellers reworks Max Weber's thesis on the spirit of capitalism: Protestant rules for living guided the New World's "pious adventurers" into market activity and helped them find "worldly success by equating Christian virtue with the market ethos of self-disciplined effort." Calvinism worked on the American mind by liberating it from traditional religious scruples; it was the "spiritual medium of capitalist transformation," and it worked its magic on capitalist sentiments by "sanctifying worldly work as religious duty and wealth as fruit of grace." One major legacy of this cultural Protestantism was its contribution to the nation's achievement of political democracy, wrote Sellers, for it "championed humble plainness over pride and luxury" and so served as a check on the powerful commercial interests embodied in the founders' strongly worded federalism.[31]

Capitalist culture in market revolution exploded with that Protestant energy in a Second Great Awakening of faith that swept the nation during the Jacksonian era. Like the first Great Awakening of the 1730s and 1740s, which guided the nation's founders and animated their translation of natural religion into natural law jurisprudence and finally into enumerated doctrines of natural rights, this second burst of enthusiasm in the 1830s translated Calvin's *Institutes of the Christian Religion* into language more consonant with the populist element of the rising culture program—the *demos* immanent in democratic republicanism. The *Institutes* had been an effective program for modern living in the light of God's everlasting love, a complex of divinely powerful knowledge laid out in practical rules for self and society.

Likewise the U.S. Constitution was an updated program for setting up modern life—a powerful complex of enlightened knowledge laid out in laws for the formation of a government designed to institute a nation on if not equivalent, at least harmonious, principles, first among which was individual conscience inseparable from public duty. When in the 1830s evangelical Protestantism validated market behavior and lent it the enthusiasm of religious conversion, the market system as nationalist motor gained an integral source of emotional energy and psychical investment.

Combining these insights, we can see the culture system that colonized Humboldt in 1850 as a place-making extension of a nationalist social movement powered by an institutional motor of market revolution, the labor energy of which was set in specific motion by constitutionally organized concepts of rights, race, sex, and religion, among others. The nation, in other words, was a people-making project with psychosocial economic coordinates in the sense that its movement—its expansion into new spaces and its (re)production over time—had an unconscious cultural, social, institutional, and thus increasingly psychically embodied dimension: a dimension uniting the concepts of nation, family, gender, race, property, and god in a definitively American style of practice that gained the authority of a natural attitude. Capital culture's logic of continuous revolution, described by Marx in the *Manifesto* and traced in its American adventure by Sellers, did in fact melt everything solid, but what it produced was more concrete than air; it made the nation as a feeling for legitimate authority—especially for property—in every new place that the system incorporated.[32]

"A nation is a fantastic thing," write Roger Friedland and Richard Hecht, and "making one involves an imaginary fusion of a people, a land, a history and the authority of a single state." The people, they continue, must "believe there is an identity between them and the land" and "understand themselves to have a history," while making that history requires that "they understand themselves as distinct, with an essential nature" and "a character worth reproducing across generations."[33]

But how were the people distributed across the vast new world to achieve such a feeling? As a collectivity of belief, identity, understanding, and aspiration, a nation must be manufactured, and the process of newspaper integration that linked Humboldt to San Francisco and thus to the eastern seaboard and the world shows how: print media constituted a national collective consciousness—a national public—by dint of providing a continental system of public address. If a nation in fact is a psychical entity, an imagined community, then the integrated media of transportation, communication, and print were its original technologies—its conditions of possibility and the actual material of its symbolic constitution. Indeed, these tools constitute the principal record we have of the nation's inscription in California and the bay redwood region.

In the 1850s, transportation and communications technologies were quickly bringing this national formation of market revolution in cultural practice to the land, labor, and ecology of California's North Coast, collapsing the distance that had long shielded the place and flooding it with people seeking gold, land, and profit. These pioneers were the libidinal-economic ambassadors of an Americanized but still Anglo-European capital culture. Their identities were cultural storehouses of the most efficient order. Each was raised in a far-off world and brought into Humboldt a gamut of cares, desires, and inclinations to act that carried the trace of their respective origins. From this perspective, cultural practices cannot be conceived as disembodied forces, ideas floating in air. On the contrary, they inhabit and draw force from the living bodies that bear them. Migrants to Humboldt were bearers of social and cultural relations into the redwood region from the outside, and their bodies fueled the energetic operation of these relations on the landscapes they reached. Subjects themselves, it should be repeated, are psychical engines running on bodily energies organized by external forces of culture and environment. They harbor potential energies that—when addressed, for example, by the nation(al imaginary)—are bound to reply. Thus called up and responding as national subjects, they mediate the forces of culture *and* environment—through their labor they produce and reproduce culture and environment, transforming the land and themselves in the process. And they came into Humboldt, as we will see, somewhat set for success.

Owen C. Coy examined the field notes from which the statistical tables of the Eighth U.S. Census of 1860 were compiled.[34] He reported that while 460 of Humboldt's total official population of 2,694 (Eureka, 581; Arcata, 524) had been born in California (this number not reflecting the indigenous population), the remainders were immigrants, of which New England furnished 315, New York 201, Ohio 123, Maine 113, Pennsylvania 111, and Massachusetts 107. The so-called border slave states of Missouri, Kentucky, Tennessee, and Arkansas together provided 233. Four hundred ninety-seven were foreign-born, with 59 having departed from England and its colonies, 97 from Ireland, 26 from Scotland and Wales, 118 from British America, 13 from other British colonies, 85 from the German states, 22 from France, 12 from Denmark, 5 from Norway, 5 from Sweden, 11 from Portugal, 34 from China, 4 from Mexico, and 2 from the Sandwich Islands.

Among these were a number of individuals familiar with lumbering, men called to attention by certain characteristics of redwood ecology. They saw redwoods as the largest trees that really did grow money. In 1860, 168 men were directly linked to the lumber business, of which 114 were native-born and 54 foreign-born, the largest portion of the foreign-born hailing from Maine (27 percent). Eighty percent, forty-three in number, of the foreign-born lumbermen were from New Brunswick and Nova Scotia—among whom the great future lumber barons of Humboldt, John Vance and William Carson, must be noted.[35]

Likewise, gold called both green and seasoned prospectors to the Trinity River mines of northwest Humboldt. Humboldt Bay called land speculators who founded merchant and seaport towns on its shores. Superproductive rivers and seas called fishermen to the docks and shipbuilders to the timbered bay. The rich alluvial lands called stockmen, dairy men, and farmers to the fertile valleys of the Eel, Van Duzen, and Mad rivers, among others. And in addition to skilled lumbermen, the forests called up migrations of unskilled itinerant woodsman, a transient population that would swell the deep-woods lumber camps during cutting season and the towns on weekends and off season.

It was not inconsequential that this external culture wave swept newspaper technology up the West Coast into Humboldt as well as these skills in blasting value from land and environment. Nothing impacted the region's development more than the information channel to the nation's eastern coast center of semiotic gravity. The prospectus of California's first newspaper, the *Californian,* shows just what the papers imagined for people. Its initial edition on August 15, 1846, presaged the production of America in Humboldt:

Prospectus

This is the first paper ever published in California . . .

We shall maintain freedom of speech and the press and those great principles of religions toleration, which allows every man to worship God according to the dictates of his own conscience.

we shall advocate such a system of public instruction as will bring the means of a good practical education to every child in California.

we shall urge the immediate establishment of a well organized government and a universal obedience to its laws.

we shall encourage imigration and take special pains to point out to agricultural imigrants those sections of unoccupied lands, where the fertility of the soil will most amply repay the labors of the husbandman.

we shall encourage domestic manufactures and the mechanical arts as sources of private wealth, individual comfort and indispensable to the public prosperity.

we shall urge the organization of interior defences sufficient to protect the property of citizens from the depredations of the wild indians . . .

This press shall be free and independent; unawed by power and untrammeled by party. The use of its columns shall be denied to no one, who have suggestions to make, promotive of the public weal.[36]

Taking this prospectus as representative of the desirous project of all newspaper editors on the California frontier, it reads as if it were a mini–Pledge of Allegiance to the federal Constitution. The civil rights of free speech, press, religion, and property come first, but the authors quickly beg for their state promulgation through public education and call out through these symbols for inclusion in the Republic, for California at this moment was not yet a member state of the national union. Then the call of its commodity imaginary goes out to labor, encouraging immigration by promising land and its sacred protection as private property, by force of communal law. By envisioning free land as "unoccupied space," it erases native presence and signifies their place as outside the coming market bonanza. In this vision, private wealth fosters collective wealth when the institution of private property is promoted by state legal authority and backed by military force. Indian depredation, according to this booster rag primer, is not good for business and will not be tolerated.

The newspaper idiom on display here made a radical invitation, a capitalist call into productive labor that provided a spectacular conceptual engine for the production of wealth. It seems hard to imagine a more condensed expression of the core American project of colonization than that which the newspaper culture of rights discourse fomented in burgeoning western publics by mirroring the nation in its continuous display.[37]

California's second newspaper furthered the project. It began publication in San Francisco on January 9, 1847, merging one year later with the *Californian* to become the famous *Alta California,* the *Californian* having moved to San Francisco in April 1847. The gold rush was on, San Francisco had changed its name from Yerba Buena, and the territory was well on its way to statehood, with the U.S. flag raised in victory over Mexico in the summer of 1846. The *Alta California* promoted "the public weal" and published a "steamer page" covering the shipping news that would later feature events in Humboldt, for example, alongside news of world-system markets and European connections to culture and history. Its prospectus stated that the paper "will contain the current news of the day and a general synopsis of political, commercial and literary intelligence. Its columns will always be open to contributors and to the fair discussion of every question of public importance."[38]

News of Europe and Ireland dominated the front page of its first issue, and succeeding issues advertised the activity and solicitations of vessels at the city's great port. The issue of April 6, 1849, for example, was typical in pointing toward the North Coast and what would someday be Humboldt:

For Trinidad Bay—to sail, positively on Saturday, 6th April, the ___ fast sailing, coppered and copper fastened schooner Col. Taylor, 200 tons berthen, Jas. Hogan master. This vessel, having superior accomodations for cabin and steerage passengers, offers a most desirable opportunity to parties about to proceed to the new diggings, with speed and comfort. For freight or passage, apply to al 6 Weiss & Pearce, Sacramento st.

On the morning of Wednesday, April 10, 1850, just one week before the history-making Laura Virginia Company sailed into Humboldt Bay, the *Alta California* was actively promoting coastal exploration of the mining region, explaining how "speculators and capitalists fanned the flame of this new enthusiasm, until vessels were *lead* for Trinidad [in Humboldt] almost without number." Here we see the newspaper spectacle helping Anglo-European colonizers sailing for Humboldt out of San Francisco to self-identify as capitalist ambassadors of liberal society and cultural nationalism.

The federal government took an active, material interest in advancing newspaper integration on this expanding frontier. Under the direction of president James Madison, the U.S. Congress passed the Post

Office Act of 1793, promoting the exchange and circulation of news by offering free post for any news publisher to send its newspaper to any other newspaper.[39] The act officially sanctioned a "cut and paste" reporting technique, by which local papers variously summarized, edited, and reprinted stories from other papers as they came in by free post.

Colonial Humboldt's weekly *Humboldt Times* reported world and national news almost exclusively by this method. Schooners and steamships brought world papers up the coast and returned to San Francisco with Humboldt newspapers, which themselves had been cut and pasted from this or that distant paper and from East Coast serials and even wide-ranging novels—as well as anything heard on the grapevine or duly reported by citizens in letters to the editor. The paper's office was in Eureka for the first year, then moved to Union Town in 1855. The type was laboriously set by hand throughout the week in these early years, finally printing the two-sheet weekly for release every Saturday, folded in half to make an eight-page format.[40]

After gold was discovered in 1848, the U.S. Post Office contracted the Pacific Mail Steamship Company to keep California connected, shipping mail by sea from New York to Panama, where it was transferred over the isthmus by rail and then back onto ship for passage to San Francisco. Rarely did the service meet its three- to four-week delivery target, but when the mail did finally arrive on the West Coast, it was farther still to Humboldt Bay and the redwood regional *Humboldt Times,* for it had to sail up the coast via irregular steamship and schooner routes.

The first transcontinental telegraph finally reached California on October 24, 1861, patched through from St. Joseph, Missouri, to Sacramento, but the connection was not extended to Humboldt until 1873, when the first message came through from Petaluma. For a brief period in 1860, the fabled Pony Express ran from St. Joseph to Sacramento, but it never braved the mountainous overland paths through Indian country to Humboldt, and the telegraph quickly put it out of business. Nor could news reach California by quick railroad mail, for that avenue also stopped at St. Joseph until 1869, when the golden spike was finally driven at Utah's Promontory Summit, linking California to New York. Until the rail system connected Eureka to Willits and thus to all California lines in 1914, major passage into the world had to pass through the treacherous mouth of the great Humboldt Bay.[41]

Humboldt's penetration by this modern print culture followed a pattern repeated across the western frontier in the nineteenth century.

Wherever a town sprang up, its growth and survival depended on attracting immigrants. Good news of economic opportunity became a first principle of civic success.[42] Across the Pacific Northwest, towns relied on local papers to send word out by post that life could be made and made well, in this or that region. Papers were used as civic boosting devices, usually by touting the riches to be made from local resources. But as noted, getting the word *in* was a concern of at least equal consequence.

For the first four years of settler life, Humboldt's paper-culture connection ran through relatively urbane San Francisco back to the East Coast and then further back to old European sentiment; and it channeled those sentiments back into Humboldt through San Francisco city papers, for example, and all the world papers that accumulated in San Francisco, with their shipping news and national reports from world-system markets, alongside poetry, jokes, and literature. For its part, Humboldt sent back statements and letters for publication in those very San Francisco papers—but the redwood settlers' own public sphere had no local print culture medium until the *Humboldt Times* appeared on September 2, 1954. Its first edition ran the following mission statement:

The Humboldt Times
An Independent Newspaper

Dedicated to Science, Literature, Arts, Foreign and Domestic News and Markets, to the great features of National and State affairs, to the domestic economy and home improvement and all matters of general interest: its chief aim being the advancement of this section of the state in general and Humboldt Bay in particular.

Then, in the high fashion of the burgeoning newspaper culture's self-conscious stewardship of the public sphere:

Introductory

Knowing as we do the responsibility devolving upon the conductor of a news paper of the die and style of ours, it is with diffidence that we, unaided and alone, present ourself in that capacity before *the public*; trusting to the kindness and forebearance of our friends and patrons, we to-day issue our first number. An independent newspaper has been for some time the *desideratum* with the people of this county. . . . The people of the present day have "progress" inscribed upon their banner and "upward and onward" as their motto and expect a public journalist to keep pace, if not in advance, with the march of intellect, the spread of religion and civilization and the extension of Republicanism. . . .

In this seminal address to redwood country, the *Humboldt Times* called up a local public using terms that carried a national mandate, imbuing this public with a national imagination—an extension of Republicanism. The paper's public extended the nation form into Humboldt.

We have no need to speculate about how deeply the emerging redwood imaginary of Humboldt ultimately embodied the newspaper culture's nationalist address. We have read the media reports of the Wiyot massacre and found its colonial discourse speaking a spectacular language of fictive ethnicity that reveals an ambivalent civic culture torn between claims of property right and public-trust duties; and we know how market revolution was a modernizing, institutional transformation that swept into Humboldt on a wave of international migration that carried in newspaper culture. But the words of the Humboldt pioneer E. H. Howard give the point sharp relief. He was a native of New York, a graduate of the University of the City of New York, a discoverer of Humboldt Bay who arrived on the *Laura Virginia* in 1850, and the superintendent of the Humboldt school system from 1858 to 1860.[43] On May 12, 1881, he addressed the Society of Humboldt Pioneers at its annual meeting. I quote him at length for the image of the modern social imaginary inscribed in his words:

Our New England *forefathers* were pilgrims of persecution from their homes and pioneers to that inhospitable shore, who sought more elbow room, so to speak, for their consciences and to have the enjoyment of their *religion,* with none to molest or make them afraid. On account of some supposed flavor of sanctity in their character they were called Puritans. On the other hand, the forefathers of the Pacific—for it does not take long to become forefathers here—came in quest of more elbow room for the pursuit of wealth and so have come to be known as Argonauts. Somehow, pioneers, in their history as in their lives, have always had a pretty hard time of it and been exposed to the charge of doing some very hard and wicked things. Our Puritan ancestors, no doubt, wrestled stoutly for the blessing of Heaven and each other's spiritual welfare and made vigorous use of means at hand for converting of Natives. But when they failed to fetch the red man to his knees by preaching and praying, the persuasive measure of powder and ball was a most effectual one in completing his conversion on the battlefield. We read that they converted heretic Quakers and Baptists into exile from their communities and witches, fastened to the stake and fenced in with blazing faggots, they converted into innocent ashes. As between the Argonauts and the Puritans, I don't know but that in many respects the Argonauts never charged anyone with witchcraft; neither did they persecute in the name of the Lord, nor in any other name, any man or woman for religion's sake. . . .

[Humboldt] was a land abandoned to savages, with its untold treasures waiting for the coming *race,* for the Argonauts of '49 and '50, to break the seal under which they had lain for ages. And here we stand—we can go no further. . . . we must acknowledge that the plucky and energetic Anglo-American, in settlement and civilization, has eclipsed every other people in the elevation of his race and the grandeur of his territorial possessions.

He loves his *country* for itself—he loves it all the more for the vastness of its solitudes. They present new fields whereon to impress the stalwart heroism and virtues of individual character, where, as founder of new communities, he can contemplate from his primitive cabin the multiplied homes and industries that owe their beginning to the experiences in which he has borne a part.

He has warrant for his possessions in the primal command which bade our first parents go forth and cultivate the earth. In olden times armies were accustomed to be in the van of settlement and colonization, but on this continent, so far as our country is concerned, the settler has always been a long way from the soldier.

It is by *institutions* molded to protect the natural *rights* of man and at the same time preserve the peaceful policy of an enlightened civilization, that he has built up new commonwealths and compelled the agencies of intelligence and matter to yield their tribute to his prosperity. . . .

In short, the triumphs of the pioneers have not been lighted up by the firebrands of war, but throughout the steady march of the century, as now, the *schoolhouse,* the *pulpit* and the *press* have illuminated his pathway and achieved his proudest victories.[44]

Doubtless the massacre on Indian Island in 1860 stands first among "the hard and wicked things" around which the pioneer Argonaut Howard molds his apology for the new culture system. The atrocity can be heard as a resounding silence in his colonial discourse of legitimation. He speaks of it without mentioning it. Total violence and Indian genocide are sanctioned in one sweeping statement of racial cultural superiority. Humboldt's fictive ethnicity is here in all its emergent hubris; it is not accidental that the little problem of redwood country—its Indian trouble—must be explained away before the idealized version of history is presented. The complicit experience of primitive accumulation was still fresh for Howard.

Colonization and domination, he said, were driven by institutions that "protect the natural rights of man," first among which are freedom of religion, speech, and individual property. They preserve "the peaceful policy of an enlightened civilization," they energize the construction of new commonwealths, and they compel each new subject of national, historical destiny to pry open the storehouse of nature and convert it into prosperity—into Providence—that is, into commodities.

For Howard—and for us he is speaking for Humboldt's nineteenth-century modern redwood imaginary—natural rights, by dint of God's design of nature, are the literal motor of colonial history. They drive the social system; they call individual agencies of intelligence into world historical labor on the land and its resources, on the whole of the world. They yield the tributes of treasure and power from this natural repository, and they rely on the free spectacle of public media address to constitute community and supplement its feeling for the national culture of property with a total image, a complete representation, not just for racial but for a complex of religious, patriarchal, and political market identifications.

Indeed, under the institutional compulsion of which Howard speaks, the wealth of capital accumulation in Humboldt and especially the lumber economy grew massive. The concept of property—inspired, inhabited, organized, and powered by the U.S. Constitution's people-making legal system—entered Humboldt through the culture system of private land tenure, public school, Christian religion, and spectacular public-sphere print capitalism, producing an archive that fundamentally structured and continues to structure the agency of the people whose labor made and still makes today the place of the redwood imaginary.

Land System, Psyche, and Frontier Authority

Having tuned in to Humboldt's colonial discourse at the point that contemporary social memory in the region indicates was its single most traumatic moment and listened in on its public-sphere voices speaking of conditions that led to the massacre, we discovered a public struggling to assume control of Indian lands. Making connections of locale to nation, we traced origins of the redwood imaginary to the social, cultural, and ecological conditions confronting the settlers: as libidinal-economic ambassadors of market revolution and media spectacle, they entered the redwoods and made them their own through genocidal displacement of Indian life. Then, in pioneer Howard's address to his peers, we found the ideological coordinates of this colonizing culture supercondensed in the voice of one of the region's celebrated fathers—his words bear the silent mark of the signature event, as would every story of Humboldt to come. Now we return to the beginning to look more closely at the institutional setup that pioneer Howard described in the ethereal poetry of his nationalist myth. The land system psyche and social institutions of school, press, and pulpit he cherished took on specific coordinates in redwood ecology that can help us understand the emerging redwood imaginary.

Consider the discovery of Humboldt Bay not by a man or even a company but by a libidinal-economic culture system whose emissary to the region was a man and his company. On April 20, 1850, Lieutenant Douglas Ottinger wrote a letter home to his wife. He was on furlough from the U.S. revenue cutter *Frolic* and serving as captain of the onetime slave ship *Laura Virginia,* recently around the Horn from those bloody Atlantic waters. Ottinger's letter recalls his discovery of the entrance to Humboldt Bay, linking his dream of land acquisition to the sentimental profit quest and heroic ideology of pioneer settlement. In his words we hear more than a man speaking privately to his wife. The culture system is here, speaking to us through his colonial idiom:

Trinity Bay, Upper California, April 20th. My dear good wife;—I am yet aboard the Sch'r Laura Virginia and have explored this coast for two hundred miles faithfully. My object was to find the nearest point to the gold regions of the Trinity River and the party with whom I am connected are to take up land for a trading post as well as for farming purposes. I had the good fortune to be the first to land a party at this point where *we took up four quarter sections which covers all the land that can be used* for commercial purposes and since that time it has been shown that it is valuable for trade and is only forty miles from the rich diggings. A party came here with horses, mules, etc. for supplies on the sixth day after we had taken up the land and there have since that time some three hundred persons arrived and many of them located on our land in opposition to us, but we feel satisfied that *our claim must be acknowledged as good and just* and have given legal notice warning them off. Others of them have taken leases and some purchased of us already at the rate of 3,200 dols. per acre. It appears to me that I can realize some three or four thousand out of this affair, but may be disappointed. We have also taken up some 12 or 14 quarter sections on a beautiful bay about 20 miles south of this. My vessel was the first to discover and enter it. . . .

I must tell you that the land is so beautiful and the soil so rich that I was almost fascinated. . . . In addition to the good qualities of the land, the waters produce clams in abundance as well as fish; and geese, ducks, snipe, plover, etc. are about as numerous as wild pigeons at Erie in spring. The wood is not less productive than the waters and droves of elks and deer, with a goodly number of bears always to be found. . . . — signed Douglas Ottinger.[45]

Here, in its first exchange with the region's indigenous communities of labor and environment, the nation was leading Ottinger's vision and he its installation in Wiyot country. The universal logic of colonization repeats itself in the first letter home that recounts the first day of action in the bay. Ottinger came representing a company of investors to take up land for farming and a trading post near the gold mines; he claimed all the commercially viable land at Trinidad Head on the first day; he then held claim against the white horde at his back by appeal to the law. Soon he found nearby Humboldt Bay and took up twelve to fourteen quarter sections there. But how could he not know precisely how many quarter sections? If the answer is simple, its implications are complex. There were no quarter sections on Humboldt Bay when he arrived. His claim created them. The land was surveyed after the claim. What was it before? Free land, empty space, idle capital. When we listen to his claim on the land and hear its appeal to the national authority of legal property culture, we feel the first rumblings of the deep culture drive of right-based juridical institutions in the bay redwood region that down to this day still shapes public life in the redwood imaginary. The knowledge of property driving his claim and calling his labor into land exploration and profit speculation was a practical understanding of the U.S. legal system for land distribution, and the authority of his claim lay in the power of that system to establish social relations of ownership between people.

The federal government facilitated the nation-building project of settling the West by juridical means—by acquiring land for the public domain, naming it property, and then selling it off repackaged as a fungible commodity. Before being claimed, deeded, and sold, its only possible value was use, but entering the culture system expanded its range of possible values to include exchange. Through its land office business, the government invited land companies to explore and make claims, installing the nation in each new location by means of its state-sanctioned naming of place as commodity—a process that guaranteed that the concept of property would inhabit the center of local cultures and struggles, at least for as long as the republic would stand. In every new place, this is the debt that is owed to the nation—a crucial dimension of the national imaginary.

Legislative acts governing private land claims, credit sales (1800–1820), land auctions and cash sales (1820–40), preemption (1841–), and homestead (1862–) were juridical expressions of the government's objectives of expanding the sphere of the republic while at the same time initiating a proliferation of opposing interests, thereby diluting the potential threat of majority faction. Land sales privatized the public domain and stimulated local economies while helping pay the national debt. The monies generated helped fund the creation of the U.S. Bank, which was then put to work funding corporate development of communications and transportation networks. The end product envisioned, and largely achieved, was a vast social class of staked owner-citizens with a feeling for property that embodied the nation.[46]

U.S. founding fathers saw private ownership as constitutive of enterprise; productive forces unleashed by land acquisition would fertilize the soil with a burgeoning entrepreneurial population. Alexander Hamilton, for example, had in 1790 encouraged the sale of western lands at twenty cents an acre, suggesting that profit from sales to speculators could help retire the nation's debt.[47] He reasoned that cheap land would mobilize the labor of speculators with the promise of profit, and we have seen how newspaper editors repeated this address to their new reading publics. Land was used as a national symbol of material action.

Jefferson, on the other hand, having written to James Madison that "small land holders are the most precious part of a state," envisioned not a nation of big-business speculation but a yeoman nation of virtuous small farmers. In the decades before Andrew Jackson's presidency, Jefferson struggled to bolster small ownership against the rising tide of corporate land aggregations by eastern commercial and mercantile capital, represented by the federalists Madison and Alexander Hamilton.[48] But the idea that laboring on the land civilized the owner *and* wild nature was common to both positions: land system as technique of subjectification was elemental to the national imaginary and remains so today. If so-called empty land would weaken the nation, the psychosocial treatment was an ownership stake.

To get a sense of how deeply this land system psychology permeated the continent and shaped the emerging nation's feeling for property, pause and recall that half of the total U.S. land area was at one time part of the public domain. Privatization and use of these public lands were a prevailing concern for the nation during the nineteenth-century market revolution and Jacksonian Indian removal, as it was in Humboldt.[49] It was geophysical discipline for a wild continent— a legal-rational resignification powered by slave accumulation and contingent on Indian removal. It was also psychical discipline for the settlers whose labor it called to action. Privatization, in other words, was a technology of subjectification that accomplished much more than the Constitution's objectives of unleashing productive forces, expanding the sphere, proliferating interest, and diluting faction. If we want to understand its effect on the redwood imaginary, we have to look back at the historical context of institutionalization of property in the bay redwood lands.

Much can be learned from the land laws that preceded the colonization of Humboldt. After the early period of experimentation with credit sales, the federal Congress ended its settler-friendly credit system and set a minimum up-front price for all cash sales at $1.25 per acre in 1820.[50] The highest bidder at public auction entered into a contract with the state binding him to occupy and improve the land by labor to perfect his title. This type of policy produced steady income for the federal government and provided a modicum of orderliness to the process of frontier settlement while ostensibly civilizing the frontier subject.

By the time California achieved statehood in 1850, the federal Preemption Act of September 4, 1841, had replaced direct-auction cash sales as the most important mode of land distribution to settlers throughout the expanding territory. The guidelines of the federal act of 1820 were reinforced in the act of 1841, and provisions were made to allow acquisition for settlement on surveyed lands before being opened to public auction—the idea being to more duly protect the settler's natural right to any real property his labor improved. This so-called Log Cabin Bill also legalized squatting on the public domain, but without extending these rights to recognized Indian lands or to large areas of public domain in the West that had yet to be surveyed. It further supported the idea of property right by labor but could not fully guarantee it and rationalize the process owing to ambiguities concerning Indian lands, uncompleted surveys, and fraudulent speculation.[51]

All public lands of California were brought under the jurisdiction of the federal act of 1841 and opened to legal preemption by the federal act on March 3, 1853, under which legal compulsion a great deal of land in California went directly to settlers.[52] Preemption legalized acquisition of up to 160 acres per claim, for the minimum price of $1.25 per acre, to any citizens or persons willing to declare their intent to become a citizen, as long as they were either head of a family or a single man over twenty-one. Single women need not apply, and so they remained at a disadvantage in the acquisition of land and the social autonomy and authority that such status increasingly granted.[53] Native Americans of either sex were not citizens and so ineligible for acquisition at this time but would later be granted citizenship by the Indian Citizenship Act of 1924. Blacks, of course, were slaves in half of the country and largely deprived of equal protection in the other half. Under these laws, great portions of Humboldt were transferred to white pioneer men.

In Humboldt and throughout California and the West, settlers continuously outstripped the land surveyors, making life on the frontier insecure in terms of property right. The act of March 3, 1853, corrected this by legislating preemption settlement of unsurveyed land under slightly different rules; claimants would now be required to file declaratory statements within the first three months *after* the plats of survey were filed at the district land office.[54] Title would then be perfected within thirty-three months, by personal appearance at the land office to present proof of inhabitation and improvement of the land and to pay the requisite fee. By law two witnesses had to be present and swear under oath that the claimant was true to his claim of personally settling the land, that no others were making the same claim, and that residence had been maintained throughout the specified period. Labor of plowing, fencing, and cultivation was required in some measure. With the local land office satisfied, all the papers were sent to the U.S. General Land Office, which would then issue the land patent. The text of these laws compelled people to work on the land, thus translating the nation's natural law jurisprudence directly into the landscape *and* the people themselves. The Anglo-Protestant nation's Enlightenment social engineers were using land laws to make places in the West and to fill their expanse with properly nationalized subjectivity.[55]

In these early years, metes and bounds were necessary terms for description of claims, owing to the absence of official surveys laying out township and section lines. The following preemption claim that Owen Coy found in the archives of the Humboldt County Recorder illustrates the work of signifying property:

> Julius W. Graham made a clame of Preemption on one Quarter Section of Land. on The 12th day of May A D 1850 Situated on the warters of Humboldt Harbour and State of California and bounden as follows. viz Commencing at Wm H Sansbury's. S. E. corner. runing Thence one half mile N. to a read wood tree. Thence one half E. to a Matharone tree. Thence one half mile. S to a spruce tree Thence once half Mile W to the place of beginning

> Said clame of Preemption was runn of and marked in presence
> Arter Graham
> Andrew. Worford[56]

This is the language of lay citizen knowledge inscribing land into the culture system. First in this naming and then by improvement, the settler established his right by positive law to a share in the public domain. Such activity mingles the force of state law with the bodily labor of subjects bound up in the dreamy expansion of republican providence. But this national call into labor on the land was not addressed to women; rather, it provided a legal incentive for women to submit to marriage in search of a home and the vaunted ontological security of its private domain. In Humboldt the institution of white patriarchal land was a legitimate system of state-driven authority that sexed up the logic of capital accumulation in the very real terms of fee simple title.

And the state's use of the land system did not stop at legislating this racialized and gendered accumulation of moral, material, and symbolic authority in land ownership; it served other directly disciplinary and subjectifying functions as well. The Preemption Act of 1841 provided 500,000 acres of federal public domain to each new state admitted to the union, to be offered for sale with the proceeds to be used for internal improvements.[57] The Constitution of the State of California, adopted in 1849, earmarked the proceeds from the sale of these lands to the establishment of public schools. The state received its 500,000-acre internal improvement land grant at its annexation in 1850.[58] The federal act of March 3, 1853, then granted further lands to the states for the establishment of schools. The sixteenth and thirty-sixth sections of each township were granted to the state for the benefit of its public schools, while two whole townships, or seventy-two sections, were granted for a university.[59] By this act alone the state government took ownership of one-eighteenth of the public domain in California and earmarked it for production of universal schooling.[60]

The implications of this school land policy were decisive in Humboldt. The first public school was organized at Union in 1852. The county itself was created in 1853 by its division from Trinity County to the east, being 3,507 square miles in size at its birth.[61] The county created a combined office of assessor and school superintendent—with D. D. Williams elected and presiding over the first schools. That year three schools in the county taught thirty-five students. Additional acres were surveyed and sold by warrant through the General Land Office of the State to fund the expansion of mandatory public education. The combination of assessor and superintendent in one office confirms the imbrication of land system morality and state education.[62]

In his inaugural address of January 7, 1854, California's third governor, John Bigler, articulated this imaginative connection of the public domain to the public school system, illuminating its role in the people-building, place-making project of nationalism:

> The people of California, though greatly absorbed in the development of her unequalled mineral wealth and in preparing for a more full enjoyment of her vast commercial advantages, have not been unmindful of other great interests. . . . Congress . . . has donated lands to the State, which . . . will yield not less than eleven millions of dollars for school purposes. This sum, judiciously applied in fostering a system of Common Schools, will be found amply sufficient to educate all the children of our State and thus give a high character to our civil institutions. The education of the masses, is justly esteemed the ground work of free institutions and the enduring basis of constitutional liberty. . . . While from our earliest boyhood, there has been instilled in our breasts, love and reverence for the free institutions of our country, we have also been taught to respect

and obey that country's laws. . . . Let us hope, then, as we have
every reason to expect, that not only our mountains and river sands
will continue to yield their rich treasures to the hand of industry,
but that our vast and unequalled commercial and agricultural resources may be so developed by the enterprise and perseverance
of our people, as to elevate California in the scale of nations and
render her that populous and powerful sovereignty of the American
Confederacy, for which she was so manifestly "destined by nature
and by nature's God."[63]

The free institutions Bigler celebrates in these words are the same
ones that pioneer Howard lauded in his speech to the Pioneer Society
and championed by California's first newspaper editor—institutions
that protect the civil rights of property and so compel the agencies of
both intelligence and matter to yield their treasures to accumulation.
Manifest Destiny in Howard, Bigler, and the daily *Californian* signified religious legitimation for what at its core was a state policy of land
privatization that institutionalized in Humboldt a landed formation
of white patriarchal authority that still inhabits the redwood imaginary. This is modernity, American style—a modern social imaginary
uniquely instantiated in an irreducibly local constellation of reciprocally constituting institutions.

In 1862, the federal government made another land-system law that
legislated subjectivity in redwood country during the Indian trouble:
the Homestead Act of May 20. It had been defeated in Congress until
after the Civil War, largely because of opposition from slave states
figuring that a policy of free land grants to individuals would be an-
tithetical to the concept of a society of slave-driving plantations and
large family ownership. Under the new act, for a mere ten-dollar fee,
any head of household or citizen over twenty-one could acquire one
quarter section on any public land that had been for sale under the
preemption laws, if it was guaranteed to be used for an abode and
cultivation. Title could be perfected only after five years from the date
of homestead claim entry; again, two witnesses were required, and
only U.S. citizens or individuals of sworn intent to become so were
eligible, but the restriction on women was dropped from the language.
From enactment through 1865, 1,116 homestead entries were made in
California for a total of 170,031 acres. Thus had one barrier to wom-
en's attainment of landed authority fallen. Another five decades would
pass before women won the right to vote in 1920. Native Americans

also remained outside the language of juridical privatization, losing ground with every land claim filed. When the Immigration Act of 1924 granted Indian citizenship, the actual space for homesteading in Humboldt had almost entirely closed, because the forest land had largely been privatized and aggregated for industrial forestry in the closing decades of the nineteenth century. Before 1865, however, settlers had managed to claim 4,773.6 acres under homestead.[64]

In that year, Humboldt's County Board of Supervisors authorized A. J. Doolittle to cooperate with the U.S. Surveyor General's Office in San Francisco and draw up an annotated map of the county. Using U.S. Land and Coast Surveys, "as far as they have been run," and "personal observations" of his own alongside those provided with "intelligence and politeness" by "very many citizens throughout the county," Doolittle produced a compelling visual and narrative document that captures the drama of Humboldt's property struggle in the closing moment of its opening scene—published in the final year of the Indian wars. Statistics compiled and charted quantify the transformation spatialized in the map and narrated in two columns along its right side. Between 1860 and 1864, the number of acres enclosed grew from 10,975 to 24,052; those cultivated rose from 3,457 to 7,078. There were now twelve schools in eleven districts with 762 children attending. "These facts," wrote Doolittle, "only serve to illustrate the energetic, fixed and invincible spirit, so eminently possessed by the early pioneers."

The greatest difficulty for settlers in the period our mapmaker surveyed was, of course, the Indian trouble, with which he chose to end his narrative account. The Indian war commenced, he said, in 1861, and by 1862 the citizens had been "driven entirely from the grazing portion—Bald Hills, or Prairie—which is as two or three to one of the farming lands and from ten to fifteen miles back" from the bay. "The destruction of stock by the Indians, in 1862, was 9,000 from 26,000 head," a loss, as he described it, "desolating the condition of the county for protracted Indian war enlistments for the same and [hampering] the great rush to the northern mines," a situation that "affords ample cause for the moderate assessment and population of the county." In other words, from the mapmaker's urban view, Indian resistance in Humboldt was retarding the white horde, who for their part contributed $2,500 "by tax for soldiers employed in ridding their county of the presence of Indians." Nevertheless, by his count, "fourteen thousand acres of land have been located in the county; 5,000 by

Detail from A. J. Doolittle's "The official township map of Humboldt Co., Cal.: carefully compiled from U.S. land and coast surveys, with personal observations" (San Francisco: Grafton T. Brown Lith., 1865). Courtesy of Bancroft Library, University of California, Berkeley.

School Warrants." Primitive accumulation for white male enclosure and national expansion meant Indian removal, concentration, and re-education, on pain of death; and it already presaged environmental destruction.

The richness of these lands is incomparable, wrote Doolittle: "Deer, elk, etc. . . . are yet abundant in portions of the county," indicating that a limit to the taking of animal bodies might already be approaching—but "salmon, bass, herring, smelts, crabs and clams abound in the waters. . . . Ducks and geese, in great variety, swarm alternatively in the waters. . . . Seals and Sea Lions pack and howl on the Rocks." And so even though "no wagon road connects this Co. to the rest of California, the mails will suffer little," and things are improving, he concludes, because "the native Digger Indian quits his pertinacious hold, his long enjoyed haunts and with them the ready fruits of the bow and the chase—[and] the Battalion Mountaineers [of volunteer Indian killers] is disbanded." Progress has finally arrived, in other words, with an "influx of money for produce—Oil Lands—returned miners—regular steam communication—[and] Overland Road and Telegraph, by which to communicate with the outer world." And this wave of progress "will forever pass the Indian to his inevitable doom and the settler to his wonted thrift, in so susceptible a clime. A. J. D." As surely the closing initials sign this author's name, so do his colonial language and commodifying visions sign that of his culture system.

Perhaps the most illuminating data in Doolittle's map, from which I have taken the detail reproduced here, are the personal names of the ranches lining the waterways and valleys where Humboldt's cattlemen lived and whence the energy, the manpower, for the massacre is said largely to have originated. These names put an intimate face on the land and suggest just how precious it must have been, not just to those whose names are signed but to all the unnamed, Indian and other, whose status as not the owner of this forest and not the owner of this salmon stream was enacted in the selfsame appellation.

A Spectacular Invitation

An editorial-advertisement from the *Humboldt Times* of November 4, 1854, the paper's third month of publication, illuminates the irreducibly libidinal economics of cultural colonization in Humboldt and helps us connect this idea to the names on Doolittle's map. It shows how colonization was already shaping up like a modern advertising campaign. The headline pleaded "WE WANT MORE POPULATION," and the editors quipped that "California requires the same means used to build up our waste places and to establish permanently those already begun, to make them prosperous and happy, that were used by that benevolent-hearted woman, Mrs. Chisolm, who accomplished so much to settle and make prosperous the people of Australia and whose plans will apply to California." Waste places? At the risk of making too much of what might be a humorous aside, we should recall how instrumental the mass media were in addressing the subjects of said colonization and how they provided a continual supply of cultural narratives that situated the reader as a national subject. Here they situate settlers on a clear open plane—empty spaces. A clean slate. Native presence is once again written out of the story.

The body of the article is a citation cut and pasted from *Harper's* magazine. Lauding Mrs. Chisolm for her "keen insight into human nature" and for her "knowledge of the wants of the colonies" in Australia, the editors concur with her finding that "it is not good for man, or woman either, to be alone; and that a virtuous society can be reared only upon the basis of family state." The state, says Mrs. Chisolm, would do well to relieve this frontier condition of "anxiety . . . for virtuous wives" and "reverse . . . the impossibility of their obtaining them." It should take heed of "the anxious question of the stockman, 'When were they to have a governor who would attend to matters of importance like that?'"—a query, she says, that "embodied more wisdom than the Colonial office was aware of."

That the colonial land office in Mrs. Chisolm's Australia and the U.S. land office in Humboldt were both in the business of trading native lands for nationalist loyalty suggests something universal in the politics of colonization. Everywhere modern capitalism would be, there moves ahead of its subject performers a specter of the law, whose shadow is cast in front of the settler by the light of spectacular media like this, as if its light were shining at the settler's back and illuminating his way but leaving the space of his immediate forward reach outside its sphere and thus obscuring his view. In lighting the way without formally determining his next step, the law opens a space for activity without forcing any action. And everywhere its illumination is cast, an invitation is extended, and a singular set of lacunae appear that pose the questions of life, love, and labor in a synchronous fashion: "Something else is requisite for a flourishing state than fat cattle and fine-wooled sheep," said Mrs. Chisolm to the world through the medium of *Harper's:* "To supply flock masters with good shepherds is a good work," she continued, and "to supply those shepherds with good wives is a better," "to give the shepherd a good wife" is to make "a gloomy, miserable hut, a cheerful and contented home." Yet grounding men using women does not quite complete the circuit, which requires one more point of investiture to constitute a full unit of household production. "To introduce married females into the interior," she wrote, "is to make squatters' stations fit abodes for Christian men," because "all the clergy you can dispatch, all the school-masters you can appoint, all the churches you can build and all the good books you can export, will never do much good, without 'God's police,' wives and little children."

The whole power source of colonization is assembled here: its complex imbrication of modern constitutionally rights-based institutions is given allegorical relief, its figure arrested from the continuous effervescence of editorial production and projected through a single figure into Humboldt's new public sphere. Mrs. Chisolm was an English woman colonizing Australia—but by newspaper transfer, her words in a sense helped colonize Humboldt, a place where colonization was no doubt unique, but all the same modern in its structure. Signifying land as empty space for commodification, it erased native presence here as it did around the globe. Primitive accumulation meant violent appropriation everywhere the national systems of property culture spread. In 1860 the advance technique of market revolution worked largely through newspaper integration that situated pioneers in its gathering spectacle as a people with feelings for nation, god, family, gender, and property. Its knowledge culture bound land system to psyche in masculine feelings for property—and laws enacting privatization were among its principal agencies. Coming generations of workers and environmentalists would find this formation at the core of the social order that advancing capital would compel them to rise up and challenge.

By quoting at length from Capt. Ottinger's letter of 1850, the *Humboldt Times* mission statement of 1854, assistant editor Hart's and proprietor's Wiley's 1860 editorials on the Wiyot massacre, the postmassacre letters of "Exodus," Major Rains, and "Anti-Thug," Doolittle's 1865 map with statistics and commentary, and E. H. Howard's address to the Pioneer Society in 1881, I have tried to let speak the voices of Humboldt that participated in the lawgiving violence by which the tribes were reduced to the tiny portions they inhabit today: At Table Bluff, around 35 homes and 50 registered Wiyot occupy 88 acres. At the Trinidad Rancheria, about 70 of the tribes' approximately 150 local members, which include Yurok, Tolowa, and Wiyot, live on 47.2 acres of coastal property at Trinidad Head. About 1,100 of the Yurok tribe's 4,374 registered members live on approximately 60,000 acres, much of which is not in tribal trust status but privately owned. Perhaps 100 Indians live at the Blue Lake Rancheria, created as a reservation for homeless Indians from across the United States. On the Rhonerville Rancheria, the Bear River Band of Wiyot live together with remnants of the Mattole tribe. Between 20 and 30 Indians live on 20 acres at Big Lagoon Rancheria, a federal reservation of Yurok and Tolowa Indians; and about 2,600 Hupa, Yurok, Karuk, Whilkut, and Tsnungwe share the 85,000-acre Hoopa Valley reservation, where members of all tribes from the bay redwood region were forcibly concentrated when the reservation was established during the effort to conclude the Indian wars in 1864.[65]

Like the labor and ecological resistance mounted by oppositional subjects in the twentieth century, nineteenth-century Indian resistance fundamentally changed the landscape of Humboldt. The reservations exist as a constant reminder that the land had to be divided up if the hostilities were ever to end. They continuously project native presence in the county, inciting historical consciousness to remember and investigate, an intellectual passage that inevitably leads back through the litany of crimes to the murderous rampage on Indian Island and the symbolic project of enclosing the land—both by means of fences for cattle and by township surveys that readied the land for market exchange and formalized the institution of capital culture in the redwood imaginary.

Whereas the massacre settled the immediate matter of conquest for the Wiyot, completing the reduction of their original population of an estimated two thousand to perhaps two hundred, in various ways the other tribes continued to put up fierce resistance. From the spring of 1860 well into 1865, it was all-out Indian war, with local militia working together with federal troops to take care of "the problem." In this moment of primitive accumulation—accumulation by force, in other words, that set the system in place—colonizers made themselves and their culture of modern institutions by making the Indians an outside identity, by treating them as outsiders in their own territory, by putting them out of bounds, concentrating them on the outside, on Indian reservations.

These traces of native presence in Humboldt show how the redwood imaginary emerged from a practical cauldron of colonial discourse as a racialized, patriarchal, and propertied machine with the power to name and control not just the land but the labor it needed to grow its accumulating power. Redwood capital had flowered in the collective recognition of indigenous place as property for commodification, blasting the redwood forest out of its quiet nest and channeling its value into that culture's commodity circuits, interminably linking the coming labor unrest back to the Indian wars. Power over land that was taken by force and signified as legitimate property grew into power over labor for redwood production. It anchored the first contradiction in Humboldt, setting the conditions for an epoch of labor trouble. Violence constituted that system's character—in the symbolic sense that it archived the colonizers' racial abjection of its Indian other, for example, in their narrative self-identifications, in their editorials, letters, histories, myths, and dedications. These being history, they have a way of returning like this, by way of staying right where they are.

I have described the period of Indian trouble as an integral moment in the formation of the redwood imaginary. From the moment that Captain Douglas Ottinger sailed the nation into the Humboldt bay region, to that in which pioneer Howard lauded its achievements and narrated its character in his mythopoeic address to the Society of Humboldt Pioneers, the nationalist project of land-system subjectification of both people and place was colonizing everything, lived, symbolic, and built in the region, remaking that landscape per its own image. Everywhere we turn we discover its name, especially condensed in Doolittle's map, which so elegantly articulates the quantitative, spatial, and narrative registers of the colonizing cultural imaginary.

The horrific intentions, actions, and field of effects of some of its subjects are impossible to speak, and at any rate I am no judge of them or the loss that we have all suffered by their work. It is equally impossible not to always already be speaking about them when one speaks of the history of capital in the redwoods. By beginning from the massacre, I have wanted to acknowledge the debt that contemporary historical consciousness owes to that terrible night. It does so much work for the writer today, by gathering together the social relations that prevailed in that moment of colonization. Its symbolic power over the perception and interpretation of redwood colonization is to always return in historical consciousness—in the thinking and writing of what came before and conditions the present. It always takes us back to its traumatic origin. It archives the trace of that collective subject, that perpetrating collective—and its modern imaginary—in a more precise image than we could have expected: namely, an incomplete image, imperfect, not completely recoverable, just as it would have appeared in the moment. Did redwood pioneers have a better grasp of the colonizing culture as a system of knowledge and institutions than we do today? Or rather, than we *can* today, after engaging this archival material? To answer yes would be to condemn every future to a fate of attenuating knowledge and even the ultimate extinction of the past as significant to the present.

The case of Humboldt seems to suggest the opposite. Over time, the continuous return of historical consciousness to the traumatic scene tends to accumulate material. Archives tend to grow, collecting connections that continuously expand the possibility for producing present meanings. This is not just because the field of effects of particular processes and events, like colonization and the massacre, continuously grows wider as time goes on, although that is inarguably the case, suggesting that things just keep getting more and more meaningful. It also has to do with how records are kept and so with the technologies of reading and writing that give different societies their own unique characteristics of historicity.

The colonizing capital culture that wrote itself over Wiyot territory in 1850 was and still is a newspaper culture, a culture obsessed with writing itself into massive archives. It was the culture of social memory that arrived—a people who came in on the text, producing more text and saving every text in the interest of future history. How else can we fathom why not a single copy of the *Humboldt Times* was lost? Every word published every week in the beginning and then every day since has been saved and at present stands available for use by historical consciousness. The American nation established in the redwoods was already a library nation.

In 2001, as I neared the end of my fieldwork in Humboldt, I visited the Table Bluff Reservation for an interview with tribal chair Cheryl Seidman. By way of beginning, I told her of my struggle to comprehend the timber wars and how my attention was drawn increasingly back to the arrival of capital culture here on the bay. Could she explain my failure to find Wiyot sources? Did she like the idea for starting my book with the Wiyot word for the bay, Qwa-la-wal-oo? "No," she said, "I wouldn't do that. As far as I understood, it was not the name of the bay. I understand that it is a greeting." Is there a word for the bay? "Wigi," she said, "W-I-G-I." Would it be right for me, in your eyes, to begin my book with this word? "Yes, because that's what the name of the bay is. That's what it was called before it became Humboldt Bay." I've been reading every story of Humboldt I can find, I said, have they been told by anybody from your point of view, by someone from the Wiyot tribe? "Not to my knowledge, no." No Black Elk. No Counting Coup. No Lame Deer. No Book of the Hopi.[66] No great book of Wiyot wisdom. No Lucy Thompson, the Yurok author of *To the American Indian*. No comparable Wiyot archive—at least not in the conventional sense. In Humboldt a quietude punctuates history right where it begins—in the stories of Indian trouble and the signature event of the Wiyot massacre.

I told Seidman of my collection of articles in which she had been quoted on matters concerning the Wiyot. One was a front-page story of the *North Coast Journal* headlined "The Terminated Tribe." "Ya," she said, "that's what we were—terminated. We became non-Indian Indians. That was the terminology they [the state] used for us after 1959. . . . Couldn't call us white." As individuals, she said, they regained their status as Native Americans in 1976, but not until 1990 was the tribe officially recognized. "But that's federal," she said. "We have never lost our own recognition. We have always known who we were."

The reservation's original land base was twenty acres, she said, purchased first by nuns in the early 1900s, after which the federal government gave it tribal trust status and the name Table Bluff Reservation. But "when that was terminated in the late fifties," she said, "the land was deeded over, allotments, to . . . all of the Wiyot people who were living on the reservation at the time." One-acre lots were created, and "everyone got the deed to their land and paid taxes on it." In 1990 the tribe acquired eighty-eight acres for the reservation in a federal lawsuit, she said, "and the first houses started to be built, I think, at the end of '92, and people started moving in at the beginning of '93, and . . . I didn't move until October 30, 1993."

I was startled to feel the proximity of these events, to sense how near we are to the surface of history. How could this struggle have remained so contemporary? "They wanted us to be white, that's the bottom line to any of this," she said. "It was assimilation, it had always been assimilation. From the very get-go when the first people got here to this country; if you don't look like me, [then] look like me! Wear clothes like me! And do what I do because I am better than you are. You speak gibberish. You don't speak English. . . . I think the termination act was just one more thing of trying to get rid of Indian people. . . . Termination was to allow us to have the land that we lived on all of our lives, and I think part of it is allowing all of us to be land-owners. But, again, another part of it is that *we never owned land. We were caretakers.*"

Are you still fighting to regain the center of the Wiyot world, the sacred site? The site of the massacre, where the tribes gathered to dance in the Wiyot village on Indian Island? "Yes," she said, we now "own 1.5 acres out of 275 . . . so, no, we have not accomplished what we have set out to do. I don't see it ever being accomplished in my lifetime." Since that time, however, the Eureka city council has transferred to the Wiyot tribe first forty acres in 2004 and then another sixty acres in 2006.[67] What is your mission at the sacred site? "To bring it back to what it was, pre-1860. Making it into a village again. A pseudo-village, so people can see it. Make it an educational tool. Making it to where we can come and dance, and sing, and become the universe, you know, the creating of our universe again. That's what I see." In one of the articles you said that you had not danced there or done that ceremony since 1860. "Correct." Are you waiting for that, to do it there again, where it is supposed to be done? "True. Yes . . . I've always said that the Wiyot people were the buffer to all the other Indians. We got hit. We got hit hard and first, and everybody else went to hide. And that's why they kept a lot of the traditional culture alive, because we took the brunt of all the ugliness that happened. I'm not saying that they didn't get hit either, but at least we got hit first and hardest."

In the spring of 2001, I attended the Wiyot Sacred Site Fund art auction at a theater in Eureka. For a fee that contributed to the Sacred Site Fund, I purchased the right to make bids on art donated to the cause by local artists, Wiyot artists, and others. On one side of my handheld sign was the number used to signal my bids, on the other was the Wiyot Sacred Site Fund logo. I had seen it before—at the mu-seum of Indian trouble in the old hospital at Fort Humboldt, in a glass case at the end of a walk-through display that cataloged a decade of broken treaties, hostile confrontations, volunteer extermination cam-paigns, and finally the massacre. An interpretation of the logo pro-vided by the tribe explained how "our logo was designed by Leona Wilkinson, who is a multi-media artist, and a member of the tribal council. The logo depicts a woman who represents the tribe. She is shown with no mouth, indicating that the Wiyot people did not have a voice in what happened to them in the taking of our land. The child in her arms represents the only infant found alive following the massacre of 1860, still nursing at the breast of his dead mother. The child also represents the future generations. The redwood tree symbolizes our strength, and the water our on-going dependence on the sea. The basket design gives honor to the memory of the last full time weaver of the tribe, Winnie Buckley, who died in 1945."[68]

If this logo, this auction, and the Sacred Site effort itself are any

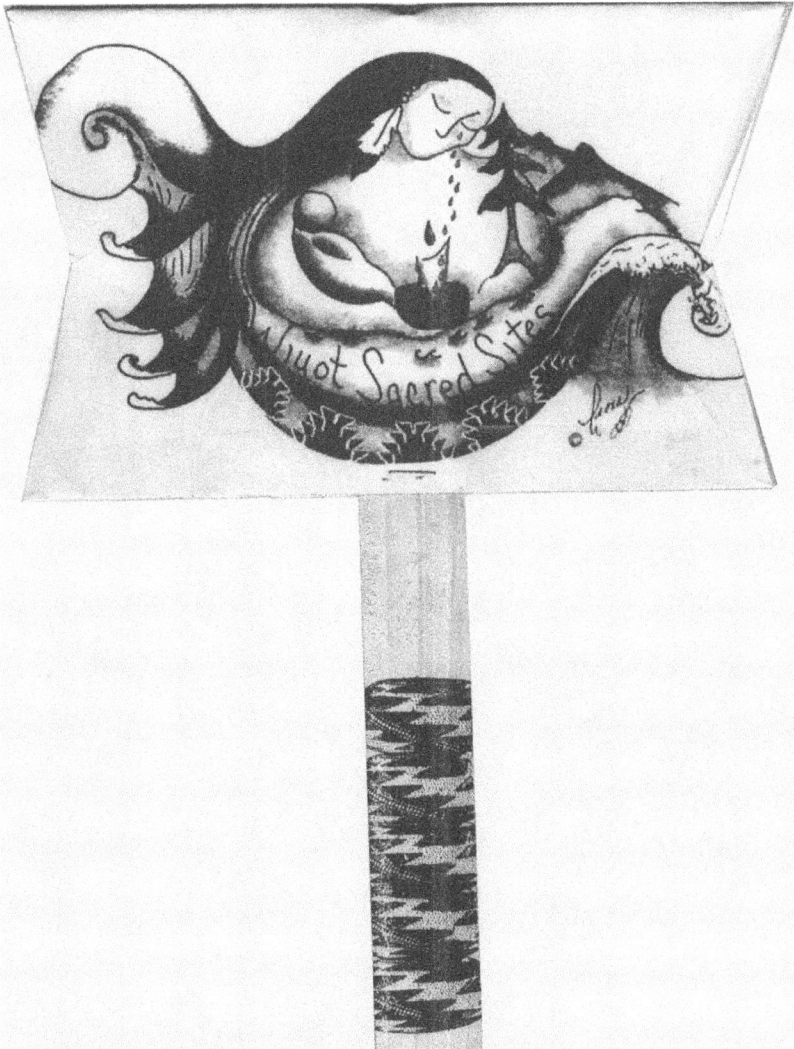

Wiyot Sacred Sites Fund handheld auction sign, Eureka Theatre, April 7, 2001.

indication, the Wiyot tribe is regaining its voice. The tribe is moving forward aggressively on the dream of reestablishing a tribal center on Indian Island. But Indian Island has already been established as a psycho-geographic landmark in the redwood imaginary—part of the symbolic order within which the people of Humboldt are compelled to live their lives, make their claims, and construct for themselves a historical consciousness. Traces of the massacre everywhere structure this landscape of possible meaning-making culture.

5. Labor Trouble: Capitalism, Work, and Resistance in the Redwoods

> The masters, being few in number, can combine much more easily . . . [and] can hold out much longer. . . . Many workmen could not subsist a week and few could subsist a month and scarce any a year without employment.
>
> —*Adam Smith*

> Every wage cut reduces the power to buy. This creates idleness and lowers the purchasing power of others, thereby creating more idleness. How anyone can advocate that a wage reduction is a benefit to the community is a mystery to us; we cannot see it and it is our hope that it will not be tolerated by any community.
>
> —*Harry C. Breit, Secretary, Cooks and Waiters Local 220 of Eureka, 1929*

From all over Humboldt the strikers came, and by 6:30 a.m. on June 21 some two hundred had massed at the gate of the Holmes-Eureka redwood mill. It was day forty-three in the Great Lumber Strike of 1935, and mills were shut down from Puget Sound to Humboldt Bay. Across the road and up on the bluff, spectators gathered at Fort Humboldt looked down on the growing crowd. Strikers blockaded the entrance with scavenged boards. They argued with special officers hired to protect the mill. Some went against the union's directive and harangued nonstriking lumbermen with rocks as they arrived for work. Soon the police arrived. The crowd gave them trouble but began to withdraw. Witnesses looking down from the bluff said they saw the police chief pull a gun and start firing at the ground. As more police and vigilantes arrived, the scene heated up. Authorities fired a tear gas canister into the growing fight. Strike leader Mickey Lima saw a woman go down bleeding and thought she had been shot in the back with a gun. At his cry some strikers rallied to him, and they charged the police. The law, company men, and vigilantes then opened fire on the crowd. A machine gun used by police jammed after just a few rounds went into the swarm. Woods cook William Kaarte was killed instantly, shot through the throat. Paul Lampella, his eye shot out of his face, went insane but lived until August 7. Pacific Lumber Company tree faller Harold Edlund was shot in the chest and died on the night of June 24.

Holmes-Eureka gate, original location, Eureka, circa 1935. This site is now occupied by the Bayshore Mall. Courtesy of Humboldt Room, Humboldt State University, Arcata, California.

Ole Johnson's wounded leg later had to be amputated. All three who died were Finnish American. Many others were wounded, including five officers of the law.

At the height of the violence, perhaps one dozen mysterious so-called G-men arrived. J. Edgar Hoover later denied that they were government men—roving federal agents policing the boundaries of the labor movement. They helped in rounding up everyone they could link to the scene, nearly two hundred in all. By 1:00 p.m. there were 166 prisoners in city hall. For three months the town was consumed in mass trials of strikers, but no convictions were won against the spirited defense put up largely by radical lawyers sent by the International Labor Defense, a left-wing legal organization with known communist members. The public spectacle of the trial ended on September 25. The prosecutor called it quits; his major witnesses had been discredited, and no more citizens were willing to serve on strike trial juries. Whenever the talk turns to sawmills and unions in Humboldt, some version of this story punctuates the conversation.

Today the Bayshore Mall occupies the site of the Holmes-Eureka labor massacre. The impressive gates that guarded the mill and witnessed the killing have been moved to the entrance of nearby Sequoia Park. In 1995 the Central Labor Council of the AFL-CIO and the Building Trades Council of Humboldt and Del Norte counties placed a historical marker at the entrance to Bayshore Mall reminding shoppers that "no one was brought to trial for the deaths of the three strikers."[1] At the ceremony in which the marker was dedicated by local unionists and police, when the chief of police started explaining how different law enforcement is today from those troubled times, Alicia Littletree and other Earth First! redwood forest defenders questioned the chief about the use of pepper spray against peaceful activists protesting Maxxam's destructive logging practices.

Humboldt's Central Labor Council holds its annual Labor Day picnic in view of the Holmes-Eureka gate at Sequoia Park. The grassy spot is near a small stand of ancient redwood trees and is hemmed in on three sides by working-class homes. Cutover timber production zone (TPZ) surrounds the park with miles of mountainsides stripped of ancient forests and scarred red with herbicided clear-cuts, a patchwork thrown into sharp relief by groves of two-thousand-year-old giants more than two hundred feet tall, recently protected by the Headwaters deal.

Holmes-Eureka gate at Sequoia Park, 2001.

To keep the human side of the story alive, the Labor Council distributes a pamphlet that pictures the gates and describes the massacre at Holmes-Eureka: *The Redwood Lumber Strike of 1935: A Tragedy in Humboldt's Labor History*.

Seventy-five years after the massacre on Indian Island in 1860, capital culture in the redwood timber extraction zone had inscribed another psycho-geographic landmark in the redwood imaginary. Whereas the signature event of the Indian trouble archives an image of the culture system in its local first phase of primitive accumulation, as it was directly expropriating First Peoples in the name of institutionalizing modern property in the bay redwood region, this new signature event of the labor trouble records an updated image of that culture system, now in its moment of historical transition from nineteenth-century competitive industrial capitalism toward twentieth-century consumer capitalism. This change embodies and drives the rise of internal contradictions in Humboldt's instantiation of capital culture as it runs up first against the limit of laboring communities to absorb and suffer the culture system's incessant demands for cheaper bodily labor energy

The Redwood Lumber Strike of 1935

A Tragedy in Humboldt's Labor History

Central Labor Council of Humboldt and Del Norte Counties, AFL-CIO
1707 K Street
Eureka, CA 95501

Pamphlet distributed by Central Labor Council of Humboldt and Del Norte Counties, 2001.

and more productivity, and second against ecological limits to its infinite expansion.[2]

The image of this historical moment that the signature event of
the labor massacre archives and transmits to the present is necessarily
incomplete, as is that transmitted by the massacre in 1860. But the
newspapers, oppositional literature, unionist records, publications,
interviews, and landscape of memories, which includes the Holmes-
Eureka gate and the plaque at the mall, do form an archive on which
we can draw—on which we *must* draw if we want to form an impression of the emergent redwood imaginary in its moment of fixation by
labor trouble. This is the stuff of collective memory—not the sponta-
neous, alive, and internal experience of memory "in the head," so to
speak, but the *place* of social memory that serves memory in the head
by structuring it and providing it with materials available for narrating and performing the present as historically meaningful.

Sign of the Times: First Contradictions in Capital Culture

On the morning after Friday's violence, the *Humboldt Times* reported
on June 22, 1935, that "one of the bloodiest outbreaks of mob violence
in Eureka's history was replaced by peace and order last nite, after
federal state, county and city officers combined to round up nearly
150 rioters and suspected radicals. Although considerable tension continued, officers believed their action had broken the back of a terrorist
campaign launched by communist leaders yesterday morning." Under
the circumstances, it is easy to speculate why the editors did not compare the event to what had in fact been the town's bloodiest moment,
the massacre of the Wiyot at Indian Island. The moment was intense
and reflected new concerns. The Wiyot massacre played no explicit
part in this new public drama, but the mill did sit on what had been
Wiyot land—directly between the island and Fort Humboldt—and
the newspaper's overt silence on the Indian question opened a structuring space of discourse within which this new violence could be
described in terms of the new culture system. With these new public
voices crying out "terrorism" and "communism," we suspect that the
core stakes of the current struggle must be near at hand. An editorial
in the *Humboldt Times* on the same date, June 22, 1935, put it directly: state violence was necessary to defend the nation in every locale
where its fundamental institutions are openly challenged.

Yesterday's Rioting

The evidence is overwhelming that yesterday morning's tragic riot in Eureka was deliberately invited and caused by the radical element which had attached itself to the lumber strike movement. . . . It was a "now or never" situation, according to the public declaration of the communist leaders on the day before the rioting occurred. The strike itself was clearly lost, after the workingmen of Humboldt had been given a fair opportunity of more than one month to decide their views on it. A great majority of them had made it clear that they wished to remain at work. It is to their credit that they were determined to do their best to support themselves and to reject the attempted leadership of class hatred. . . . But the communists among them would not accept the free verdict of these home-owning, thoroughly American workers.

This language says a lot about the changing conditions of struggle—social radicals, terrorists, anarchists, and communists had replaced the diggers, those poor fellows, the savage Indians, redskins, bucks, and squaws included, as the obstacle and other against which capital culture must struggle to grow—at least according to the hegemonic voice of Humboldt's major media organs, which claimed to speak for all but in fact gave privilege to corporate interests and rather short shrift to workers and Native Americans.

Expressing regret for the tragic event, but defending the law as a bastion of freedom for "home-owning, thoroughly American workers," the *Times* editorial puts the symbol of property in play—property in land, in the home, is equated with American feelings for nation, establishing a chain of equivalence between freedom, home ownership, and work that suggests how deeply the concept of ownership had permeated constituted authority. Having delivered property in land to white male immigrants during colonization, this feeling for nation was growing synonymous with its privilege. Property had been made a naturalized symbol of liberty, a gift for which all owners are indebted to the celebrated nation and an inviting promise to everyone else. And here we see the *Times* deploying the symbol to call up the citizenry's passionate obligation to defend the nation's law.

But the strikers' demand was also for property—property in their labor and what it produces—and so the strike should be seen precisely as a crack in the consensus on the symbol, a challenge to the established idea that owning the land and the capital needed to buy labor and use it confers absolute right of control over labor's conditions and all of its product. The strikers had made the juridical definition of the power over labor that constitutes property as the receptacle into which the actual flow of alienated values accumulates to one party or another into the point of contention. The public struggle over redwood unionism is shown to be rearticulating the core values that identify the nation as a structure of feeling and really challenging the reigning culture system.

How deep was the fissure? One column asked "why we have been so slow about dealing with the communistic and Bolshevik element that has been growing in our midst by leaps and bounds." In taking action against the strikers, the mayor and police chief indicated their sweeping concerns by jointly dispatching the police to the rail yard with orders to burn down the hobo "jungle."[3] At 1:30 that afternoon they warned the occupants to leave and then torched the shanty village, but not before discovering "several copies of Communistic literature and copies of the journal *Man!,* a periodical of the anarchist idea and movement."[4] Tipped off that an incoming Northwestern Pacific freight train would be importing even more vagrants, police greeted the train and promptly ordered unknown folks out of town on the next southbound train. With these two acts, the *Times* reported, the law has ensured that "no longer will the odor of mulligan stew mix with the fetid odor of the nearby swamps," for this "rendezvous of tramps and hotbed for radicals" had been rooted out. "It is believed that the city was purged of much of the undesirable element that has contributed to the unrest leading to the riot yesterday morning." The witty headline put it this way: "Eureka's Picturesque 'Jungles' Only a Memory."[5]

The June 1935 issue of *Man!,* presumably among the issues confiscated in the jungle raid, included an article titled "Rampant Fascism in America." Its author reported on the emergent student and professorial movement that the issue of May 1935 also described. In an unsigned column, "The Youth of the Country Speak," *Man!* asserted that a wave of socialist antiwar demonstrations had occurred on April 12.

> In colleges of the reactionary south, in fact, from Maine to California, over one hundred thousand students as well as professors voiced in no mistaken terms that they are against any contemplated wars for exploitation and legalized plunder. The brutal attacks by

the "law and order" brigade upon some of the militants in various parts of the country is but another added proof as to how fearful the misrulers of our lives are of such demonstrations and to what means they will resort in order to suppress these. . . . The answer to a call for war by and in the interest of perpetuating the present order, should be answered by a General Strike and followed by a Social Revolution that would end every form of oppression and rulership.[6]

These are the fighting words of anarchists who saw in the state a looming specter of fascism. When business and government resort to vigilante violence against civil organization, as they did in Humboldt's Great Lumber Strike, such public charges by radical elements of a broad social movement gain credence—at least we can understand why such changes were being made. In 1935 state support of business corporatism, in its struggle against the insurgent populist corporatism expressed by unionism, formed a basis of grievance among workers and organizers.

But mainstream unionism also fundamentally challenged redwood capital culture. If redwood capital can aggregate property for power over labor, it would seem only just that workers be allowed to aggregate labor for counterpower over capital. The question raised concerns about the role of state intervention in this competition of rival collectivisms—not intervention itself. Federal and state governments had historically demonstrated willingness to use the law to structure economic growth however government saw fit, usually promoting economic expansion and corporate control at the expense of other public interests. Resource conservation, for example, always came second to exploitation and extraction, and workers had to fight anew for relief from every conceivable deprivation that capital might newly conceive in its interminable internal struggle to lower its costs.[7] Labor's basic demand for the right to associate and bargain for control constituted its challenge to business and the state; this was its only hope, because the association of property had already been juridically programmed into the capitalist system. We will have cause to revisit the developing language of labor law later, but first let us listen a bit longer to the redwood strike public and gain a better sense of its direct concerns.

Not only lumber was agitating in June 1935. "All Quiet on the Fishing Front," proclaimed a Saturday headline next to the strike violence report, explaining how the "fishermen express[ed] themselves as satisfied with the new arrangement by which the recently organized Eureka Fisheries [will] act as the sales agent of the striking fish union." For five weeks the fishermen had been demanding a higher guaranteed price for their catch, but the processing companies refused to produce a price schedule, so the fishermen combined against them. At the same time, train crews and merchant mariners sporadically halted work in solidarity with redwood strikers.[8] Also on May 14, the night before the strike, the International Longshoreman's Association pledged to go out with the strikers, but over the next few weeks the docks operated normally. However, after the lumbermen called a nationwide boycott of redwood products on June 5, the dockworkers refused to handle any of the lumber stamped "hot cargo" until after the violence ended the strike.[9]

Meanwhile mass-media editorials channeled local attention into the national political public tuned in to Washington politics. For example, on May 27 the Supreme Court ruled that the mandatory code section of Roosevelt's National Industrial Recovery Act (NIRA) was unconstitutional, thereby nullifying the act's Codes of Fair Competition, including the Lumber Code Authority's "Code of Fair Competition for the Lumber and Timber Products Industries," which in 1933 set redwood lumber industry wages at thirty-five cents per hour, limited labor time to forty hours per week, and set elaborate price controls for product.[10] On June 11, the *Humboldt Times* carried the conservative syndicated columnist David Lawrence's "From the Battlefield in Washington": "This has been a historic two weeks," he wrote, "and the country knows to a certain extent what has been the effect of the Supreme court's unanimous decision." As he put it, the court ruled that it takes "an actual amendment to bring about a centralized government." If it had instead ruled in favor of the NIRA, it would have driven the stock market down, first for "fear of the disintegration or confiscation of private property through government management," and second because of "the certainty of reduced earnings." Lawrence argued that stretching the Constitution's commerce clause to allow the National Recovery Administration to set mandatory production codes ultimately meant that every single transaction would fall under federal authority, weakening state and local control. But "the entire economic structure of the United States has been built up on the preservation of equities and property rights under the Constitution," he protested, "and once the federal congress is given authority to impose cost items, whether for labor or any other purpose, the management ceases to be individual and becomes public."[11] This national debate over the legal relations of capital to labor touched the core of Humboldt's public-sphere struggle for control over labor. "Which side are you on?" had become a local question in which national and worldwide concerns were imploding.

The *Humboldt-Standard,* Eureka's other daily since 1892, put out an evening edition on the day of the killing, featuring a special front-page editorial. Beneath banner headlines reading "114 Arrested in Riot; One Dead; Many Others Hurt," the editors called their public to order in terms that bear examination:

Law and Order Must Prevail
(editorial)

One thing must stand first and foremost in the minds of all fac-tions and all interests and that is that law and order must prevail in Humboldt County. . . . The right to strike is unquestioned, but that right does not connote the license to interfere with a single individual worker who desires to work. . . . In all probability, when constituted authority has had time to fix responsibility . . . it will be found that the instigators and perpetrators are not loyal Humboldters, not legitimate members of any labor union, but pro-fessional agitators and undesirables who flock to any locality where they may use an existing labor trouble as the excuse for projecting their own peculiar doctrines into the situation. . . . Decent labor must purge itself. Eureka must join with other states and with law enforcement organizations to drive from every community and from the state and nation this trouble-breeding element that recog-nizes no law but under the protection of our laws would overthrow our form of government.[12]

From the sound of this authoritative public address, which arrogated the task of defining for all what is decent on one hand and "trouble-breeding" on the other, and associated the nation of laws with the de-cent, one might think that the strikers had little support in the county. That impression is belied by circumstances surrounding the funeral of the first picketer to die in the violence—William Kaarte, shot dead at the scene.

Three days after the signature event, the front-page headline of June 24 dispelled at least part of the editorial myth: "2,000 Attend Funeral of W. Kaarte." In Eureka, fifteen hundred union members from all labor sectors joined one hundred carloads of friends and relatives in a "five block solid phalanx followed by the body" of the former woods cook. Each local union formed a group and marched in formation, with local assemblyman M. J. Burns out in front. A large American flag taken from the Labor Temple was carried high in a stiff wind by J. B. Williford, president of Timber and Sawmill Workers Union Local 2563.[13] Local 2563 was in fact a "hotbed for radicals," but the funeral of Kaarte suggests that they were hardly pariahs. There were only 16,000 people living in Eureka that year and 48,000 in the county.[14] It was quite a turnout to bury an associate of disreputable communists and terrorists intending to "overthrow our form of government."

Unfortunately the only known collection of strike literature put out by Local 2563 has been lost, but we do know they produced the *Redwood Strike News* and used it to counter what they understood to be a dominant press bias in favor of capital. Nor were the strike-time meeting journals of the local among those collected by labor historians—they are missing, without explanation, from local archives and unreported in any treatments of the strike. But public voices of

Labor Temple, Eureka, 2001.

redwood unionism can be heard elsewhere. We have, for example, the *Official Yearbook of Organized Labor: Humboldt County 1929,* published by the Federated Trades and Labor Council of Humboldt County. W. E. Gladstone is quoted on the cover ("Trade unions are the bulwarks of modern democracies") alongside Abraham Lincoln ("Capital is the fruit of labor and could not exist if labor had not first existed. Labor, therefore, deserves much the higher consideration").[15] Redwood unions were explicitly claiming the patriot's mantle. The nation here is contested terrain in which civil and political rights are provinces to conquer and defend. Workers want democratic participation; owners defend their property rights. Both claim the mantle of American tradition. Who has the right to control production? For their part, labor made mechanization and capital's control over constantly increasing productivity into sites for contesting the hegemonic status of national capital culture.

In the 1929 *Yearbook,* for example, American Federation of Labor (AFL) president William Green spoke to Humboldt about "the Attitude of the American Federation of Labor toward the machine displacement of Industrial Workers." Although the industrial revolution and mechanization brought displacement and conflict, he wrote, and mass production, standardization of output, and specialization have become "a substitute for individual skill and training," nevertheless everyone benefits because "human drudgery has been relieved, social well being has been advanced . . . knowledge has been more widely disseminated . . . wider opportunities for enjoyment of leisure have been created . . . and spiritual values have been enhanced in correspondence to the enhancement of material values." The AFL and "the great mass of working people [have] come to understand" these benefits, and though unions in some countries have "set themselves in opposition to the introduction and use of mechanical processes in industry," the AFL "has accepted it, has adjusted itself to it and will be found co-operating with management in the extended and efficient use of mechanical technique and mechanical improvement." But "common justice demands that the wages of workers shall increase in accordance with their increasing power of production," and "in a corresponding way the hours of labor can be reduced so that the great working mass of working people may enjoy higher wages, short workdays and shorter work weeks through the introduction and installation of machinery."[16]

The union ideal is common justice, an appeal in which we cannot fail to hear an echo of natural rights: workers should share in the

In Hoc Signo Vincens

FEDERATED TRADES AND LABOR COUNCIL

HUMBOLDT COUNTY.

Insignia of Federated Trades and Labor Council, 1929. Courtesy of the Humboldt Room, Humboldt State University Library, Arcata, California.

values produced because it is their right to own the value produced by their labor. But this appeal to the natural, the obvious, is supported by an equally powerful claim: sharing the benefits of mechanization equitably is a matter of public good—it is in the public interest to create a consuming class by justly rewarding the producers of value. Unionists understood, in other words, what the New Deal would over decades prove to the world—that a mass culture of highly paid workers could be forged, that an emerging army of consumers could be raised from the standing army of the unemployed. Harry C. Breit of Eureka's Cooks and Waiters Local 220 theorized this relation of wage labor to capital with equal foresight in the *Yearbook,* presaging the prosperity that New Deal laws would have a large hand in shaping.

> So much has recently been said unfavorably about the demands of the wage-workers for increased pay, a shorter workday and better working conditions, that it becomes necessary to refute them or correct them. The general impression is that to get things cheap, at a bargain, cheap labor, at a knocked down price, is good business. The question, however arises, "is it?" . . . We contend that cheap labor is a positive injury to the community and a detriment to all classes alike. We further contend that a general reduction in wages is a public calamity. Cheapen labor and the incentive that spurs men on to effort and improvement is destroyed. Reduce wages and the workers in the performance of their duty are disheartened; they become careless and sometimes indifferent; they have no ambition in their work and do not care whether it is done right or not. The poorly paid worker cannot be depended upon to do the work required and in the manner it should be done; he seldom gives satisfaction and is rarely employed steadily. . . . It should also be remembered that the purchasing power of a people creates employment and that only well paid workers possess that purchasing power.[17]

In these dueling public claims of capital and labor, all parties are striving for control over laws that create and maintain relations of property. Each voice we hear is attempting to channel attentions from the mass of available loyalties and investments into its own respective camp and call up a public with specific definitions of the labor situation, struggling to build both individual and collective consciousness of law; and each voice works with the concept of nation, claiming to embody the sacred values that make America great. One side stresses liberty, and the other equality; both claim the nation's law as their principal ground.

Editors and timber capitalists proclaim that individual ownership rights are in the nation's public interest, while unionists claim that their right to combine also serves that interest and thus requires protection equal to that accorded private capital. The social space laid out by these claims is a sign that the modern social imaginary is here, as it is locally performed under conditions of redwood production. Each position is divided by this much-remarked-on private-public fissure in political consciousness, which we have learned to recognize as a mark of the founders' design. This is precisely the pattern they built into the public sphere when they enumerated individual civil rights of property over against political rights of democratic participation in using the texts to which juridical institutions continually return in their role of regulating the political body of the nation—it is, in other words, a cultural performance of perpetually conflicting rights.[18]

Humboldt's labor trouble, like the Indian trouble of decades before, was a broadcast system for this discourse of rights. It disseminated the nation form by performing its mandate in a free public spectacle. People entered this discourse by speaking to everyone and no one in particular, everyone of *us,* that is, we Americans who animate and obey the authority—law and order—embodied in the concepts of free speech, press, and property that collect the force of the nation's liberal institutions and archive its display of the symbols that drive it.

Noticeably absent from the voices of labor trouble we have heard from so far are direct concerns about race and religion. These familiar nineteenth-century styles of conflict over property, omnipresent in the discursive production of Humboldt's Indian trouble, are altogether less prevalent in the manifest content of redwood labor trouble. But it would be a mistake to conclude from this silence that Humboldt had somehow left all of that behind. On the contrary, its dominant institutions had been deeply racialized by decades of Indian removal, the effects of which were regularly expressed in the redwood imaginary. Several incidents confirming this colonial legacy bear mentioning.

From the earliest days and increasingly in the 1880s, successive waves of antiforeigner sentiment swept across California and the West into Humboldt. In 1885 Eureka became the first of many California towns to expel their Chinese. Although they numbered only 1.5 percent of the county's total population (whereas in 1880, 16.3 percent of San Franciscans were Chinese, as were 8.7 percent of all Californians), the Chinese were seen as competing for scarce jobs and were singled out for a variety of cultural differences. Out of 2,700 people living in Eureka, only 101 were Chinese. In the whole county there were only 288 Chinese employed, and only 6 worked in the lumber industry. Nevertheless Humboldt was swept up in California's antiforeigner sentiment when an event occurred that prompted local people to retaliate dramatically.

On February 6, 1885, Eureka city councilman David C. Kandall was accidentally killed by crossfire in a shootout between two Chinese men who were never identified, and a local white youth also took some errant lead in the foot. In the racial hysteria that followed, six hundred white men gathered downtown within twenty minutes. After rejecting the first proposal made by a would-be community leader—to massacre every Chinese person in Eureka's Chinatown—a committee was appointed to contact each Chinese person and inform them that they were to leave the county within twenty-four hours.[19] The next day *all* of Humboldt's Chinese were forced onto ships and sent south to San Francisco, with many tons of their belongings heaped in big piles on the docks as they waited to depart. Two steamships were employed for the job—the *City of Chester* took 175 Chinese, and the *Humboldt* took 135. This collective expulsion was as much a community affair as the Wiyot massacre, but it was done without murderous violence and so is less well remembered in Humboldt today.

Another revealing incident occurred during a wave of lumber strikes in 1903. A number of mill workers at Scotia objected when the company brought in a crew of blacks and Filipinos. The lumbermen would not allow the newcomers to eat at their cookhouse table. When several of the offended men were fired, things went back to normal quickly, but the foreign workers remained and with them a source of aggravation that functioned quite well for the redwood operators.[20] The managers and owners well understood the racial situation and used it to their advantage.

Pacific Lumber Company and Hammond, for example, sought out ethnic difference in general and put it to work in their favor.[21] They transported ethnically selected workers from the East and Midwest who were willing to sell their labor at rates cheaper than those of local men. In one case, local men fiercely resisted a group of Italian workers imported by Pacific Lumber. The companies very often charged such workers for transportation to San Francisco by rail and then to Humboldt by ship, where their pay would be docked to cover the cost of their delivery to the new land of opportunity.[22] The strategy was doubly successful: cheap, imported labor divided worker solidarity along racial and ethnic lines, helping to secure control and thus even cheaper labor. Institutions of work and management were deeply racialized in these years leading up to the big labor trouble.[23]

We can enter this racial dimension of the redwood imaginary through the prism of an easily overlooked item in the *Humboldt Times* of May 14, 1935, the same day that Local 2563 announced in the press that it would join the West Coast lumber strike.

Scotia Redmen Accept Invitation

The Scotia Red Men have received and accepted the invitation from the Garberville Chamber of Commerce to take part in the forth-coming redwood highway celebration scheduled to take place at Garberville on June 9th. Attired in attractive Indian costumes the Scotia Red Men will add considerable color to the event, featuring an Indian pageant in the big parade. There will be a news reel photographer on hand to make pictures of the affair.[24]

Weott tribe 147 of the Improved Order of Redmen was an active part of Scotian civic life in the 1930s. The Redmen say they are "the oldest secret society of purely American origin in existence, a claim which rests on the fact of its being the virtual continuation of the Sons of Liberty, formed prior to the American Revolution and the secret societies which it gave birth."[25] The Sons of Liberty famously performed the Boston Tea Party, in which American revolutionaries, dressed like Mohawk Indians, revolted against the commercial oppression inherent in English tax policies—in this case the Tea Act of 1773.

Established at Baltimore in 1834, the Redmen modeled their government on that of the Odd Fellows, which itself is of the Masonic pattern.[26] *The Official History of the Improved Order of Redmen,* compiled from the order's official records by Carl Lemke in 1964, describes in detail the descent of the Redmen from the era of patriotic societies and champions and their mission as "conservators of the history, the customs and the virtues of the original American people." The Redmen's self-styled mandate is "to perpetuate the memory of this, the noblest type of man in his natural state that has ever been discovered"—but only whites were accepted into the order. Local chapters called themselves tribes and awarded members the chance of achieving several degrees: an "Adoption Degree," exemplifying Indian firmness and the power of endurance; a "Warrior's Degree," illustrating the hunt and the chase, wherein sustenance is provided for the tribe and the manners of the warpath are conveyed; and finally a "Chief's Degree," which "illustrates the religious ceremonies of these primitive men, they being firm believers in the Great Spirit. Their beautiful Legends showing unbounded faith in a future life of immortality of the soul."[27] For women there was a separate order and a Degree of Pocahontas, the character of which can only be surmised.

At the time of the big parade in which the Redmen would dance to celebrate the new redwood highway, Scotia was a wholly owned lumber company town in the midst of a bloody regional industry strike. Though one of its tree fallers would soon take a bullet and enter the historical consciousness of Humboldt's labor trouble forever, the company itself was completely unorganized; it operated normally throughout the strike. Redwood corporate paternalism reached perhaps its most eloquent achievement of social control in this impenetrable fortress of goodwill and fair treatment. The workers' houses were owned and maintained by the corporation. Their children were given college scholarships and hiring privileges. Garden competitions and picnics were organized. Scotia was paradise in the redwoods, according to many. But little dissent from its goodwill was tolerated. And as the empire of Pacific Lumber continued to grow through the turbulent years, generations of workers channeled their labor through Scotia and into the forested landscape. No doubt workers benefited from the company's legendary good citizenship, but control over labor was the defining question of the historical moment, and Pacific Lumber achieved it with a company town strategy and forceful repression of union organization.[28] The Scotia Redmen who danced on June 9, 1935, were white company men living and working on company land in what had been Wiyot territory.

Polk's Eureka City Directory, 1935, a complete "Buyers Guide and Classified Business Directory," listed twenty-nine additional "Secret and Fraternal Societies," including chapters of the Masonic Order, Ancient Order of Egyptian Sciots, Order of the Eastern Star, Ancient Order of Foresters, Benevolent and Protective Order of Elks, Knights of Columbus (Catholic), Knights of Pythias, Loyal Order of Moose, Native Daughters of the Golden West, Order of Herman's Sons, Fraternal Order of Eagles, United Order of Ancient Druids, and two additional chapters of the Order of Improved Redmen—the Hupa tribe 146 and Pocahontas Ewa-Yea 56. We also learn from the *Directory* that the town had fifteen churches serving a population of 16,038, in a county of population 43,189, up from 27,000 in 1900 and 2,694 in 1860, of which 162 were "colored," meaning "Indians and Negroes."[29]

While this is not the place to excavate the full range of relations these fraternal orders had to the cultural life of Humboldt, it should at least be said that their common structures and functions suggest a shared style of life. Invariably the spiritual requisite for membership was belief in a higher power and a willingness to submit before it and the guidelines of the order in a unified program for rational Christian enlightenment and patriotic nationalism. It is well known that such orders had financial and other concerns as well, not the least important of which was simple entertainment, but the basic assumption was Christian spirituality and an explicit commitment to national pride that was usually expressed as "working for the public good." In the Scotia Redmen, economic concerns and right rules for Christian living were fused with patriotic nationalism and racial identification in fun public displays of masculine civic pride, as on this occasion in which the cause of celebration was the completion of Highway 101.

The prevalence of these orders in 1930s civic life communicates something that is perhaps difficult to conceive: Humboldt, Eureka, and the region's principal company town of Scotia were brilliant flowers of patriarchal, Anglo-capitalist hegemony in the redwood region. Their schools, churches, fraternal societies, libraries, theaters, railroads, chambers of commerce, and market base of natural-resource commodification formed a complex institutional machinery of civic authority and cultural (re)production.

As a further reminder that this was a cultural order not merely in consensus but riven by conflict, *Polk's Eureka City Directory* of 1935 also listed twenty-five labor organizations, each of which met at the Labor Temple in Eureka. These included the Brotherhoods of Locomotive Engineers, Locomotive Firemen and Enginemen, Painters and Decorators of America, Railway Conductors and Railway Trainmen, as well as the Bakery and Confectionery Workers, Cooks and Waiters, Electrical Workers, Journeymen Barbers, Laundry Workers, Machinists, Musicians, Plasterers, Plumbers, Printers, Textile Workers, Brewery Workers, Carpenters and Joiners, and finally the Saw Mill and Timber Workers.[30] Certainly these union folk were members of fraternal societies as well, indicating that a lot more was at work than opposing class interests.

In fact, what we have is a network of interlocking unions, churches, timber corporations, and fraternal orders that formed an intricate structure for articulating redwood politics and identities, something like a linguistic and racial community of the order described by Étienne Balibar's concept of the nation form.[31] The nation form works, as he put it, through "a network of apparatuses and daily practices [in which] the individual is instituted as *homo nationalis* from cradle to grave, at the same time as he or she is instituted as *homo oeconomicus, politicus, religiosus*."[32] The discourses and practices of religion, unionism, and fraternal brotherhood each called their members into *the people*—and each had its eyes as well as its laboring hands set on the public good. That is what united them—the public form of discursive practice in and through which they entered the perpetual conflict for control over property, for example, in labor.

In the nation's dominant collective public form and space of communication—the public sphere, a legacy of the framing—each individual is compelled to champion the public, common, and universal good. It is important to note the underlying consensus—a commitment to rational public discourse whose singular focus is universal human rights. It is here once again that the record seems most clear: the public sphere's continuous spectacular display of constitutional claims carried out a powerful invitation to collective action. The words of the Constitution are a symbolic drive, both providing spokespeople for the status quo with a rationale for authorizing claims of legitimate ownership and encouraging working people to identify with each other in a common effort to wrest from their masters a legal concession of right to control their own labor values. Both were public, symbolic demands on the nation to fulfill its promise, and where these demands fell back on constitutional claims, we encounter the legacy of the framers, in all its real, material, symbolic, and imaginary effects—a culture of rights in perpetual conflict.

Symbols are powerful things. As we have seen in the case of Humboldt, the symbol of liberty embodied in property made struggle the business of thousands of people, each fighting for what has always already been promised by the text. When the Constitution claimed nature's law and thus God's legitimizing approval in formulating the basic rights to be granted in the new nation's compact, it programmed the colonizing culture system to produce in the people precisely this feeling—a nation in mind. This nation is an image—a social imago—whose original unity can be traced to a text that largely entrusted to property rights the promise of liberty that gives America its sacred name.

What we are learning is that the deep cultural drive of rights-based institutions produces individuals who embody a tendency to fight about life, liberty, and equality in terms of their juridically constituted objective manifestation in property, which again is not a thing but a social relation. The social relation of property exercises a patterning effect on people's passionate connections, such that property must come to be seen first as *libidinal,* because the psychical energies of the people involved become deeply identified with, and therefore bound to, the beliefs and values promulgated by those institutions; second as *economic,* because those institutions shape the way that labor energy is objectified as value in exchangeable commodities and determine how and where they accumulate (for example, to owners or workers in various proportions); and finally as *sociopolitical,* because the forms of those institutions are always already outcomes of struggles in the field of spectacular public power.

In the 1930s, after six decades during which redwood capital succeeded in exercising great power over labor and extracting cost savings while combining in industrial associations and refining the labor process, mechanizing production, and expanding in scale, organized labor began to succeed in getting better wages and conditions by refusing to work. These are strong local signs of internal contradictions—the first contradictions in capital culture that economic sociology has always understood as the motor of history, but which environmental theory has also established as the structural source of its external, second contradictions. Under intense competition, firms transform themselves in defense of profitability by means of continuous application of science to production, innovating cost-cutting measures and increasing scales of production that externalize costs that ultimately tend to impair the social and ecological conditions of production, raising costs on capital in general. Communities of labor and nature are made to absorb every cost that firms can figure out how to avoid. But these costs accumulate, sparking worker revolts and finally environmental resistance.[33] And because no institution stands alone in society but rather is nested in the full constellation of modern institutions, this process, in driving labor's challenge to capital, had effects not just in labor law but also in the practice of those religious, racial, sexual, gender, media, public-sphere, and ultimately environmental institutions, the forces of which contextualize modern law in every domain.[34]

New Deal for the Body?

These traces of the redwood strike massacre in the public-sphere archive of Humboldt's labor trouble remind us that the year was 1935. It was a moment of turmoil across the globe. Decades of monopolizing capitalism were realizing their effect, producing the conditions for systemic change. Efficiency gained in Fordist assembly was ramping up output across the industrial sector but imperiling workers by overproduction. When the market crashed in 1929 and world economic depression set in, production started shutting down, driving millions out of work. Labor organization was reinvigorated in proportion to workers' new misery under a failing world system.[35]

Nazi power emerging from Weimar Germany channeled recovery through racist state corporatism into total war, daily gripping the attention of Humboldt's news-reading public. Word of Mussolini's program for Italian recovery through fascist aesthetic corporatism, expansion, and war also passed continuous review in local media. Japan in Manchuria for a decade of Sino-Japanese war similarly made militarism an economic policy that redwood papers daily perused in Associated Press coverage and syndicated punditry. Russia's Bolshevik revolution, moving through the hopeful horrors of Lenin toward the grim-faced terrors of Stalin, offered its totalitarian image to both conservative capitalists and welfare-state liberals, who used it to contextualize radicalism in the redwood labor movement and to cope with their own local specters of communism.

In the United States, the dialectic of capital and labor socialization was proceeding at a quick pace. By 1935, 0.1 percent of American corporations reporting to the Bureau of Internal Revenue owned 52 percent of U.S. corporate assets. Concentration on the same scale also occurred in employment: 0.8 percent of all manufacturing firms employed 27 percent of U.S. wage earners engaged in manufacturing.[36] Trouble was brewing, even boiling over, in the great melting pot of American labor.

Gains were made by workers in proportion to the lengths they were prepared to go in challenging the established order. Agitation and organization exacted concessions industry by industry, region by region, transforming public consciousness and creating a social context in which the political will to institutionalize labor rights was effectively emerging. A brief look at some of the laws that relegislated the dynamic relation between labor and capital in the years leading up to the Great Lumber Strike will help to establish just what was at stake and how the labor trouble in Humboldt embodied the crisis the nation was facing.

As far back as 1842, one state supreme court ruled that labor combination was not in itself a conspiracy if its purposes and methods were consistent with the law. The court ruled that seeking a closed shop for the maintenance of working conditions satisfied the requirement of lawfulness.[37] And the Erdman Act of 1898 attempted to limit retaliation against railway unions, stating in section 10 that "any employer subject to the Act . . . who shall require an employee, or any other person seeking employment as a condition of such employment to enter into an agreement . . . not to become or remain a member of any labor . . . organization; or shall threaten any employee with loss of employment, or shall unjustly discriminate against any employee because of his membership in such a labor . . . organization . . . is hereby declared to be guilty of a misdemeanor."[38] But even with precedents and laws such as these, capital maintained an extraordinary advantage over labor, winning a whole host of laws that promoted its interests and facilitated its combination.

Laws legitimating corporate personhood may be the most important forces structuring this class competition. In 1886 the U.S. Supreme Court issued a novel interpretation of the Fourteenth Amendment that extended equal constitutional protection to corporations—it gave them constitutional status as *fictitious persons* and thus shelter under the Bill of Rights.[39] In *Santa Clara County v. Southern Pacific Railroad Company* (argued January 26–29, 1886, filed May 10, 1886), the court ruled in such a way that the court reporter felt justified in writing a headnote to the decision: "The defendant Corporations are persons within the intent of the clause in section 1 of the Fourteenth Amendment to the Constitution of the United States, which forbids a State to deny to any person within its jurisdiction the equal protection of the laws."[40] The effect of the headnote was that the case would later be regularly cited as grounds for holding stare decisis that private corporations are natural persons under the Fourteenth Amendment, even though the ruling itself made no such explicit determination. From this sketchy beginning, the concept of corporate personhood has grown, through numerous applications of the precedent, to be a well-accepted tenet of U.S. jurisprudence and legal culture.

The Santa Clara precedent exemplifies a tradition of judiciary knowledge, statute law, legislated policy, and constitutional norms that together did more than simply recognize the legal aggregation of capital—it invited it, amplified it, and made it authoritative.[41] This was the juridical matrix in which labor conflict grew.

By legitimating capitalist combination in the rights of corporate personhood while refusing to fully recognize and guarantee the rights of labor to combination in unions, the state was intervening in the competitive struggle, not exactly setting the price of labor but programming the system for capital accumulation and financial socialization and so for the socialization of labor and all the other productive forces, including natural environments, as would become evident later.

Late-nineteenth-century capitalism of the Gilded Age should be seen as both the cause and effect of such state intervention, for it was the rising power of corporate accumulations that energized the ultimately successful push to institutionalize such law. Equal rights for corporate personhood, in other words, were part of a national recipe for class conflict. The laws that finally legalized unionization explicitly represent the state's realization that this bias in the competition for control over labor and thus property in its product needed correction, if not reversal, and that it must do so for the common interest and the public good—that is, in the interest of both labor and capital. New Deal labor law envehicled labor's radical critique of the state's legal maintenance of capitalist advantages over labor.

The National War Labor Board set up during World War I, the Railway Labor Act (1926), the Norris–La Guardia Act (1932), the National Industrial Recovery Act (1933), and the Wagner Act (1935) each include language recognizing that the state's historical intervention on the side of capital exacerbates the crisis character of capitalism, which had long been described in the labor movement's most radical literature, for example, Marx's "Wage Labor and Capitalism" and even the journal *Man!* Consider, for instance, a crucial passage from the Norris–La Guardia Act of 1932, which had the effect of limiting the power of any court to issue injunctions against striking workers; such sanctions had become a weapon of major effect against labor organization after the wartime truce had been broken, helping to stifle the rebound of labor organization in 1920s.

> Whereas under prevailing economic conditions developed with the aid of governmental authority for owners of property to organize in the corporate and other forms of ownership association, the individual unorganized worker is commonly helpless to exercise actual liberty of contract and to protect his freedom of labor and thereby to obtain acceptable terms and conditions of employment, wherefore, though he should be free to decline to associate with his fellows, it is necessary that he have full freedom of association, self-organization and designation of representatives of his own choosing, to negotiate the terms of contracts of his employment and that he shall be free from the interference, restraint, or coercion of employers of labor, or their agents, in the designation of such representatives or in self-organization or in other concerted activities for the purpose of collective bargaining or other mutual aid or protection.[42]

This language is telling; capital accumulation is described not as the natural function of laissez-faire markets but rather as an effect of legal signification of aid for property owners, which deprives workers of actual liberty of contract and freedoms of association and labor. Workers must henceforth be allowed to associate and use representatives to negotiate the conditions of their employment. The state was recognizing that it had grown to disproportionately represent private property, accumulated in corporate ownership, and that the workplace must be democratized to correct this injustice.

In other developments, the election of Roosevelt in 1932 registered a major victory for the Democratic Party and its somewhat muted representation of progressive causes. The administration quickly offered and won passage of the National Industrial Recovery Act (NIRA), creating the National Recovery Administration, an agency empowered to enact the controversial market interventions of mandatory wage, price, and production schedules. To many capitalists this looked like pure socialism—a leap into centrally managed markets. Though its mandatory Codes of Fair Competition were ruled unconstitutional in 1935, its historic section 7[a] legitimated the unions' cry for rights to association and jump-started widespread unionization.[43] Section 7[a] held

> (1) That employees shall have the right to organize and bargain collectively through representatives of their own choosing and shall be free from the interference, restraint, or coercion of employers of labor, or their agents, in the designation of such representatives or in self-organization or in other concerted activities for the purpose of collective bargaining or other mutual aid or protection; (2) that no employee and no one seeking employment shall be required as a condition of employment to join any company union or to refrain from joining, organizing, or assisting a labor organization of his own choosing; and (3) that employers shall comply with the maximum hours of labor, minimum rates of pay and other conditions of employment, approved or prescribed by the President.[44]

The libidinal engines of labor trouble embodied in this language of state signification are unmistakable—they bring the question of national political economic compromise formation down to the level of bio-psycho-social bodily production by proclaiming the *right* to free participation in decisions that control the conditions of labor and the full flow of values it blasts out of nature, values that in a more concrete language would be names for the movements a body makes in the world. Texts on economic history and theory seem to harbor an instinctive fear of these terms, but a coal miner, choking on dust, black lungs infected, struggling to reach one more time for his mate, does not need neoclassical economic theory to understand the stakes in his struggle. His embrace uses energy in too short supply in a body that has been spent in a life underground. By comparison, the lumber-jacks' trade was considered even more dangerous, and logging the redwoods was hardest of all. Were a body's labor somehow distinct from its more intimate proceedings, it might be possible to substitute some other term for libido in this economics, but the voices of labor envehicled in New Deal legislation and permeating the public-sphere conflict in Humboldt suggest it has just the right temper for describing what was at stake.

Section 7[a] speaks of rights to association in a radical unionist idiom of choice, freedom, self-organization, collective bargaining, mutual aid, protection, compulsion by force, time, the price received for time given in the body's intimate labor and control over every con-ceivable condition this body might encounter in the activity of work—these are coordinates of life and death both on and off the job. Where the new labor laws reach this bodily level, legislating the time a body spends at home and the power over the conditions of its vital well-being, the law decides who will control the division of the working life's quality, energy, and value and who will accumulate these trea-sures. In other words, when these new labor laws delegate the power to name what part of life will be in the market and what part will stay out of it, and when they decide who will set the price at which these goods will enter the commodity circuit, they reach all the way back to that constitutional engine on which so much has already been found to rest and whose authors built their case for setting up the law as a public-sphere conflict over property on early modern ideas about the need to control an unruly body. Just how much labor and its psycho-physiological energy this new law channeled can be multiplied in the following terms: when Roosevelt was inaugurated in 1933, fewer than 3 million U.S. workers were unionized; under the influence of New Deal legislation, that number increased to 15.5 million by 1947.[45]

Workers organizing in redwood country in the months leading up to the Great Lumber Strike were excoriated in the press for views and for language no more socialistic than what had already been made law under Norris–La Guardia and section 7[a] of the NIRA, indicating that the juridical structure organizing the flow of labor value was already changing and that the discursive, symbolic struggle in the streets, workplaces, and public-sphere press of the bay redwood region was lagging behind.

When the Wagner Act, officially named the National Labor Relations Act, became law in July 1935, it reaffirmed the achievements of section 7[a] and concretized the New Deal transformation of the American social contract by establishing the National Labor Relations Board, a panel of three experts mandated to control labor-management relations. The board had powers to issue "cease and desist" orders against noncompliant corporations, and circuit courts gained the power to recognize the authority of individually elected worker representatives to speak for the whole, to decide the suitability of bargaining units in supervised elections, to proscribe unfair labor practices, to rule on reports of management retaliation against workers who complain, and to sanction employers who refuse to bargain in good faith.[46] In line with the Railway Labor Act of 1926, Wagner maintained and the Supreme Court concurred in establishing rights to selection of bargaining agents, eliminating some minority rights but radically democratizing the workplace.[47]

Much has been written about the effect of the Taft-Hartley Act, otherwise known as the Labor-Management Relations Act of 1947, on the gains made by unionism culminating in Wagner, showing how the later legislation had been passed at the recommendations of the National Association of Manufacturers and how it substantially eroded the right to collective bargaining. Taft-Hartley represented the successful reaction of capital to the advance of labor, adjusting labor practices to the desires of the employers, who, for example, used the act for outlawing closed shops. As the tool of reaction to New Deal progressivism, its passage reaffirms my conclusion that the gains up through Wagner really did embody the radical critique of capital.[48]

From these remarks several conclusions can be drawn. The Railway Labor Act, the Norris–La Guardia Act, and the Wagner Act (the National Labor Relations Act) represent a compromise formation between two public forces of collective interest vying for control over labor and the patterns in which its psychophysical energy accumulates as property value to one class or another; they rerouted history, where history is the word for labors whose energy flows into environment and transforms its value into forms of both use and exchange. Legal control over recognition of labor rights represents the power of knowledge to signify property and therein determine its practical existence—it creates property.[49]

At the national level, this legal naming of labor as equal in right to capital was a momentous change in the modern constitutional state's juridical address to its national public of productive activity, its libidinal economy: it changed the world, in other words, and it happened just in time, perhaps, to determine a middle path between communist revolution or full fascist militarism, which today might sound farfetched but at the time was a matter of immediate concern. In 1935 other nations had already turned toward totalitarianism and fascism when faced by the same crisis in the world system of capitalism that was driving the labor trouble in Humboldt to violence.

On this local scene, labor's rising legal status entered the landscape of memory as it transformed the place-making colonization of the redwoods by capitalism. Scotia, for example, and the other company towns of Crannel, Samoa, and Korbell were architectural programs for containing this rising culture of labor resistance. They were built around commodity circuits that unions succeeded in forcing to pay living wages to thousands of families that, carried on over generations, helped to establish the mass attitudes of psychosocial solidarity with an industry that offered laborers not just an opportunity to work hard and live in the present but also a future of family home ownership in community with nation and God. These are great achievements of capital culture in Humboldt, representing decades of work objectifying and sublimating the bodily energies of labor into the redwood imaginary, into every dimension of its living institutions, its symbolic and textual archives, and its built environments of architecture, ecology, and physical geography. They came at the price of worker revolts that signaled accumulating internal contradictions, the memories of which are inscribed very deep and organized around the signature event of extraordinary violence that archived an image of prevailing social relations in this historical moment.

Redwood Capital and Labor

Having tuned in to Humboldt's discourse of labor trouble at the point that contemporary social memory in the region indicates was its single most traumatic moment, and listened to its public-sphere voices speaking of conditions that led to the massacre, we discovered a public struggling to cope with first contradictions in capital culture and poised to embark on a massive expansion through which new contra-dictions and new forms of revolt would soon make their mark.

In the decades after its Indian trouble and before the Great Lumber Strike of 1935, a small class of extremely powerful so-called timber barons came to dominate Humboldt's burgeoning redwood indus-try. Having arrived in the redwoods relatively short of means, John Vance, John Dolbeer, and William Carson, for example, joined Pacific Lumber in coming to embody Humboldt's local gilded age of baro-nial capital. Their empires held strong through World War II, after which they were diminished by the combined effects of the War Labor Board and the long lumber strike of 1945–47, during which the large operators held out for so long that a whole host of upstarts, spurred by unmet demand, entered the market and broke the monopolies.[50] A brief look at what big timber had accomplished by 1935 gives a sense of what the nascent union movement was up against when the signa-ture event occurred.

The timber industry had been central to Humboldt's economy since the 1850s, but employment was greater in agriculture until the 1880s. In 1865 Vance and Company was milling at a rate of 5 million board feet per year, Dolbeer and Carson at a rate of 3.5 million.[51] By 1866 eight mills were operating with 1,000 men at work in the Eureka area,[52] and by 1870 $700,000 of lumber was produced in the county, exceeding the output of agriculture by $150,000. By 1890 80 percent of export-generated revenue was lumber, the total figure being $20 million in that year. Whereas in 1870 40 million board feet were cut, production had tripled by the end of the decade, when twenty mills were operating and 2,000 men were employed in the cutting and mill-ing of 120 million board feet. By 1935 there were 4,000 men work-ing and a total of 5,400 in the tri-county area including Mendocino and Del Norte. Pacific Lumber, Hammond Lumber, and Dolbeer and Carson had emerged as the biggest operators. Holmes-Eureka was strong, but not of such scale. Little River and the Arcata Barrel Company were significant enough to complete the list of five compa-nies explicitly targeted for strike action by Local 2563 in the Great Lumber Strike of 1935.[53]

The Pacific Lumber Company, incorporated in 1869 with about 10,000 acres, had two mills putting out 500,000 feet per day by 1905.[54] Then a wave of strikes motivated the company to begin long-term plan-ning and engineering of its company town at Scotia in 1910—a program of corporate paternalism that disciplined labor and locked out unions. For union organizers, company towns were like fortresses.[55] By 1914 Pacific Lumber had acquired a total of 65,000 acres, making it Humboldt's largest operator.[56] In 1921 it quit using steam and electrified its Scotia mills, hiring 150 new men to handle the increased capacity, for a total of 1,500 employees.[57] In 1950 it bought Carson and Dolbeer and ran way out in front of every competitor.[58] Finally in 1958 it acquired the Holmes-Eureka Lumber Company, adding lands in the Van Duzen River valley and Eureka to its holdings.[59]

The story of the Hammond Lumber Company parallels that of Pacific Lumber. In 1900 Andrew Hammond, a Montana timber capitalist, banker, and owner of the Oregon Western and Pacific Railroad, purchased the Vance Lumber Company from the pioneer lumber baron John Vance. Until his death in 1945 he bought up companies and land all over the region.[60] Five to six hundred men were working out of two lumber camps run by Hammond in 1901, with another four to five hundred men working in his mills, cutting up to 300,000 board feet of redwood per day in two shifts. In 1912 he dropped the name of Vance and called the operation the Hammond Company.[61] Hammond died in 1934 at the age of eighty-five, so he never saw the large protests that would greet the Hammond men as they returned from Samoa on the ferry named *Antelope* to the docks of Eureka each day during the Great Lumber Strike of 1935. His men signed a loyalty pledge and never stopped working—but that meant crossing a picket line at the foot of F Street in downtown Eureka.

When John Dolbeer, that other great pioneer lumber baron, died in 1902, there were one hundred men working in his mill on the Eureka waterfront and another one hundred working out in the woods. All told they produced 100,000 board feet of redwood per day for the Dolbeer and Carson Lumber Company (189–90). William Carson died a decade later in 1912; he was the last of the original pioneer lumbermen (201). He had started the mill in Eureka and ran it until his death, becoming famous for his benevolence, high wages, and ostensible concern for the workingman. The story of his mansion, called the "Castle in Fairy Land," reflects the legendary high regard with which the community is said to have held him; during a downturn in the industry, instead of laying off his men, he employed them for two years to construct the mansion. In 1950 the company sold all its assets to Pacific Lumber.

Former home of William Carson, "pioneer lumberman," Eureka, 2001.

Smallest among the big five operators at work in 1935 was the Holmes-Eureka Lumber Company, created in 1903 by J. R. Lane of Arizona. It owned 680 feet of Humboldt Bay shoreline just below the bluffs on which perched Fort Humboldt and just north of the town of Bucksport on the road leading into Eureka. This land was later acquired by Pacific Lumber and ultimately sold to developers who built the Bayshore Mall, which occupies the site today.

The rise of these great lumber empires reflected the power of eastern capital flowing west, consolidating ownership in land, mills, and markets. The industry's payroll continuously expanded in the process, growing from 3,500 men in 1916, receiving an annual sum of $2.5 million, to 7,000 men receiving $10 million in 1922. Business continued to be good until the depression of 1929, whereupon the mills were forced to dramatically curtail production, and many chose to shut down completely. During the ensuing consolidation, Hammond bought the Little River Redwood Company and its company town of Crannell in 1931 (207). The resulting firm, named the Hammond–Little River Redwood Company, owned ten billion feet of ancient redwood. The Little River Mill was shut down soon afterward, and production was transferred to Samoa, a move resulting in two hundred men being laid off. This pattern of accumulation and consolidation during moments of crisis set the stage for the strike.

This short outline of Humboldt's complex history of capital accumulation would be incomplete without taking a brief look at associationism among the redwood capitalists during these formative years. In 1854 redwood operators formed their first combination, the Humboldt Manufacturing Company, which set prices and controlled output. A redwood glut combined with a general recession in 1854 ended their effort.[62] After that early experiment, the large operators tried to stay independent, but by 1896 they recognized their need for a common plan. Peculiarities of the isolated redwood geography, difficulties in manufacturing, and competition with lumber production in other regions drove them to combine to clear the lumberyards of the excess inventories that new technologies of logging and milling were constantly producing. They also needed to open foreign markets, so they colluded on conditions and prices to avoid being destroyed by their own competition.[63]

The California Redwood Association was founded in 1916, locating its offices in San Francisco. At its inception there were thirteen mills in active membership, a marketing effort that controlled 80 percent of the industry. Then the companies combined again, at another level, creating the Redwood Export Company under the auspices of the Webb-Pomerene Act of 1918, which invited firms to join forces in the interest of penetrating foreign markets, as long as competition at home remained robust. In 1931 the Redwood Export Company was reorganized and designated the California Redwood Association—a combination of combinations that allegedly achieved monopoly power, allowing the industrial redwood giants to operate legally while setting prices and breaking strikes.[64] Formal antitrust charges were brought only once, in 1941, after the major operators combined again in what they called the Redwood Lunch Club. This group had a joint bank account and successfully fixed identical prices at mills. The Hammond Lumber Company, Pacific Lumber, Union Lumber, Dolbeer and Carson, Holmes-Eureka, and the Redwood Manufacturer's Company were each fined $2,500, with the club paying an additional $5,000 for violations of federal laws proscribing price fixing.[65] The club was nominally disbanded, but the association carried on, and by 1950 it was able to determine single-handedly who would receive redwood shipments.[66]

Many in Humboldt saw the great financial empires as anti-American, and from the 1870s through the Progressive Era organized dissent to ascendant capital ran strong. In the field of political consciousness, such oppositional sentiment registered as early support for the California Workingman's Party, the Greenback Labor Party in the 1870s and 1880s, the International Workingmen's Association in the 1880s, the Knights of Labor in the 1880s, and the Humboldt Populists in the 1890s. In the twentieth century, populists, socialists, and ultimately communists and Wobblies carried forward this local tradition of direct-action union and party opposition.[67]

Common to these political subjects was the understanding, informed by the culture and literature of radical criticism, that monopoly power distorts markets, concentrates wealth, and produces excess exploitation. Labor opposition took the form of public movements advancing some version of the labor theory of value and envisioning egalitarian democracy as a bulwark against rising corporate power. Socialist critique of capital often explicitly guided their programs, but local workers also had firsthand knowledge that redwood logging had the highest accident rate and lowest pay of all logging populations in the country. They had direct experience of the industry's long history of collusion among owners and fraudulent consolidation of title, which compelled aggregation and favored capital, and of how the unique commercial properties of redwood trees, the region's geographic isolation, and the physical difficulty of transporting redwood to market contributed to this dynamic.[68]

Humboldt's workingmen's parties, for example, were the direct offspring of Karl Marx's International Workingmen's Association, which had been founded in 1864 only to collapse in 1876. In the following year, the Workingmen's Party of the United States formed out of its American remnants. At a mass meeting called in San Francisco in 1877, the men railed against both capital and the Chinese who competed for jobs.

In 1883 Charles Ferdinand Keller established the first International Workingmen's Association in Humboldt.[69] During Keller's tenure, and in league with the San Francisco division, the IWA put out a four-page pamphlet titled *To the Laboring Men of Humboldt County: Men and Brothers,* from which we can draw some sense of the critical cultural resources the oppositional subjects had available.[70] It began with conviction: "There has been a systematized plan and continued effort, on the part of mill and timber owners of Humboldt Co., to degrade labor and to bring wages down to starvation figures." They continually advertise for more and more men, the pamphlet explained, driving down wages by producing an oversupply of labor, while the local papers are scared to criticize for fear of losing corporate patronage. Meanwhile the California Redwood Company is stealing timberland at labor's expense, using a scam in which $50 is paid to hire men to make fraudulent land office claims that the company then acquires. "Are you aware that you are in the same boat with labor as regards this campaign?" wrote the IWA.

At the base of their charges, we find a critical labor theory of value: "We have associated ourselves for protection only; we have no wish to interfere with the action of any man or his business, so long as he or they do not trample on our just rights, among which we count freedom of speech, independence as regards our dealings with our fellow man and honest and sufficient wages for an honest day's work. We understand well the truth, that labor creates all wealth and we have determined that the man who produces, shall enjoy more fully the wealth his labor creates. Who will join us?" That the group operated in secret—the pamphlet is signed not by names but by division secretary I-41, sub-division secretary M-81, and division secretary M-831—says a lot about the hostility to unions at this early stage.[71]

It is unclear what effect this call had on the workingmen of Humboldt, but in quantifiable terms it met with little success; they were unable to substantially raise wages, ameliorate conditions, or wrest

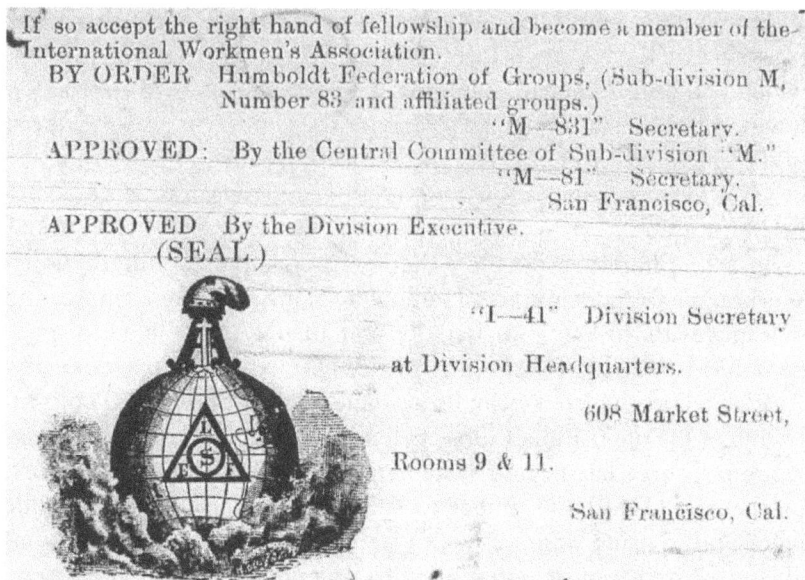

If so accept the right hand of fellowship and become a member of the
International Workmen's Association.

BY ORDER   Humboldt Federation of Groups, (Sub-division M,
Number 83 and affiliated groups.)
"M—831"   Secretary.
APPROVED: By the Central Committee of Sub-division "M."
"M—81"   Secretary.
San Francisco, Cal.
APPROVED   By the Division Executive.
(SEAL.)

"I—41"   Division Secretary

at Division Headquarters.

608 Market Street,

Rooms 9 & 11.

San Francisco, Cal.

Signature page of International Workmen's Association, *To the Laboring Men of Humboldt County* (San Francisco: International Workmen's Association, c. 1880). Courtesy of Bancroft Library, University of California, Berkeley.

control from the operators, let alone achieve union recognition.[72] But
these early stirrings of labor consciousness cannot be ignored, for not
long afterward, another, perhaps not unrelated effort surfaced; the
Knights of Labor established a Eureka Local in 1884.[73]

Events in nearby Fort Bragg, Mendocino, illustrate the power still
wielded by capital over nascent unionism. The Knights of Labor called
a strike for May 1, 1892, demanding a reduction in labor time. In reaction, mill owners fired every member they could discover, a tactic
repeated whenever the Knights appeared—any worker determined to
be a Knight would be terminated.[74] Wherever they demanded a "closed
shop," owners fought back with "right to work" campaigns, arguing,
as they did in this case and right up to the Great Strike of 1935, for the
"right of every man to labor as he pleases and in a free and independent action."[75]

The next significant wave of union activity occurred in 1905, when
the International Brotherhood of Woodsmen and Sawmill Workers
founded the first international union of lumber workers anywhere
in the world. At the time redwood labor was still largely transient,
with lumber camps forming and then moving, determined as much by

rhythms in the market as by feasibility of cutting timber, for the era of company towns, railroads, and tractors was yet to come. Only Scotia was looking like a permanent settlement when the International began organizing. By 1907 the Brotherhood had up to ten locals, including one in Scotia.[76] They launched a series of strikes—which failed—and then simply faded away. But their strike of May 1, 1907, had an impressive start. The strike vote was overwhelmingly in favor, and the action was well organized, exempting plants in Eureka that already had favorable conditions and not going after Hammond at Samoa either, for that mill's practice of antiunion repression had been so successful that the Brotherhood had nothing to gain from calling out the few men it did have on the rolls. On the strike day, 740 men walked out at Scotia, as did 650 at the Northern Redwood Company and 400 at Hammond.[77]

The corporate reaction was swift and sharp. Men were evicted from the camps and from their homes in the towns, and the union had scant funds to back them up. Frustrated and hurting for sustenance, hundreds left the county. The companies would not agree to a closed shop under any circumstances. The strike dragged on for weeks until a referendum on May 31 sent the men back to work. While labor had not succeeded in closing the shops, management had succeeded in developing a relentless, industry-wide, right-to-work open-shop campaign that has kept the shops open through to the present. After lingering for a while, the AFL finally pulled the Brotherhood's charter in 1911.[78]

Around World War I, as redwood unionism made strong gains owing to wartime labor shortages and shipbuilding booms, the International Workers of the World moved more forcefully into California. The IWW established an office briefly in Eureka in July 1917, but it closed within two months and failed to generate much interest through the early 1920s.[79] Their Timberworkers Union made some early gains but suffered harsh judicial repression. By 1919, for example, it was illegal to be in possession of an IWW card.[80] On December 30, 1919, the *Humboldt Times* reported that a Pacific Lumber employee was arrested on precisely that charge.[81] From these beginnings, the owners made the 1920s a period of stability in which they renewed their commitment to paternalism with improved conditions, shortened hours, better services, higher pay, and company towns built for better provision of needs, but with no quarter given to union activity.[82]

The AFL had better luck in the 1920s than did the IWW. Membership in their International Union of Timberworkers (IUTW) ranged between 800 and 3,000 in 1918. They pressed for the eight-hour day and achieved it in March of that year, with help from the War Department in convincing the operators. But one year later, timber capital reinstated the nine-hour day without notable resistance. Then the IUTW struck against Hammond in the fall of 1919; five hundred men went out demanding time-and-a-half for overtime in excess of eight hours. But Hammond held out and successfully received an injunction that broke the strike in October.[83] Soon afterward it was revealed that IUTW leadership had relations to the IWW, after which a number of lumbermen destroyed their union cards, much to Hammond's delight (213).

There followed a difficult period for lumber organization. Wages were cut between 10 and 15 percent in 1921, and some operators returned to a mandatory ten-hour day, again without significant resistance, and finally the IUTW dissolved in 1923.[84] Describing Humboldt timberworkers in their journal *Industrial Pioneer* in 1924, one Wobbly writer said that Pacific Lumber and Hammond victimized labor leaders and were responsible for major deforestation in this desperate period, but for the next decade relations between redwood capital and labor were largely quiet. Capital had the upper hand.[85]

When the 1930s brought depression, only the three largest mills were able to operate continuously—Pacific Lumber, Hammond–Little River, and Dolbeer and Carson. A 10 percent wage cut was adopted industry-wide to keep the mills moving wood.[86] Other operations were mothballed or disappeared. Recovery began in 1933 under NRA guidelines, although some have argued that the renewed growth was independent of the new Keynesian policy. Around 2,000 men were working in Humboldt's redwood industry when the Lumber Code Authority published its first Code of Fair Competition for the Lumber and Timber Products Industries, a massive state intervention in the collapsing market: 700 at Pacific Lumber, 700 at Hammond–Little River, 200 at Holmes-Eureka, 175 at Dolbeer and Carson, and 50 at the Northern Redwood Company.[87]

This brief survey of redwood labor history suggests that Marx's radical critique of capitalism must not be underestimated as a cultural force shaping the emergent redwood imaginary. His analysis of competitive markets was continuously part of the public discourse—part of a countercultural practice working to modify the structure of property relations. Granted, the resistance movements failed to gain their immediate demands in almost every case, but it remains a fact that they built a tradition of dissent into a counterpublic archive and managed to raise wages and provoke paternal management and company town reforms that did in fact improve the conditions under which they worked. In this way it helped build a working-middle-class logging culture, the remnants of which are still the most powerful political force in the region, although its authority has come under enormous attack in the era of globalization that witnessed the purchase of Pacific Lumber by the Maxxam corporation. Labor opposition in the years before the Great Lumber Strike constituted a diffuse network of counterpublics, communities of subaltern practices, and underground knowledges. These were connected across barriers of time and space by narratives, symbols, meanings, and critiques and together constituted a growing oppositional archive whose collective address to the working people of Humboldt helped gather up their attentions and energies into that history-making public that finally did strike back at hegemonic capital first in 1935 and then again from 1946 to 1948.

Solid communities of working-class Humboldt emerged from this cauldron of property struggle—a compromise formation between commodity circuits formed under constitutional address by the national commodity imaginary and the radical resistance guided, at least in part, by the critique of capitalism embodied in socialist labor unionism and IWW Marxism. The lumbermen of Humboldt made claims on the nation of law and of right, forcing the hand of capital to improve conditions for the body of work and changing its relation to property accumulation. If this is the case, then the specter of Marx will always be haunting Humboldt's landscape of memory, visible to those who have knowledge of its history and embodied, for example, in the town of Scotia, the Vance Hotel, the Castle in Fairy Land, the plaque at the mall, and the Holmes-Eureka gate that now stands guard at Sequoia Park in Eureka, but also in the whole of Humboldt's psychosocial achievements, as they have been concretized in the living, symbolic, and built institutions that capitalist colonization produced in the redwoods.

Having gained a sense of how much had been built in this epochal struggle of labor and capital, we should pause briefly to consider its impact on the redwood environment. According to the U.S. Forest Service Survey of 1921, conducted by the California State Board of Forestry, 90,000 acres of ancient redwood had already been "completely denuded and devastated by logging." Another 785,000 acres were approaching this condition. Operators were cutting the timber two or three years in advance of yarding it and then burning the slash, precluding reforestation. That left 1,069,192 acres of uncut ancient redwood forest. Of Humboldt County's timberland, redwood and otherwise, 543,340 acres were privately owned, and 245,800 were owned by the government, there being a total of 2,325,760 acres of timber originally available. Twenty-six percent of the total acreage in Humboldt had been cut by 1921; to the south in Mendocino that number was 28 percent, and farther south in Sonoma it was 62 percent. To the north, in Del Norte, 9 percent had been taken. Pacific Lumber owned 67,800 acres at this time and had cut 26,620; Hammond owned 87,280 acres and had cut 17,640; and Dolbeer and Carson owned 25,720 acres and had cut 11,600.[88]

The next survey was conducted in 1934, by which time some eight billion additional board feet of redwood had been cut and milled, fully 10 percent of the total standing inventory tallied in the 1921 survey. By 1946, a mere twelve years later, a total of 49 percent of the timber had been logged off—half the redwoods were gone. By 1948 Humboldt had become the second-highest county in lumber production on the West Coast, producing at a rate of 816 million board feet per year. By comparison, Lane County, Oregon, cut and milled more than 1.3 billion board feet. The success of the redwood production circuit had made deforestation an issue.[89] Thus did consumer society materialize in Humboldt's redwood commodity culture, its continuous expansion driven by internal contradictions in redwood capital accumulation toward increasing scales of production and cycles of consolidation that ultimately made advertising necessary to clear inventories and in the end stripped the land both of its trees and of its unions. The era of manifest second contradictions in capital culture was now close at hand. Soon the trouble with redwood production would arise not merely from labor on what had been native commons but also from defenders of remnant forest and salmon ecologies.

Spectacular Humboldt

By the 1930s, the ongoing colonization and occupation of Humboldt by the knowledge-culture system of capitalism was producing a local instantiation of what Guy Debord later called the "society of the spectacle." According to Debord, spectacular society runs on a fusion of politically shaped commercial and media interests, creating an ethico-political, libidinal-economic display that transforms the social and psychological conditions of production. The very success of industrial production had created a new problem for society as a whole, as well as for specific industries and firms—how to dispose of the immense accumulation of commodities?

When redwood capital experienced the need to expand and produce more desire for its products—more demand for its commodities—it embodied this fusion of commercial and media imperatives and entered the age of mass-media advertising. By substituting innovative machinery for labor and increasing output, individual capitals grew more competitive and in so doing were forced to expand market share or find new markets altogether. Competitive advantages gained in this way could be sustained only briefly, because each new round of technological innovation spurred another round of market entry, more competitive pricing, and renewed incentives to innovate further in search of still lower prices, lower wages, more mechanization, higher output, and further market expansion. In response to the new problem of overproduction generated by this cycle, the advertising system emerged. Competition had driven production into the realm of signs.

Debord wrote of these changes with critical attention to the human consequences, developing Marx's analyses of alienation and commodity fetishism to account for these new conditions. "Whereas at the primitive state of capitalist accumulation 'political economy treats the *proletarian* as a mere worker' who must receive only the minimum necessary to guarantee his labor-power, and never considers him 'in his leisure, in his humanity,' these ideas of the ruling class are revised just as soon as so great an abundance of commodities begins to be produced that a surplus 'collaboration' is required of the workers." What he means is that capital can no longer act with *total contempt* for the worker "built into every aspect of the organization and management of production." By its own imperative—to sell, to realize profit—as soon as the working day is over, capital must start treating workers "with a great show of solicitude and politeness, in their new role as consumers," and finally must attend "to the workers' 'leisure and humanity' for the simple reason that political economy *as such* now can—and must—bring these spheres under its sway."[90] The age of sign production was at hand, in other words, as soon as the capitalist masters realized the new imperative to produce desire for increasing consumption. The new economy demanded and rewarded advertising for taking on this explicitly psychological project. Consumer society—the society of the spectacle—emerged from this changing matrix of rights-driven political, economic, and mass public-sphere media institutions.

The rise of the society of the spectacle thus follows directly from the first contradiction in capitalism and gives rise to the second contradiction. It represents the system's own attempt at self-correction but assumes infinite economic growth is the horizon. And as it turns out, the economic system's imminent realization crisis can be deferred by the spectacular production of desire only as long as the external conditions of production—namely, social labor and planetary ecology—hold out. But faltering cities and communities of labor and nature give rise to sociopolitical forces—largely social movements—that signal, in fact, how such limits began being met in the twentieth century. The labor trouble went critical, and the changes it wrought in world-system capital are with us today embodied in both the New Deal and the spectacle, with its ideology of consumerism and the feverish psychical energy it channels into globalization, both by driving capitalist firms to seek foreign sources of labor and nature to remain competitive, and in offering consumers cheap imported products to alleviate their suffering of stagnant wages and job growth at home.

Labor's struggle to combine in self-defense led to labor market regulations that redistributed the flow of values associated with accumulation in general. New forms of political consciousness were legitimized. Income redistribution helped reproduce the willingness of labor to contribute anew to capitalist expansion. This amounts to internalizing what had been social costs. Social security, workers' compensation, medical benefits, and a growing system of social welfare also contributed to this socialization of the market culture. Of course, socialization of this order is the mirror image of privatization—the latter being synonymous with externalization of cost. To privatize is to isolate, singularize, make separate, alienate, and withdraw. To socialize, on the other hand, is to combine. In advanced economies, this logic is continuously driven to higher and more complex expressions, as when capitals combine to externalize costs from their sphere of production onto workforces and environmental carrying capacity but thereby add impetus to the very grievances that are driving labor into tighter association and everyone into closer identification with environmental interests. During the process, capital increasingly turns to the spectacle to organize mass attitudes in its favor by systematically employing media technology to propound ahistorical, uncritical, labor-friendly, and ultimately eco-friendly and sustainable images of itself. It uses the spectacle as a veil of illusion.

The spectacle is not merely an effect of the first contradiction. Though it develops from within as a necessary fix to the crisis-ridden system, it turns out also to be one of its causal conditions. In my description of the continuous public-sphere spectacle of the law as a crucial technology of colonization in Humboldt, I traced the spectacle all the way back to its constitutional call, where the performative texts of the national project wed the rights to free speech, pulpit, property, and assembly to the printing-press technology of public display. In that opening moment, the technology of the printing press was already exercising its formative power over human attentions, massing them together, channeling them, focusing them, and shaping them. The framers multiplied the force of this machine by anchoring mass democracy in its capacity for the free creation of publics. Tom Paine's "Common Sense" and *The Federalist* exemplified the identificatory potentials of popular media, as did the whole of emergent eighteenth-century print capitalist pamphleteering, newspapering, postering, and Bible-pushing literacy campaigns. And the great spiritual awakenings in America were democratic in the sense of literally publishing the values of mass literacy, free public education, state schools, and free press. But perhaps most importantly for the case of Humboldt, the production of California and the colonization of the redwoods were a fully mediated, spectacular process for the productive expansion of the nation form—that structure of feeling that the cross-continental and intercontinental culture channel of newspaper exchange made possible. Media revolution was integral to the westward-driving market revolution, and the first contradiction must be analyzed accordingly.

One riveting display of Humboldt's arrival in the age of spectacular consumer society appeared in a half-page advertisement that ran in the *Humboldt Times* on May 15, 1935, coincidental with the day that the Great Strike began. It reveals the psychological coordinates of the mythological nation form in an articulation of commercial and media interests with the advertising system. Reading closely, we discover that its explicit purpose was to solve the society-wide problem of production—to produce the desire that alone could dispose of the immense accumulation of commodities. It is an advertisement for advertising itself, which it portrays as a patriotic, national duty or activity that stimulates mass markets for the hardworking landowners' agricultural products and likewise brings world commodities to their deep rural doors. It mobilizes the most powerful symbols of the national form of fictive identity, which the case of Humboldt has shown to be fully colonized with all that is modern in the modern social imaginary (some emphases have been added).

Duke of the Plowshare

The slow, plodding water buffalo and the tireless tractor stand for more than the old and the new beast of burden—they are symbols of man's conquest over the soil, marks of the tiller's power.

Since time began *our great conflicts have been waged for land.* Wars have been fought for fertile valleys. Battles have been waged for protective hill tops. The first kings were those who could control the most land. Dukedoms were created for those feudal landowners who threatened the king's reign.

Yet all this bloody struggle made little difference in the lot of those who *worked* the land. In every country in the world, until the opening up of America, the actual *producers* of food held no power of their own. They were regarded as the lowly peon class, the serfs. The sons of soil were the sons of toil.

The Farmers Were Serfs

Landlords of old furnished their vassals with little more than the most meagre needs of life and the poorest make-shift tools. They gave no thought to the back-breaking labor of producing grain which, for the most part, paid only for the peon's right to exist. Even today, in many parts of the world, water buffalo or oxen, hitched to wooden prongs, furnish the only power—other than human backs— for tilling the soil.

Farming Becomes a Business

The broad, fertile fields of America did far more than offer *freedom of religious thought.* They yielded bountiful harvests and a high return for the effort spent. Unfettered by generations of servitude, American sons of the soil developed a freedom of thought and action unknown elsewhere in the world. The *science* of farming was born. New methods and machinery were invented. Tilling the soil became a pursuit worthy of the best.

Today's oriental *coolie* farmer has little in common with the American Duke of the Plowshare. Both work the soil—but one is still a slave of toil, the other a master.

The Duke of the Plowshare wields his power as the ancient duke never dreamed. Thanks to his own ingenuity and the aid of modern business he is lord and master of his own domain. From his broad acres he *feeds* the world and the world *serves* him. Idaho potatoes, Kansas wheat, Iowa corn and *Alabama cotton* have reached the farthest corners of the earth. The world of industry is ready to trade the products of its craft for the products of his field. From industrial centers everywhere come his motor cars, his tractors, his home conveniences, clothing. World products have been brought to his door *through advertising.*

Mass Production and Advertising

Furthermore, his time has been devoted to his task as a specialist in *mass production* of raw food and clothing materials. Industry and business serve as his middle men, his agents, in stimulating a *mass market* for his produce. Every convenient sales outlet—the advertising that you see in the newspapers of packaged and canned foods—helps to sell the farmer's product. The greater the demand created for wool, cotton and leather clothing *by advertising,* the more surely does the Duke of the Plowshare find a demand for the growth of his soil.[91]

The illustrated collage shows a tractor climbing uphill toward the reader-viewer, with a family farm in the right background and rolling hills of orchards and farmland stretching to the horizon, beneath which (below on the right) stands the diminished Oriental coolie as negative signifier of the rising American cyborg farmer of enlightened and machine-augmented modern progress.[92] The tractor and water buffalo, says the text, are symbols marking the American modern's growing domination of nature—its conquest of the soil and triumph over toil.

The text envehicles the entire unconscious grammar of the American liberal nation form in its explicit call for the citizen-subject to gear up and get busy in the people- and place-making world historical action of consumption—and of advertising for even more consumption. Consider how it mobilizes symbols of nation, God, liberty, equality, progress through science and industrial mechanization, and the patriarchal authority founded on individual property for sparking up shoppers—he who answers this cultural call is, of course, the Duke of

Duke of the Plowshare

PLOWSHARE

The slow, plodding water buffalo and the tireless tractor stand for more than the old and the new beast of burden—they are symbols of man's conquest over the soil, marks of the tiller's power.

Since time began our great conflicts have been waged for land. Wars have been fought for fertile valleys. Battles have been waged for protective hill tops. The first kings were those who could control the most land. Dukedoms were created for those feudal landowners who threatened the king's reign.

Yet all this bloody struggle made little difference in the lot of those who worked the land. In every country in the world, until the opening up of America, the actual *producers* of food held no power of their own. They were regarded as the lowly peon class, the serfs. The sons of the soil were the sons of toil.

The Farmers Were Serfs

Landlords of old furnished their vassals with little more than the most meagre needs of life and the poorest of make-shift tools. They gave no thought to the back-breaking labor of producing grain which, for the most part, paid only for the peon's right to exist. Even today, in many parts of the world, water buffalo or oxen, hitched to wooden prongs, furnish the only power—other than human backs—for tilling the soil.

Farming Becomes a Business

The broad, fertile fields of America did far more than offer freedom of religious thought. They yielded bountiful harvests and a high return for the effort spent. Unfettered by generations of servitude, American sons of the soil developed a freedom of thought and action unknown elsewhere in the world. The *science* of farming was born. New methods and machinery were invented. Tilling the soil became a pursuit worthy of the best.

Today's oriental coolie farmer has little in common with the American Duke of the Plowshare. Both work the soil—but one is still the slave of toil, the other a master.

The Duke of the Plowshare wields his power as the ancient duke never dreamed. Thanks to his own ingenuity and the aid of modern business he is lord and master of his own domain. From his broad acres he *feeds* the world and the world *serves* him. Idaho potatoes, Kansas wheat, Iowa corn and Alabama cotton have reached the farthest corners of the earth. The world of industry is ready to trade the products of its craft for the products of his field. From industrial centers everywhere come his motor cars, his tractors, his home conveniences, clothing. World products have been brought to his door *through advertising.*

Mass Production and Advertising

Furthermore, his time has been devoted to his task as a specialist in *mass production* of raw food and clothing materials. Industry and business serve as his middle men, his agents, in stimulating a *mass market* for his produce. Every convenient sales outlet—the advertising *that you see in the newspapers* of packaged and canned foods—helps to sell the farmer's product. The greater the demand created for wool, cotton and leather clothing *by advertising*, the more surely does the Duke of the Plowshare find a demand for the growth of his soil.

The Newspapers of the United States

This historical material from the *Humboldt Times* during the great lumber strike of 1935 shows how the economic conditions of societal transition from industrial to consumer capitalism found expression in local public culture. An advertisement for advertising, it explains the social function and lauds the national benefits of advertising in stimulating mass markets. Produced by the American Newspaper Publishers Association and printed in the *Humboldt Times*, May 15, 1935.

the Plowshare. You, the reader, are, or can be, the duke, the sovereign, if you will only project yourself into the position provided for you here.

But the message functions as such only by way of complete mystification of the actual relations of production that delivered the goods it is trying to sell. Silence on the Indian question haunts the statement that "our great conflicts have been for land." Native American opposition to the privatization programs that made American agriculture as well as Humboldt's great lumber empires bastions of white male capital accumulation are elided for the romantic vision of revolutionary fathers escaping feudal bonds in ancient Europe. The slave question is similarly buried; slave labor alone was supplement enough to drive the engines of market revolution during hundreds of years of colonization that finally made "Idaho potatoes, Kansas wheat, Iowa corn and Alabama cotton" into profitable export crops to be grown for world-system markets. Indian genocide was the origin of the mythical duke's property, and cotton slave lives of cruel human bondage largely built the foundation of American capitalist grandeur that this advertisement treats as a hypnotically natural expression of national virtue. But what red-blooded, freedom-loving, home-owning American would not respond to this image, feeling its inviting call to drive the tractor of progress right up the shining hill? The picture's connotative dimension lies just ahead, for the tractor is aiming to plow the ignorant, backward, and servile Oriental right into the rich soil graveyard where his corpse will commingle with millions of other savage colored bodies, his Indian brothers. Given Humboldt's record of Chinese expulsion, the image acquires additional resonance when displayed in the redwood public sphere.

What are we to make of the idea, proffered here, that America's wide-open spaces gave its progressive sons room for "freedom of religious thought"? The text invites us to forget the laws by which Native American religious practice was instrumentally proscribed in the process of taking the land with as little resistance as possible. It invites us to take part in that law, to participate in it, to take up its position, to be a part of it, to project ourselves into the social myth that it elevates, to identify with its image and ultimately to forget. It beckons us into the modern church—the secular church of ownership. The religion is capitalism, which by this time was fully inhabited by state-powered media institutions and thus the philosophical enlightenment that engineered the internal tension between civil liberties and political rights in the modern American social imaginary. You too can be a duke. Join us on the inside.

On Saturday, June 22, 1935, the *Humboldt Times* addressed its local strike public with this same lexicon of symbols when it spoke to the people just after the signature killing of redwood strikers at the Holmes-Eureka gate. In naming the rejection of radical unionism "the free verdict of these home-owning, thoroughly American workers," the editor's message was clear: This is America, land of the free, home of the owners whose owning is their freedom. This national, spiritual "law was not to be defeated so easily," they said, and the "threat of class government was met promptly and decisively . . . tragically, too and that we all regret." Could there be a more definite statement of class government than this claim itself? Government by the owners is the overarching message, and that is what the Duke of the Plowshare signifies as well: the symbolic structure of American authority is reproduced here in continuous, historical, practical, communicative, and spectacular public-sphere struggle over how private, individual property shall be construed and thus distributed, always in the interest of promoting the public good.

This advertisement, partaking in the continuous display of business enterprise, narrating its mythological, triumphant American exceptionalism, reveals once again the function of public-sphere technologies in driving the ongoing colonization of Humboldt. The ability of consumer society to clear the massive inventories of ascendant industrialism increasingly relied on efforts like this, just one among countless cultural calls on American subjectivity made in that classic vernacular of public-sphere discourse that media analysts have described as a force constitutive of imagined communities: it is addressed to everyone, by no one in particular, or rather by just another one of us. Such rhetoric works by forming chains of equivalence between the subjects addressed—for example, by eliding social differences of race, class, and gender, as well as the historical details concerning disproportionate access to everything promised by the national myth. Thus does the spectacle situate each reading subject as the duke, aspiring to ownership and willing to forget by remembering the script. The society of the spectacle, from this perspective of critical historical sociology, is an archive on fire with the economic incentive to forget the cost of these things, to stop being so negative and take the gifts you are offered in return for your work.

Having begun by tuning in to Humboldt's public-sphere discourse of redwood industrialization, and having listened to the struggle over property at the point when socializing capital and labor collectives brought the system to its knees in a signature spasm of violence, we can now understand how at that moment an image of reigning social relations was archived in the redwood imaginary. In this record we encounter, again, the idiom of rights, this time in the defensive rhetoric of property and freedom thrown up by supporters of redwood industry. On the other side, oppositional subjects demanded inclusion, claiming the right to participate in the law that subjects them. In the 1929 *Yearbook,* local labor leaders and the AFL president put the dominant culture of capital on notice that it well understood how the law directly determines the commodification of labor—of itself—and how it channels the value that labor blasts out of nature disproportionately to the bosses for private accumulation, at the expense of fair wages and thus the public good, whose common interest would be better served by producing a robust consuming public. In language that presaged emergent New Deal legislation, which in the end did redistribute legitimate power for control over property in labor, their claims reveal the libidinal coordinates of this epochal struggle: desire for life takes the form of demand for improvements in wages and conditions and for hours spent at home. These are the erogenous zones of redwood labor trouble, a local performance of the first contradictions in capital culture that excited the people into property conflict.

6. Trouble in the Forest: Earth First!, Redwood Summer, and the Alliance to Save Headwaters

At 11:54 a.m. on Thursday, May 24, 1990, Earth First! forest defenders Judi Bari and Darryl Cherney were headed up Park Boulevard in Oakland. Later that day they were scheduled to speak and play music down the coast in Santa Cruz at an organizing and recruitment rally for what they hoped would be the largest environmental action in history: Redwood Summer, they called it, Mississippi Summer in the ancient forest, an entire season of nonviolent civil disobedience designed to slow the elimination of old-growth redwoods. Mass protests and direct-action logging blockades would grab the attention of local loggers and world media, revealing how Maxxam/Pacific Lumber, Georgia-Pacific, Simpson Timber, and Louisiana-Pacific were cutting the remnant forests fast, knowing that if California State Ballot Proposition 130 passed, they would lose their right to clear-cut ancient trees in California.[1] Redwood forest defenders had placed the Forests Forever initiative on November's general election ballot, hoping to institute a new ecological era in forestry law.

By 1990 most of the last 1 percent of uncut, unprotected ancient redwoods were clustered in tiny isolated stands of primordial biodiversity surrounded by Humboldt's cutover timber production zone (TPZ). Forest defenders had named the largest fragments Headwaters, Owl Creek, Allen Creek, Shaw Creek, All Species, and Elk Head Springs, but Maxxam owned them all. The idea of Redwood Summer was to preserve these and other groves by physically interrupting the logging. As in Mississippi Summer, the plan required an influx of outsiders determined to break through a local monopoly on established reality, this time the dominant timber culture: corporate timber hege-mony was to ecological destruction in the 1990s redwood region as Jim Crow was to racism in the 1960s South.

Bari and Cherney never made it to Santa Cruz. A motion-activated pipe bomb exploded under the driver's seat, occupied by Bari, shatter-ing her pelvis and tailbone. Her injuries were critical. Cherney's were somehow limited to a damaged eye, broken eardrums, and multi-ple skin lacerations. Both survived the blast, though Bari would be hobbled for the remainder of her short life; weakened by the assassina-tion attempt, she succumbed to breast cancer on March 2, 1997. One thousand people attended her wake. Bari and Cherney were arrested while in the hospital for allegedly transporting the bomb themselves, but prosecutors never pressed charges because they had no evidence. Despite their injuries and the FBI's heavy-handed inquiry, the forest defenders carried on their organizing activities and made their treat-ment by federal and state authorities the object of a historic civil rights lawsuit naming members of both the FBI and the Oakland police as parties in a conspiracy to disrupt environmental organization in the redwood region, undermine Redwood Summer, and discredit Judi Bari, Earth First! and the forest defense in general by associating the movement with violence.

This widely circulated image of Judi Bari's bombed-out car was published in Judi Bari, *Earth First! and Timber Workers: Alliance for Sustainable Communities*, edited and compiled by Tanya Brannan (Santa Rosa, Calif.: Red-wood Justice Fund and Rainy Day Women's Press, 2000). Reprinted with permission of Redwood Justice Fund.

Redwood Summer endured the assault and had some success; activists sporadically interrupted the cut, engaged timber workers in dialogue, organized redwood workers under the IWW banner, and used civil disobedience to disrupt the dominant culture of timber extraction, creating a global public by using the media spectacle of forest defense. But the timber industry fought back hard, using trade associations to finance a so-called grassroots yellow-ribbon campaign that pitted the loggers' right to work against forest and species preservation. According to the rhetoric of yellow-ribbon reaction, loggers were the targets of dangerous, un-American environmental activities aimed at destroying California's treasured lumber culture.

As the fall election neared, big timber attacked Forests Forever with everything they had. They conceived, drafted, and financed Proposition 138, a counterinitiative euphemistically titled the Global Warming and Clear-Cutting Reduction, Wildlife Protection and Reforestation Act. They wrote it not to win but rather to defeat Proposition 130 by dividing the field and clouding the issue with a separate initiative disguised in eco-friendly terms. Under the measure's language, a logging plan that cut every tree on a parcel but spread the cut over three years would not be defined as a clear-cut at all. All remaining, privately owned ancient redwood groves would likely disappear entirely.[2] And because it classified old-growth forests as sick and dying while defining fast-growing young seedlings replanted in clear-cuts as healthy and valuable for fixing carbon dioxide from the atmosphere at a faster rate than older trees, the legislation's authors called it a measure for "global warming reduction." Critics, on the other hand, called it the "Big Stump" initiative because it promised to increase logging.[3]

Both the yellow-ribbon campaign and the counterproposition were the timber industry's efforts at shaping public consciousness of the timber wars. To turn the people against environmentalism and against redwood forest defenders in particular, they claimed that Earth First! "ecoterrorism" posed a mortal threat to private property and therefore timber jobs, families, and "our way of life."

Their efforts paid off. Forests Forever failed to pass by a narrow margin of 4 percent in a media climate punctuated by FBI allegations of violence by Earth First! The Oakland police and the FBI told the press that Bari and Cherney were the only suspects in their own bombing, at the same time disregarding material leads pointing elsewhere and ignoring complaints by the activists of ongoing death threats.

But the industry's counterinitiative also failed—by an overwhelming 32 percent margin. The net result was no change in the forest practice rules. Status quo old-growth liquidation logging had won the day, setting the stage for ongoing conflict. The timber wars were on.[4]

Throughout the 1990s, Maxxam continued to cut ancient forest, polarizing Humboldt and helping make the Save Headwaters campaign the most intense and long-running ancient forest defense movement in history. Hundreds of direct actions, protests, rallies, and marches in defense of the grove led to several mass arrest scenarios, helping to congeal the local subculture of ecoresistance and making the movement emblematic of modern environmental struggles.

Judi Bari died at the height of this timber war, during whose continuous display of testimony and countertestimony she emerged first as a heroine and then as a martyr for the redwoods. Every year from 1990 to 2002, on the anniversary of the car bombing, forest defenders held a rally at the federal courthouse, either in San Francisco or in Oakland, where Bari and Cherney's civil rights lawsuit against the FBI was ultimately fought. At the San Francisco rally in 2000, a fiddle-playing street carnival puppet of Judi Bari operated by two people towered twenty feet over the crowd as Cherney and friends sang protest songs including "The FBI Stole My Fiddle" and "Who Bombed Judi Bari?" with civil rights leaders, union organizers, Black Panthers, and members of the American Indian Movement standing by in solidarity.[5]

After eleven years of litigation and four weeks of trial, on June 11, 2002, a federal jury ruled that FBI agents and Oakland police officers violated the activists' First and Fourth Amendment constitutional rights by false arrest and illegal search. It awarded $4.4 million in damages to Bari's estate and to Cherney. The extensive punitive damage awards granted on the First Amendment claim signified the jury's belief that the FBI's failure to investigate and willful false claims to the media constituted an egregious violation of the activists' civil rights to free speech and political expression. Its verdict proved to the struggle's global public what the redwood forest defense had never really doubted—that the FBI had illegally attempted to discredit Earth First!, undermine Redwood Summer, defeat Forests Forever, and deflate the so-called radical element of U.S. environmentalism. Said Cherney to the crush of TV cameras and reporters after the verdict: "We're blockading the FBI from clear-cutting the Constitution."[6]

But the forest defense achieved more than formal justice; with more than nine years of hard work after the bombing, they also saved Headwaters—2,700 acres of pristine forest now at the heart of Headwaters Forest Reserve, a tiny island in a sea of cutover mountains and rivers degraded by landslides and silt from a century of clear-cuts. They had demanded that the grove be preserved at the heart of a 60,000-acre ecosystem called Headwaters Forest Complex, but instead the state and the corporation handed them this, 7,500 acres of clear-cut buffers and mature second growth cobbled around the old-growth core.

Six decades after the massacre of lumber strikers at the Holmes-Eureka mill during the Great Lumber Strike, and thirteen since the genocidal massacre of Wiyot at Indian Island, Humboldt's perpetual public-sphere conflict over property had inscribed another psycho-geographic landmark in the redwood imaginary. The Headwaters forest stands as another landscape memory that invites and enables historical consciousness, as do the plaque at Bayshore Mall honoring slain redwood strikers, the gates of the old Holmes-Eureka mill that witnessed the killing and now keep watch over Eureka's Sequoia Park, the hospital museum at Fort Humboldt that tells the Wiyot story, and the Sacred Site on Indian Island, where pioneer citizens did their most catastrophic deed of primitive accumulation. As of yet there is no public monument to Bari at Headwaters Forest Reserve, but the story of its making turns on her name and the violent blast that disfigured her body, the landscape, the forest defense movement, Earth First!, Maxxam, the timber industry, and the place of collective performance they inhabit together.

In redwood country, conversations about Headwaters and forest defense invariably lead back through the long-running struggle to protect these remnant trees, species, ecosystems, and remnant communities of labor to that signature event—the bombing—that binds them together in the region's place-bound collective unconscious, which we are learning to describe as the redwood imaginary. Recounted in books, debated on the airwaves, and tracked in miles of newsprint, this story of capitalist colonization, environmental degradation, citizen mobilization, attempted assassination, political repression, and the bracketed triumph of corporate globalization expresses the complex historical relations bound up in the timber wars.[7] In the discourse around this signature event, we discover again the deep culture drive of rights-based public spheres and property still driving the production of everything lived, symbolic, and built in the region. It brings us right up to the present moment of ongoing occupation and transformation of the Humboldt Bay redwood region by the culture system of capital—the era of second contradictions, external contradictions in which capital's struggle to maintain profitability runs up against its ecological limits.[8]

Sign of the Times: Second Contradictions in Capital Culture

The *Eureka Times-Standard* published Humboldt's first newsprint report of the bombing:

Area Activists Arrested for Blast
2 Earth First Members Suspected of Own Bomb

Two Earth First activists from the North Coast were arrested today on suspicion of possession and transportation of an explosive device after a pipe bomb blew up in their car Thursday. Judi Bari, 40, and Darryl Cherney, 33, whose organization plans a blockade of logging on the North Coast this summer were injured Thursday when the explosion tore through their station wagon, sending it careening into a car and a truck. . . . The car was apparently one of a convoy of three vehicles carrying Earth First members. Just after the blast, a man in one of the other cars jumped out and started yelling, "It's the loggers. The loggers are trying to kill us."[9]

Though the bomb exploded in Oakland, the event happened dead center in the public struggle over Humboldt's ancient forest. The local campaign to save Headwaters had begun in conversations between Cherney and Greg King, a young reporter from Sonoma who discovered the grove in 1986, just after Maxxam's hostile takeover of Pacific Lumber brought information-age global finance capital into the last cherished stands of high-priced redwood timber. Cherney and King met in the parking lot of the Environmental Protection Information Center (EPIC) in Garberville, en route to an antilogging action.[10] Sometime later Cherney suggested and King agreed to create the Redwood Action Team under the aegis of Earth First!, launching a long career of direct-action environmental challenges to Humboldt's logging culture and Maxxam in particular.[11]

For most of the century, Pacific Lumber had employed a relatively conservative forestry method called selective cutting, preserving substantial old-growth acreage for future harvests and ensuring a sustainable yield of timber revenue on its 200,000 acres.[12] The region's other large landowners had long since begun clear-cutting, stripping their old growth from Humboldt, Mendocino, and Del Norte counties and converting to even-aged management, otherwise known as tree farming. When Maxxam came to town, the Georgia-Pacific Corporation (GP) owned 200,000 acres in Mendocino county; the Louisiana-Pacific Corporation (LP) owned 500,000 acres across the North Coast; and Simpson Timber owned 300,000 acres in Humboldt.[13] Of these so-called big four redwood timber operators, only Maxxam still had ancient redwoods left to cut in 1990.[14]

Whereas for the incipient forest defense movement Maxxam in Humboldt signified the local beginning of a destructive new era of corporate malfeasance, for many lumber workers the buyout and logging speedup meant an economic boom, with more jobs and overtime pay providing new pickup trucks, new chances at home ownership, and a get-it-while-you-can approach to economic security. For some it allowed a proud family tradition to continue—and this pride suffered a symbolic assault when environmentalists rose up and asserted their irreverent, challenging counterculture of ecopreservation. After everything the logging communities had endured—the seasonal disruptions of work, the cyclical market, the physically demanding and dangerous labor, the loss of good timber and jobs to Redwood National Park, the continuous threat of regulations, layoffs, and plant closures—the intervention of these moralizing outsiders was too much for many to accept without fighting back. They saw logging as hard, responsible work and logging communities as mainstream American living, drawing strength from ideas of God, nation, family, and property that had come to define the deep structure of the redwood imaginary over 150 years of struggle. When environmentalists appeared to be jeopardizing their livelihoods, they put up a spirited reaction.

But the actual history of timber and sawmill worker opposition to redwood capital qualifies this idealized narrative of happy, traditional logging communities living out the American dream before the arrival of global capital. There had always been dissenters. Many loggers and mill workers were hostile to management, and unions gained strength in the Georgia-Pacific, Louisiana-Pacific, and Simpson redwood mills in the decades before the 1980s. But during the recession of the early 1980s, after the expansion of Redwood National Park in 1978 took land out of production and sparked a round of layoffs that weakened the workforce, Louisiana-Pacific broke the Brotherhood of Carpenters and Joiners, and Georgia-Pacific quickly followed suit, forcing a wage

reduction on its International Woodworkers Association employees, from nine to seven dollars an hour to match Louisiana-Pacific's
wage cut.[15]

Nevertheless redwood labor was and is socially and politically
conservative by comparison to the redwood forest defense. Pacific
Lumber, for example, had never been organized since the failed strike
of 1907. Its company town strategy successfully contained labor agitation for over a century. The method worked to prevent unionization
largely by treating its employees a little better than other operations in
the area. In this way, union gains elsewhere helped Scotian labor by
raising the bar. On the eve of the 1935 strike that ended in disaster, for
example, Pacific Lumber raised its wage to the level demanded by the
local strike leaders and then had its workers sign a loyalty pledge. In
the decades that followed, especially during the postwar boom, Pacific
Lumber employees benefited handsomely from company largesse:
their children received college scholarships, their pensions were generous, and the company-owned houses in Scotia were rented at below
market value to workers who had families. By the 1980s, goodwill
toward the company had grown legendary. Its company town labor
strategy, however, was not workplace democracy but paternal control
based on overwhelming power, and consequently, with the unions
broken and Pacific Lumber's more than six hundred employees largely
pacified, timber workers had little public voice when Maxxam took
over in 1986.

That the forest-defending activists appeared to be outsiders intervening in a traditional life process angered a timber folk already
stressed about long-term job security under Maxxam's reorganization
plan. When these new concerns were added to negative perceptions
of the 1970s revolution in environmental legislation, workers who already felt that they were losing ground took it hard when so-called
outsider environmental groups began raising their voices in dissent.[16]
EPIC, for example, had been founded by urban migrants to the North
Coast in the 1960s and 1970s. Earth Firsters Bari and Cherney were
both from the East Coast. And Greg King was from Sonoma, a county
to the south. But most loggers were migrants too, or their parents
were, and as for the companies themselves, Hurwitz and Maxxam
were merely the most recent wave of outsider capital to wash over
Humboldt after World War II, and the long strike of 1946–48 broke
up all the local redwood monopolies except Palco.

Understanding why timber workers saw the environmentalists as outsiders is both complex and simple. One rural teacher from northern Humboldt told me it takes only one generation to establish deep psychical bonds to timber culture in the growing child: "The father works in the woods," he said, "then his boy has the dream." Something similar occurs among forest defenders. First-wave migrant activists raise children who see logging protest as traditional. While many logging families go back a couple of generations, ostensibly giving them deeper roots in the area than some high-profile activists, the actual history of the area is one of a procession of consequential migrations by fortune seekers of one kind or another, from the original gold miners and land speculators to the settlers, loggers, and eastern capitalist pioneer timber barons—who were themselves actual colonizers engaged in the bloody ordeal of primitive accumulation—to the global corporations that finally arrived in the 1950s.[17]

The primary engine of loggers' claims to authentic tradition and a way of life authorizing it to define redwood country is the industry itself, which over decades drove the expansion of logging culture by reinvesting values extracted from nature by labor. The redwood industry is a name for this commodity circuit, in which workers and trees give each other value—it presents a seductive spectacle within the field of cultural production. Anyone who sees it on display can attach himself to it and make himself a part of the logging way of life—just by taking a job and working for pay. The tradition itself is a public story—an imaginative narrative always in circulation and on display, which people can and do seize on for telling their stories and making meaning in their lives. It is integral to the construction of self-identity because it provides a position and authorizes subjects to secure their place in their world, by which I mean their place in the timber war field of cultural politics. Maxxam actively promoted this narrative as an engine for corporate identification, using advertisements and company slogans to position every worker in its mythic tradition of community support and environmental stewardship. The idea of accusing outsiders in this land of outsiders is untenable because the actual inside to which the accusation refers is merely the established narrative reality of the hegemonic timber culture within the field of production: "outsiders" is merely a label used by insiders to signify those who stand opposed to the narrative of business as usual.

As mentioned earlier, in the 1970s a migration of back-to-the-landers or counterculture folk entered Humboldt, particularly its southern mountains around the confluence of the Eel and Mattole rivers near the town of Garberville. Judi Bari simply called them "hippies."[18] In certain respects these new arrivals were no different from the gold rushers and loggers who preceded them in the second half of the nineteenth century; they came looking for what they most desired—in this case a place to live more simply, closer to what they felt were the earth's natural rhythms, away from the pollution and corruption of urban life. Subsistence economy and rural homesteading were part of their utopian ideal. The forest was a place to build new communities. Some were part of a modern gold rush called marijuana cultivation, and others were just urban refugees. What united them all was their movement toward Humboldt, their relatively higher level of education compared to rural folk, and their experience of political unrest and social movements during 1960s and 1970s. Their generation had learned to challenge authority in struggles over labor, civil rights, Vietnam, and women's rights. From their ranks rose the true founders of Humboldt's forest defense, people for whom the bay redwood region was more than just space for utopian dreaming; it was a beautiful world besieged by corporations that were decimating the remnants of once great forests and salmon runs.[19]

Voluntary associations for forest, river, and species defense sprang up from this intersection of urban sensibility, environmental awakening, and rural tradition, challenging big timber's monopoly over established reality. Citizens from southern Humboldt formed EPIC in 1977, a grassroots group seeking to preserve ancient forests, watersheds, and endangered species, especially salmon. The Mattole Restoration Council and the Salmon Group organized themselves in the Mattole Valley of southern Humboldt in 1984. Then, in response to the penetration by Maxxam in 1985–86, Earth First! groups materialized in southern and northern Humboldt, precursors to North Coast Earth First! and the tree-sitters of the late 1990s.

But these groups were not made simply by outside individuals riding a wave of alien ideologies into the redwoods. They were local formations producing organic intellectuals in and through authentic struggles for self-determination against capitalist power on a decimated landscape. Forest defenders often point out that they would rather be doing something else, but civic duty calls when they witness destruction of public-trust resources; corporate malfeasance calls them into action. From this perspective, the spectacle of redwood industry invites people to work not just for money in the mills and the woods but for ecological and social justice. It provides more than economic opportunity bridled with nation, God, family, and property; it also gives structure for use in performing resistance that is equally freighted with historical symbols.

The physical, built, environmental world is an integral component of this generative structure. For example, the forest's sublime aspect seems to impose a feeling for the duty of stewardship and protection on the widest possible range of individuals, from the first stirrings of forest defense among the migrants, to the Earth Firsters who would make their mark on the movement in the late 1980s, to the residents, landowners, and even lumbermen who came more and more into the forest defense movement as it developed in the 1990s. The experience of being in Humboldt transforms people: citizens become loggers and activists, loggers become homeowners and environmentalists, back-to-the-landers become litigators and social movers, and at least one carpenter became a forest defender.

Judi Bari was a carpenter in the redwood region and before that an active union organizer in the Washington, D.C., area, having once led a wildcat strike at the U.S. Postal Service's Bulk Mail Center in Largo, Maryland. Living among redwoods and building homes with their wood, she witnessed firsthand both the destruction of forests and the exploitation of timber workers. She connected labor activism with environmental consciousness, and when the bomb exploded in Oakland, she had recently founded Earth First! and IWW Local 1.[20]

On the day after the bombing, the *Eureka Times-Standard* reported

Rally stage at Fort Bragg during Redwood Summer, 1990. Reprinted from Bari, *Earth First! and Timber Workers*, with permission of Redwood Justice Fund. Photographer unknown.

the arrests, repeated the FBI charges against the pair of activists, stated that Bari was in critical but stable condition and that Cherney was still in jail, and then quoted Jess Grant, "who represents the Industrial Workers of the World" (IWW):

> [Grant] believes Bari and Cherney were targeted for an attack be-
> cause of their organization's efforts against certain logging prac-
> tices. "They were combining the labor and the environmental
> issue. That is why Judi and Darryl were so dangerous to the timber
> barons. I think this action could make the summer protest even
> stronger."[21]

Grant's language, seemingly anachronistic with references to the "timber barons" and the IWW, exposed again the central dynamic of Humboldt's conflict over ancient trees; forest defenders and redwood capitalists are both after the same thing—the psychical investment of working folk. Environmentalists tuned in to redwood ecology see red-wood labor as a key ally to enlist in their struggle; redwood capital, on the other hand, relies on compliant redwood labor and a community of support to keep the trees falling, the mills running, and the money flowing out of Humboldt. Everything rests on attaching the workers. A statement attributed to Bari in the Mattole Forest Defense Calendar of 2001 embodies this focus on labor: "It is only when factory work-ers refuse to make the stuff, it is only when loggers refuse to cut the ancient trees, that we can ever hope for real and lasting change." Here was a new approach to environmentalism in general and forest defense in particular: Bari understood the structural power of labor's position in the timber wars, so she publicly challenged redwood labor to join environmentalists while calling on environmentalists to support the workers and renounce any tactics that might endanger them, like tree spiking.

In 1990 Earth First! in Humboldt was following Bari's lead, pro-moting a deep notion of ecology that understood labor communities to be embedded in communities of nature. She publicly targeted cor-porate management, not loggers, identifying the powerful interest and the systemic logic of capital with the exploitation of both workers and the environment.[22] The interests of environmentalists and tim-ber workers coincide, she argued, "because both the forests and the workers are exploited by out-of-town corporations, whose policy is to liquidate the forest and then leave."[23] When the boom-time log-ging is over, and the profits siphoned out, workers get laid off, leaving labor communities with permanent damage to the public-trust values provided by forests—for example, wildlife, salmon, clean water, and a lush, moist climate.

After the bombing, Bari wrote *The Timber Wars,* a collection of writings, interviews, essays, and recollections that appeals directly to workers and environmentalists, warning that the big four redwood corporations' continuous modernization of technology in the mills, shipment of factories out of the country, and rapid cut rate together ensure that jobs will steadily decrease regardless of forest defense and preservation.[24] Redwood capital was acting in its own interest, she said, not those of workers; the corporations are driven by short-term profit maximization, not long-term community building. She was trying to raise workers' consciousness that environmentalism only cost jobs that were already endangered by mechanization and the cut-and-run asset-liquidation business model of big timber. Later, as the Headwaters deal was taking shape, when it became clear that Maxxam would get most of what it wanted, Bari worked on an alternative called the Headwaters Forest Stewardship Plan, which called for a much larger conservation area to be set aside, but also for generous worker training and compensation. She envisioned jobs in restoration forestry for workers displaced by the preservation act.

By embracing labor, Bari pushed Earth First!, the campaign to save Headwaters, the redwood forest defense, and the broader environmental movement, which was watching carefully over the years, beyond their variously quasi-spiritual and scientific concepts of ecology and their tendencies toward antiworker attitudes. She espoused an ecological socialism based on harmony between environment and society, to be achieved through greater workplace democracy, and she also gave the local campaign a feminist edge. From her perspective, Western society's hegemonic ideology of profitable scientific mastery over nature, unlimited resource exploitation, population growth, technophilia, and mass consumption require an all-encompassing counterhegemonic feminist response, one based on sustainable, non-hierarchical, biocentric communities existing in harmony with the environment. Her social ecology sought to enlarge the concerns of forest defenders and environmentalists to include the totality of relations between organisms—the communities of human labor *and* those of the natural forest. She thought that a practical program for preserving both jobs and environment was possible if the right ideas took hold. If the forest defense movement could be pro-labor, biocentric, deep ecological, and feminist, it might be capable of developing a broader base. With this platform laid out in books, pamphlets, and journal articles before she died, Bari established herself as an innovator within Earth First! and American environmentalism.

Writing in *Ms.* magazine in 1992, Bari gave voice to a rising strain of environmental sentiment that seeks to extend the concept of natural rights to nonhuman beings, inviting animals and even plant communities into the so-called long march for social justice: "It was the philosophy of Earth First! that ultimately won me over." Its foundation is biocentrism, which holds "that the earth is not just here for human consumption. All species have a right to exist for their own sake, and humans must learn to live in balance with the needs of nature, instead of trying to mold nature to fit the wants of humans."[25]

In 1990, however, when the signature event of extraordinary violence that so deeply marked the redwood imaginary took place, Bari was planning direct-action defense, not just writing about it. EPIC had put Forests Forever on the ballot, and timber companies were cutting fast and hard in front of its threat to outlaw clear-cutting. The new law would mean little to Humboldt if the pristine groves of redwoods were all cut before it passed, so Earth First! and friends concocted the plan for Redwood Summer: blockades to halt logging and preserve biodiversity in major redwood groves, rallies to address the public, media campaigns to transform collective consciousness, and alliances between timber workers and environmentalists to direct the movement toward social ecology. It was to be a season of spectacles drawing directly on the methods and symbols of the civil rights campaign in general and Mississippi Summer in particular. It also leaned back on early Wobbly tactics of direct action, musical identification, unmitigated internationalism, and radical critique.

On Friday, June 1, 1990, big timber struck back with a spectacle of its own—a display of hegemonic logging culture it called the "Right to Work Rally," a five-hour show of "worker solidarity" in protection of "private property and our rural lifestyle," as the advertisement for the event in the *Eureka Times-Standard* proclaimed. The mills shut down for half a day, and the workers convened at a local stadium.

Presiding over the event was William Perry Pendley, president and counsel for the Colorado-based Mountain States Legal Foundation and a leader in what was then called the wise-use movement, but which had begun as the so-called sagebrush rebellion and which by the end of the 1990s was calling itself the multiple-use movement.[26]

Advertisement announcing the Right to Work Rally. Produced by Aahlberg and Cox Advertising Concepts Inc. and printed in the *Eureka Times-Standard*, May 27, 1990.

Some background on Pendley and his organization and wise use is essential for grasping the significance of the Right to Work Rally.

Founded in 1977 with funding by the politically conservative industrialist Joseph Coors, the Mountain States Legal Foundation embodied widespread and well-funded corporate efforts to oppose environmental regulation dating from the 1970s, when new laws presented capitalists like Coors with the unpalatable prospect of paying a larger share of the social costs of their operations. Corporations, landholders, and conservatives in general had been stung by the passage of the Wilderness Act of 1964, the National Environmental Protection Act (NEPA) of 1970, the Clean Air and Clean Water Acts of 1970, and the Endangered Species Act (ESA) of 1973. In the state of California, the pain was redoubled in 1973 by the passage of the California Environmental Quality Act (CEQA) and the Z'berg-Nejedly Forest Practice Act. Ronald Reagan was elected president in 1980 with strong backing from the injured parties—namely, big western capitalists whose interests were served by loose regulation of public resources like Forest Service timber, minerals, oil, and water, which lay ensconced in the vast public domain of the American West. Reagan appointed Pendley as undersecretary to James Watt in the Department of the Interior. All three men shared the property rights philosophy embodied in the sagebrush rebellion of the late 1970s—an effort by several Republican-led states to take control of federal lands and privatize them within the grazing, timber, and mining industries, of which Watt was an important ally.

In 1993 Pendley had been the keynote speaker at the "Multiple Use Leadership Conference," hosted in Reno by the Center for the Defense of Free Enterprise (CDFE). The CDFE is the organization of Ron Arnold, another wise-use leader and the author of *Ecoterror: The Violent Agenda to Save Nature* (1995), a volume I spotted on the shelf in Maxxam/Pacific Lumber spokeswoman Mary Bullwinkel's office when I spoke with her at Scotia during my field research in 2001. CDFE's first "Wise Use Strategy Conference" had inaugurated the wise-use movement in 1988. Its stated intent had been to make corporate monies available for the support of individuals and so-called grassroots citizens' groups that work to advance a legislative agenda of states' rights, which, broadly defined, includes promoting the privatization of public-trust lands and resources and supporting lawsuits filed in defense of private property against so-called regulatory takings—lawsuits in which federal, state, and local laws that protect public-trust values are framed as assaults on the economic value of private property. The battle is fought on constitutional grounds, namely, the "just compensation" clause of the Fifth Amendment. At the 1993 conference, Pendley used language familiar to Humboldters: "There is a culture war in this country," he declared; environmentalism is "socialistic," and its activists are "watermelons—green on the outside and red on the inside."[27]

Pendley's Mountain States Legal Foundation and Arnold's Center for the Defense of Free Enterprise are national antienvironmental organizations, dedicated to exposing how environmentalism is supposedly "trashing the economy," an idea that provides the title for a book Ron Arnold and Alan Gottleib published in 1994, the first line of which declares that "the American dream is in trouble." Its subtitle says even more: *How Runaway Environmentalism Is Wrecking America*.[28] Pendley and Arnold's philosophy of property rights is embodied in the work that they do. The MSLF, for example, does pro bono legal work for takings cases, and its online mission statement includes the following:

> Mountain States Legal Foundation, by seeking the proper application of the Constitution and interpretation of the law in the courts, administrative agencies and other forums:
> —Provides a strong and effective voice for freedom of enterprise, the rights of private property ownership, and the multiple use of federal and state resources.
> —Champions the rights and liberties guaranteed by the Constitution in support of individual and business enterprises and against unwarranted government intrusion.[29]

It would be an interesting rebuttal to MSLF, and to like-minded property rights philosophers, that the Constitution actually makes no provisions whatsoever for the "rights and liberties" of "business enterprises." It took one hundred years of capital accumulation and political empowerment before corporations were finally granted the juridical standing of fictitious persons, beginning a process by which the idea of corporate rights would become so naturalized that, at present, most people would notice no slippage at all between "individual" and "business enterprises" in the language of MSLF's mission statement. In fact, the Constitution and the Bill of Rights together form a unique balance between private and public interest—but where is the public interest in MSLF's free-market dream? It is entirely entrusted to the care of private property. Rather than recognize public interest in environmental commons, it pushes privatization and unmitigated private and corporate access to federal and state resources, which it advances under the disarming terms of "wise use" and "multiple use."

On the eve of Redwood Summer, Pendley roused the crowd at Humboldt's Right to Work Rally with a fiery speech: "We are in the midst of an environmental passion that has gotten totally out of hand," the *Eureka Times-Standard* quoted under the front-page headline "Thousands Rally for 'Right to Work,'" with a stunning photo of the working-class crowd standing with hands clasped over their heads in solidarity. The timber industry is "under siege," Pendley said, and concern for endangered species is now endangering people. The *Humboldt Beacon* of June 7, 1999, quoted Pendley more fully: "The loss of property rights makes the Communist Manifesto look like the Magna Carta. . . . If those environmental terrorists' hate-crimes were done to people in the cities it would be a national outrage! . . . Letting Redwood Summer activists meet in the national forest is like letting the Islamic Jihad meet at Dulles International Airport."

In his reference to endangered species, Pendley was talking about the northern spotted owl. In April, just before the bombing of Bari and Cherney, the U.S. Fish and Wildlife Service recommended placing the spotted owl under protection of the Endangered Species Act (ESA). Jack Ward, a Forest Service biologist, issued a report calling for 7.7 million acres of national forest to be set aside for the owl. Big timber spread the news through timber communities across the Northwest, framing the proposed listing as a provocation and organizing a yellow-ribbon campaign—against environmentalism. Among workers, the confluence of Redwood Summer, Forests Forever, and the spotted owl's new status felt like a massive assault. "I feel my lifestyle

The crowd at the Right to Work Rally, Fortuna, California, June 1, 1990. Photograph by Mike Harmon; printed in *Eureka Times-Standard*, June 2, 1990. Courtesy of the *Eureka Times-Standard*.

is threatened," said Gary McCutcheon, a mill worker, to the *Eureka Times-Standard* reporter at the Right to Work Rally. "Trees are like any other crop, if you harvest correctly they're renewable," he said. "That's the way I make a living, the way I pay for my house."

His fears were not unfounded. Local papers had recently reported that thirteen thousand jobs might be lost in the Pacific Northwest if the owl was listed. Public experts were predicting that increases in alcoholism, vandalism, divorce, violence, and domestic disputes would wrench lumber towns if the new rules and conservation set-asides were instituted.[30] Because a number of Humboldt mills relied on Douglas fir coming down out of Six Rivers National Forest, in the county's eastern highlands, any spotted owl conservation measures would further depress the redwood economy.

Pendley's address to the rally followed that of Bruce Vincent, a representative of Montana's Communities for a Great Northwest; Vincent primed the stadium with the claim that "the big lie is that in order to protect our environment, we have to lock man out." A forklift driver working for Simpson Timber Company echoed this sentiment to reporters: "If God didn't want us to use the trees, he wouldn't have made them able to grow back. The time is now for us to take a stand and stand up for our beliefs." Pete Porter, a truck driver, exclaimed, "That's their ultimate goal, to stop logging. They don't care about the impact it's going to have on people here." A T-shirt spotted by the *Times-Standard* reporter read, "Loggers are the endangered species." Four to five thousand people attended, wearing yellow ribbons, yellow baseball caps that read "Doing Our Job with Pride," and "People First!" T-shirts that played off the name of Earth First! A number of companies closed early so that workers could join the rally. The hats and shirts were provided as they entered the arena.

Next to the photo of the crowd ran another headline: "Coalition Working for Peace." Operating out of Arcata, the coalition hoped to head off violence between timber workers and environmentalists throughout the looming Redwood Summer. The Reverend Peggy Betzholtz of the First Congregation Church of Eureka said that each side in the controversy was out to destroy the other. What tended to be obscured in this reporting, and in the arguments of the MSLF, the CDFE, the coalition, the yellow-ribbon campaigners, and the rhetoric of the local timber managers, as well as certain of the workers, was that no environmentalists could be found calling for an end to logging and the destruction of timber communities. Rather, they called for

Right to Work Rally hat, 1990. I found the hat in this photo for sale in a bin at a thrift store in Eureka in 2000. Photo from *Eureka Times-Standard,* June 2, 1990.

ancient forest protection of the final 1 percent and sustainable forestry on the lands already converted—a program to preserve the rural life-style against the cut-and-run program of resource extraction by global capital, for example, Maxxam in Humboldt.

This is the real battle being fought—the battle over public percep-tion and sentiment that, channeled through political process and pres-sure, determines the regulatory environment for logging and therefore the relative direction of accumulating values. On one side, constant, spectacular public pressure applied through forest defense rallies, blockades, lockdowns, tree-sits, banner hangs, office raids, letters to the editor, timber harvest plan monitoring, and watchdog litigation forms a culture of resistance to corporate impunity and the established authority of big timber capital. On the other, right-to-work rhetoric and its yellow-ribbon packaging camouflage large corporate ownership with the psychosocial authority embodied over decades in the system of property, an authority whose seductive appeal is mobilized in the media spectacle of address to the public with property rights rhetoric that equates large and small ownership, establishing a chain of equivalence between Maxxam's ownership of more than 200,000 acres and working individuals' ownership of a home or a car, the three of which are presented as enjoying equal status before the law.

We know what happens when the corporation prevails in this struggle: it carries forward its project of commodification and free private accumulation. Because its victories are material and symbolic, they address the public with signs of success. Commodities are messages, and commodified landscapes call people into specific forms of labor and action. Once an ancient forest has been logged to bare hillside, for example, it no longer appeals to the forest defense public in the same way. Whereas it had formerly been an object of desire that invited protection, it now becomes a symbol of what has been lost. On the other hand, commodity circuits ramped up in corporate victory invite people into the culture of logging—by offering a job and a paycheck, delivering the goods, and providing a theater not just for masculine performance of tough physical work of which one can be proud but for building up families that repeat national gestures of American dreaming.

Throw up a lumber mill and hang out a Help Wanted sign; people will answer the call. Some will have experience and call themselves mill workers or loggers, but for others the work will be their first chance to perform and thus model themselves on the image inscribed in the commodity circuit. The culture pattern embodied in the circuit can thus be understood as an identity-producing machine that works by providing a wage and making life possible, for example, by providing for a house in which a family might be raised. Remember how quickly the dream can be established: the father takes a job in the woods, and the son is already second generation. By selling their labor for dollars per hour, workers enter a field of subject formation that runs on the value their bodies produce when directed at redwood ecology. In Humboldt, the tradition and the dream of working the woods are performative institutions that run on the power that this circuit channels in from working bodies that answer its call. Together with the world that they build, they form a symbolic system that addresses each person who enters its field of effects with a spectacular image. That image is Humboldt—redwood country—the living, symbolic, and built social world of the redwood imaginary.

To the degree that resistance succeeds in establishing limits to timber extraction, it channels timber values out of this circuit into reservoirs of public trust. This does not necessarily mean that lands are taken entirely out of the market, though that is the often case where lands are preserved in parks or conservation easements. If the owner is forced by regulations merely to avoid steep slopes and preserve a streambed, then salmon habitat can recover, and the fishing economy, both commercial and sport, can realize increased private accumulation from a healthier public-trust ecosystem—a hypothetical example that exhibits the complexity of the issue.

This internal division of the redwood imaginary into opposing private and public value accumulation is organized by parallel and opposing definitions of ecological connection. The project of corporate capital accumulation derives its character by defining environment as resource and property—as private—and disposing of it as if it has no connection to the world around it. To the extent that a particular firm acknowledges its ecosystem—the totality of connections between its property and the world—and internalizes that knowledge in its production process, it moves toward the position of the ecological knowledge that identifies and drives environmental resistance. Maxxam, for example, has agreed to numerous revisions of its forestry practice under direct pressure from environmental groups. The Habitat Conservation Plan/Sustained Yield Plan (HCP/SYP), though deeply flawed and perhaps fraudulently written by the company itself, did set aside land and adopt forestry guidelines that, had their letter and spirit been followed, might have made the company a standard bearer for responsible, sustainable forestry.[31] To the extent that they improved on the company's established practices, mountain, streambed, river, bay, and ocean salmon ecologies might recover proportionately, letting public-trust values accumulate and making Humboldt a place more fully constructed by the forest defense. But cost considerations are always on hand, and the company is under continual pressure from EPIC and others for not adhering to the plan and for cheating on the science that supposedly legitimates the agreement.

In this way, the institutionalization of ecological knowledge in public consciousness transforms the channel of value accumulation by changing the character of labor directed at the environment; it changes life in the redwoods. But again, this effort runs up against the established authority of the property culture, which we have seen to be rooted in psychical structures of feeling for liberty and nation that 150 years of colonization and occupation by capitalism have built into the redwood imaginary. Everywhere you look, the cultural unconscious—the symbolic order that governs this performance—is built into the world: it is the world! Humboldt County, as it has been achieved so far, is defined by, and built around, the authoritative institution of private property, which addresses the public through the media of Humboldt's spectacular, symbolic architecture of accumulation and redwood commodity circuits. This address comes inscribed with representations of everyone—workers, protesters, citizens, tourists—which everyone encounters as an objective structure manifest in the great installations of industrial timber production, including not just the company mill town of Scotia but all the great factories in Fortuna, Carlotta, Korbell, Arcata, and Eureka, as well as in the individual houses owned by hundreds of workers. They call on us all as American citizens; they invite us to perform in their cultural circuitry.

There are public parks and redwood preserves as well, and there are public streets and national forests—places where the institution of individual ownership might not seem to be working, but even here private property is integral to the meanings of people and place. The boundary that marks the private yard off from the street is no different from the gate on a logging road that keeps the public off a mountainside owned by Maxxam. There can be no private without a public from which it is enclosed. There could be no feeling of freedom attached to the ownership of a family home if it were not surrounded by public space.

In this sense, the place of Humboldt should be seen as a structure of action and a structure of thought—what Clifford Geertz, in describing of the theater state of Negara, called "a constellation of enshrined ideas."[32] The prevailing ideas enshrined in property institutions and commodity circuits are concepts whose meanings take form only when put into action and articulated by land, labor, and environmental laws. This is what the modern social imaginary has achieved in its colonization of the redwoods. It follows that the living, symbolic, and built place of Humboldt itself is a cultural call into the performance of public conflict over redwood labor and environment. We are learning to see the entire place of redwood industry as an imaginary institution of commodity culture, a bastion of authority whose specific manifestation or expression is a landscape of memory that situates everyone who enters the region within its powerful field of effects.

The laws of property that saturate Humboldt and form the juridical histories and conditions that structure this authority and therefore the timber wars can be treated as constitutive of three distinct but mutually determining types of property: property in land, labor, and environment. Each has been fought for in public struggles that can be taken as expressions of the redwood imaginary. As each type of property moved to the center of public discourse and contributed to Humboldt's archive of violence and landscape of memory, it recorded the local career of the nation's founding speech and sacred texts. During the land war against Humboldt's First Peoples there had been no discourse over labor law and regulations, but once the system was installed and the industrialization process began, power over labor dominated the public-sphere struggle over property for decades. In the working compromises that were formed in that second epoch of conflict, little or no attention was paid to ecological conditions and the effects of so much work on the physical world; rather, land and labor prevailed as the objects of interest, while conservation ethics were only just emerging. But as the culture system expanded its circuitry, it transformed the land and thus its own conditions of production: in the epoch of first contradictions, the struggle over property and capital accumulation produced for itself an ecological challenge—its second contradiction. This is what capital culture faces in the timber wars: the accumulation of its own history.

Challenging the Body of Property

The traces of the car bombing described earlier remind us that the year was 1990. The economy was in rapid transition from Fordist spectacle through flexible capitalism toward an informational, transnational structure. It was the age of globalization, the principal motors of which had their origins in New Deal policies of social engineering, themselves a response to the crisis character of capitalism and the massive market failures and imperial wars that were sapping world markets of all their strength and producing protectionist measures that bound up trade circuits with domestic politics. Strong, legitimate labor unionism, for example, helped set profit-hungry companies in search of cheap overseas labor and new markets everywhere. That in itself was nothing new, but combined with declining environmental conditions produced by domestic industrial capitalism and the ongoing revolution in communications and transportation technologies, the stage was being set for the rise of spectacular environmental poli-

tics aimed at transforming the balance of power between local defenders of public-trust values and the colonizing, authoritative forces
of accumulating property. It was everywhere a matter of changing
perceptions—of changing consciousness, that is, and so changing how
labor is directed at nature.

When Redwood Summer got under way, and the bombing of Bari
and Cherney made its mark, the New Deal social contract and Johnson's Great Society programs had been under relentless attack by conservatives for over a decade. The fiscal and social policy of the Reagan
era ushered in a sustained period of financial deregulation that, in the
context of direct assaults on established environmental policy and
budget cuts for EPA enforcement, helped produce an era of escalating
corporate profits and rapid disengagement from environmental protection. The wise-use movement was gaining momentum, feeding on
elective affinities between intensive resource management, property
rights activism, and the free market trickle-down supply-side ideologies driving the conservative backlash.[33] As the sagebrush rebels and
secretary of the interior James Watt followed Reagan to Washington,
seeking to transfer millions of acres of western lands from public to
private ownership, membership in the large national environmental
organizations multiplied.[34]

Helping to fuel the conservative reaction were decades of postwar growth in union enrollment and political influence, white flight
from the troubled urban cores, emergent black power, the feminist
critique of patriarchal society, and the monumental gains in environmental law during 1970s. In that decade of energy shocks, strong
unions, and purported decline in the competitive advantage of U.S.
corporations, labor and environmental regulation became targets of
business-minded elites.[35] By the end of the decade, dedicated antien-
vironmentalists began laying the groundwork for a corporate-funded
network of "grassroots" property rights organizations whose leading
ideologues addressed Humboldt at the Right to Work Rally in 1990.

Looking back on the development of U.S. environmentalism, we
can trace the reciprocal determination of declining environmental
conditions, rising oppositional consciousness, and ensuing environmental legislation that ultimately found expression in the redwood
timber wars. Having taken our bearings from the signature event of
the Bari car bombing, and learned from similar moments through
which the labor trouble and Indian wars came to inhabit historical
consciousness, we know that in following these juridical trails, we

will eventually encounter all the institutional spheres that instantiate the modern social imaginary in the bay redwood region.

Alongside concerns for aesthetic heritage, conservation of the nation's natural resources had been a central value of environmental thinkers since the late nineteenth century. In 1872, Yellowstone National Park was the first of many lands to be taken out of the developing market to better manage the provision of public goods.[36] Then Carl Schurz, a political reformer and German immigrant, was appointed undersecretary of the interior by President Hayes in 1877. Schurz opposed privatization of the nation's timber resources on the grounds that market mechanisms would fail in managing resources of such long-term regeneration and used his influence under President Harrison to establish the national system of forest reserves in 1891. When Gifford Pinchot was appointed director of the Division of Forestry in 1889, he advocated a public-private alliance in managing the nation's forests and other natural resources. "Multiple use," he argued, should strive for efficient, sustainable exploitation. This utilitarian ethic, though not completely unmoved by aesthetic values, was rather more influenced by the conservation economics espoused by Schurz. It also embodied populist and progressive political sentiments that were swelling at that time, part of a nationwide socialist challenge to Gilded Age excess. In 1907 the Department of Agriculture assumed management over these reserves, newly named "national forests," in a system that essentially socialized a large sector of one of the nation's basic industries, much to the chagrin of conservatives then and now. Theodore Roosevelt set aside much of the nearly 150 million acres of the timberland that had accumulated in the national forest system of the lower states by 1924, while another 85 million were set aside in Alaska. He also created fifty-one bird refuges, four national game preserves, and five national parks.[37] The emerging conservation movement received another boost in the 1930s when the National Forest and National Park systems were expanded under Franklin Delano Roosevelt.

These efforts officially rationalized the socialization of land exploitation with arguments that public management for multiple use would more efficiently distribute the forest's public-trust values than would open-market mechanisms; centralized government planning of the timber industry would better preserve the beauty of forests and regulate land for efficient use, reducing wastage among private timber firms.

As early as 1909, this conservation ethic was surfacing in Humboldt. In that year the *Humboldt Beacon* of Fortuna reprinted a position paper that Ms. Edith Woodridge had recently read to a local citizens group:

Humboldt County Should Replant Her Depopulated Forests

[Humboldt] originally had about 510,000 acres of this timber, but from the date of settlement to 1900 it is estimated that 90,000 acres have been cut. . . . These gigantic trees hold a weird fascination for anyone who beholds them. . . . Today . . . the hills are no longer covered with redwood forests but charred stumps and fallen monarchs. . . . The annual output of Scotia is 55,000,000 feet and that of Vance is 50,000,000. . . . At the mill [in Scotia] a great fire burns night and day to get rid of the waste lumber, the slabs and blocks and short pieces, that are only in the way. Some of these fires have burned ceaselessly for twenty or thirty years. Think of the mountains of fuel they have destroyed. This waste lumber is not saleable just now and it would not pay to ship it to market, but later on when it will be valuable and sorely needed by the people it cannot be recovered. . . . There is a grave danger threatening the younger portion of the nation unless these things are recognized and prompt action taken; there is a reckless, selfish use of natural resources which if unchecked will lead to blighting poverty. These renowned men of the west who have been busily engaged in building a new Empire toward the sun are only human, after all. Their personal ambitions are desire to make money quickly, to roll up fortunes instantly for quick success, have led to the short sighted destroying of the resources which can be most quickly and easily converted into cash. . . . Every acre destroyed is a national loss which future generations will feel.[38]

Her sentiment shared the Progressive Era's growing concern that commodification run wild on American landscapes wreaks havoc on the future, and she makes an appeal for rational usage of common resources. There is no concept of ecological systems science operating here, nothing like the mid-twentieth-century emergence of wilderness values, and no trace of contemporary Gaia imagery representing the interconnectivity of human and natural communities as some kind of transcendent power. There is no concept of biodiversity, biocentrism, or direct-action defense of nature. Her concern is not with despeciation. But there is a sense of alarm—a critique of a business ethic that willingly squanders public-trust values for quick private gain.

Ten years later the *Saturday Evening Post* published a call to save the redwoods inspired by a similar conservation ethos, but which also gave voice to emergent aesthetic and romantic values. In "The Last Stand of the Redwoods," Samuel Blythe explained that "the inevitable destruction that would follow commercial ownership" has been pointed out, "but neither the Federal Government nor the state of California has taken steps to preserve the trees, and they came into private control. Wherefore, the only way to save them now is to buy them back from the men who own them."[39] He then introduced an organization—the Save the Redwoods League, founded in 1918— that became the first significant voice for saving Humboldt's ancient forests.[40]

In the 1920s the league began negotiations with Pacific Lumber to buy large tracts of uncut ancient forest, but only after considerable agitation, including the events of November 19, 1924, when league members James and Laura Mahan discovered logging in what is now Humboldt's treasured Founders' Grove. Laura lay down in the area where the loggers were working and would not move while her husband went to Eureka to secure the paperwork to save the trees.[41] According to Pacific Lumber's 1987 account, the first agreement to transfer ten thousand acres to the league, made by handshake in 1920, was contingent on the group paying fair market value.[42] But protracted fund-raising and price negotiation delayed the purchase until 1931, when the company finally deeded Founders' Grove and Rockefeller Forest to the league, two groves around which Humboldt Redwoods State Park is presently constructed. Later, in 1969, just after the creation of Redwood National Park, the company sold the famed Avenue of the Giants to the state of California. Then another seventy-two acres were added to Rockefeller Forest in 1989, increasing the total acreage of conservation set-asides originating from Pacific Lumber/ Maxxam before the Headwaters deal to just over twenty thousand acres. According to the company, in 1987 this amounted to "considerably more board feet of old growth than remain . . . on the company's own commercial logging property."[43]

These long-running efforts by the league began in the interwar period of dramatic mechanization in redwood production, when chain saws and tractors first entered the woods on a large scale. In 1939 one forward-thinking University of California scientist commented that chain saws and Caterpillar tractors were producing a startling spectacle—the mountains were becoming barren. Still, widespread recognition of the problem by industry and science did not take hold until the 1950s. Whereas 24 billion board feet of redwood had been cut in the one hundred years up to 1955, in the seven years between the end of the Long Redwood Strike in 1948 and 1955, the rate of redwood lumber production tripled, leaving a mere 18 billion standing board feet on the tax rolls. In that year two hundred manufacturing plants operating in Humboldt were cutting virgin old growth at a rate of 1.25 billion board feet per year, while regeneration measured only 25 percent of that rate.[44] The postwar logging boom that contributed so much to the contemporary landscape of Humboldt had begun signaling trouble in redwood country.

Between 1955 and 1978, ongoing mechanization brought increasing scale and division of labor to Humboldt with a vengeance. Two-thirds of the smaller independent logging operators in the county vanished in this period, as less and less of the easy takings remained.[45] Large operators, on the other hand, could better absorb increasing costs of deeper-woods logging that took crews farther up the steep slopes into the rugged coastal range. They bought up land as the smaller operations failed, fueling a dramatic consolidation and signaling the arrival of second contradictions in redwood capital culture. In this emergent landscape, capital's new struggle was maintaining profitability in the face of its own effects, which returned to antagonize it in the form of rising costs. The growing dearth of big trees sitting close in and cheap to take was a new environmental contradiction to business as usual.

In this context, the 1950s saw large out-of-state industrial corporations, most notably the Georgia-Pacific, Louisiana-Pacific, and Simpson Timber—with Pacific Lumber, the so-called Big Four—moving in and buying up cutover land, and with it any residual and hard-to-reach old growth. These corporations built massive landholdings and consolidated mills into industrial powerhouses that only the Pacific Lumber Company could rival in size and production.

Pacific Lumber itself acquired enormous acreage in this same period, buying up, among other concerns, the Holmes-Eureka and Dolbeer-Carson companies. But whereas Pacific Lumber practiced sustained-yield forestry under the conservative management of Simon Murphy, cutting at less than the growth rate on its land and preserving some lucrative old growth, the out-of-state corporations quickly cut it all and converted their forests to even-aged management.

In these decades, two additional events that took trees out of production helped set the conditions for redwood timber war in the 1980s. First, the creation of Redwood National Park in 1968 preserved 58,000 acres, including 11,000 acres of old growth. In 1978 it expanded to 106,000 acres with the addition of 48,000 acres of mostly cutover lands. Combined with Humboldt Redwoods State Park at 53,000 acres—of which 17,000 acres are old growth, 10,000 acres of which stand in the justly revered Rockefeller Forest—that left only 20,000 to 30,000 acres of old growth in private ownership: a mere 1 percent of the original two million acres, and Maxxam owned nearly all of it. Ninety-six percent of the trees had already been cut, and now 3 percent were preserved in parks.[46] By the 1970s, big timber had run itself out of easy-take old-growth values, and park preservation had helped to set the conditions for grassroots ecological resistance.

In that decade, extensive destruction of salmon spawning grounds by sediment runoff and mass wasting (landslides) from logging operations into the rivers and creeks initially motivated the formation of voluntary citizens' associations that advocated watershed defense and the preservation of species, many of which already appeared to be permanently lost. Salmon fisheries were collapsing, and fishermen were selling their boats or converting to crabbing. Humboldt Bay itself was filling in with the massive sediment load of soils displaced by logging, and dredging was required to maintain shipping channels. Pesticides used for industrial reforestation, air quality, and wastewater pollution also became of increasing concern as Humboldt registered the second-highest cancer rates in California. Air downwind of two big pulp mills on the Samoa Peninsula offended visitors and worked against the development of tourism as a viable supplement to the diminishing-resource economies.

Over the decades in which these environmental problems accumulated in Humboldt, a national environmental movement emerged from roots in the eastern United States, starting perhaps with the founding of the Appalachian Mountain Club in 1876 and moving out west with the founding of the Sierra Club in 1892. For the first half of the twentieth century, these organizations remained relatively small and regionally based. The Sierra Club, for example, was limited to California until 1950. But under the direction of David Brower, it began fomenting national environmental debates in the 1950s and 1960s, most notably in the fight against the proposal to submerge Dinosaur Monument behind Echo Park Dam in Utah. The club succeeded in preventing that project, as it did the massive Southwest Water Plan, which proposed to construct two additional dams on the Colorado River inside the Grand Canyon. But these victories came at a price. Under Brower's leadership, the club had compromised. In exchange for saving Echo Park, it agreed not to resist the production of Glen Canyon Dam. Thus was Lake Powell born and Glen Canyon lost in 1963. The modern mass environmental movement had emerged from the battle with permanent scars of compromise and loss driving its ongoing struggle to preserve wild places and scenic beauty as treasures of national heritage.

With the Wilderness Act of 1964, environmentalists attained a landmark victory that embodied the scenic, recreational, wildlife, and wilderness values that had been rising within the conservation movement for most of the century. The act mandated the government to inventory roadless areas for consideration within a new official designation of "wilderness."[47] The objective was not merely utilitarian resource conservation or the protection of "scenic values" but transcended these while conserving them and integrating them into a more comprehensive ecological vision with roots in the new ecosystems science.[48] As a result, the National Forest Service's Roadless Area Review (RARE) got under way in 1971.[49] The initial survey (RARE I) identified nearly 1,500 roadless areas, of which 12.3 million acres were set aside for further study and 67 million acres were opened up for multiple use, but the Forest Service backed down when the Sierra Club announced it was going to sue under NEPA. A second survey (RARE II), one that survived environmental criticism and led to enactment, designated 10 million acres in the lower forty-eight states as wilderness areas, opened 36 million acres for development, and set aside 36 million acres for further study.[50] In Humboldt RARE II created the Trinity Alps Wilderness Area inside the so-called multiple-use Six Rivers National Forest, on the mountainous eastern end of the county that lies outside the range of redwood ecology.

For a full sense of how landscape functions in the redwood imaginary, it is necessary to understand two additional major divisions in land usage: the King Range National Conservation Area, on the county's southwestern shore, and the Humboldt Bay Wildlife Refuge. When combined with the state and national redwood parks, these conservation set-asides amount to significant public ownership in the county. Yet Six Rivers National Forest and the Trinity Wilderness Area are not redwood regions but mixed-conifer, largely Douglas fir forests that are intensively managed. Six Rivers sells standing timber to logging firms like Red Emerson's massive Sierra Pacific Industries (SPI). SPI, the largest landowner in California with over two million acres, is also a large owner in Humboldt, cutting mostly Douglas fir on both private and public lands and trucking logs to a sawmill located on the north shore of Humboldt Bay, just west of Arcata, that being one of the region's largest industrial facilities.

All told, North Coast forestland holdings are 88 percent private. For this reason, private corporations, not the National Forest Service, dominate the scene. In the words of one regional historian, this large proportion of absentee ownership by large corporations "imbues the region with a colony-like land tenure pattern in which timber firms enjoy significant economic and political power."[51]

In the same years that saw the Wilderness Act take shape and Redwood National Park forged out of northern Humboldt, the consumer revolution and Cold War militarism began to register across the nation in new and frightening ways that gave impetus to the era's big shift of environmental consciousness, helping catapult it beyond mere conservationism and wilderness ethics to full-blown ecologism. It is obligatory in this context to mention the publication of Rachel Carson's *Silent Spring* (1962), which has been widely credited for alerting the nation to the harmful effects of chemical poisons, most famously those associated with DDT. After this the entire postwar revolution in chemical manufacturing came under scrutiny, and suddenly the nation started awakening to environmental crises. Among so many other species, the plight of the whales, elephants, dolphins, and pelicans shocked the world. Was industrial society producing an extinction crisis? The decade's end was marked not only by the Santa Barbara oil spill in 1969, from which the images of oil-soaked beaches, birds, and sea mammals were broadcast worldwide, but by the Cuyahoga River fire of that same year in which the heavily polluted water itself caught fire and burned bridges in Cleveland, Ohio, as it had previously in 1936 and 1952.[52]

The rise of televisual mass culture and the advertising spectacle during these same years was not inconsequential to the national environmental awakening. It was the age of the nightly news, and critical eco-awareness depended on the transmission of images like those of the Santa Barbara oil spill. Advancing technologies of information and transportation coincided with the first pictures looking down on Earth—the vulnerable blue planet: what could better announce to the world the impact of time-space implosion and the global village that Marshall McLuhan had prophetically seen as heralded by the electric age? If environmentalism had a global mirror stage, that was it.

During the 1970s the use of spectacular media emerged as the signature tactic of environmentalists, who especially in those early days tended to be short on financial resources but full of ideas and genuine causes. Greenpeace was first to organize actions aimed at generating news images (mind bombs) for exposing ecological destruction. Then a Greenpeace member who wanted to push direct-action ecodefense even further founded the Sea Shepherds Conservation Society, and finally Earth First! was founded in 1980, with similar designs for high-profile public interventions, banner hangs, and dramatic stunts that would call attention to ecodestruction in hopes of transforming public consciousness and thus the collective political will to stricter regulation.[53] If there was any hope of countering the hegemonic culture of unfettered capitalist destruction of ecology, it lay in enlarging existent and creating new environmental publics using critical images to shatter the comfortable illusion that consumer society could be sustained ecologically without structural change and systematic regulation. The regulatory preservation not just of the environment but of the possibility of sustaining human life on the planet became the common issue, and spaceship earth became a symbol through which everyone concerned could henceforth identify.

On the national level the new environmental public consciousness invoked a massive federal response. Legislation established the Environmental Protection Agency (EPA) in 1970, and then, in quick succession, the nation's most important environmental laws were passed: the National Environmental Protection Act of 1970 (NEPA); the California Environmental Quality Act of 1970 (CEQA); the Clean Air Act of 1970; the Clean Water Act of 1972; and the Endangered Species Act of 1973. Both NEPA and CEQA contained language that schematized the concept of ecology into government policy: they required that every government action likely to generate environmental effects must take those effects into account in the planning stage—not just as isolated effects of disparate activities but the cumulative effects of that agency's policy taken together with all other impacts on the ecosystem in question. The holism constitutive of ecology—the science of ecosystems, of environments considered in all their relations—forms the content of the concept of cumulative effects, and it came to govern the coming conflict over forest practices and ancient forest preservation in Humboldt.

In California, logging regulations had remained essentially unchanged between the Forest Practice Act of 1945 and 1971. Under that regime industrial foresters dominated the rules-making process. Companies were required to file timber harvest plans (THPs) with the state Board of Forestry, which approved them according to the rules but also allowed for alternative plans to be submitted—an enormous loophole that allowed timber corporations to first write and then escape their own rules. The net result was an absence of logging regulations, at least according to the 1971 court ruling that changed everything.

The event that spurred the court's intervention and set the conditions for conflict over forest practices up through the contemporary redwood timber wars involved the Simpson Timber Company of Washington (recently renamed the Green Diamond Resource Company). Proceeding under the old rules, Simpson filed an alternative logging plan with California Department of Forestry that requested approval to clear-cut 147,000 acres over the next twenty years. The boldness of the proposal sparked a major debate and a court case that led to passage of the Z'berg-Nejedly California Forest Practice Act of 1973, which required private landowners to explicitly protect the productivity of public resources.[54] Drawing on rules established under CEQA and the Porter-Cologne Water Quality Act of 1969, which created a state system of regional water quality control boards, it made assessment of the cumulative impacts of human activity on wildlife populations, soil productivity, and water quality the explicit objective of the THP process. Under the new law, which remains in effect today, the governor of California is mandated to appoint nine people to the California Board of Forestry, which oversees timber production. Four of these must be representatives of the timber industry, making the industry's prevailing influence over the forestry regulation an arrangement guaranteed by the law itself. The board then decides whether or not each THP meets the cumulative-impact requirements under CEQA and Z'berg-Nejedly.

But in 1975 a state court, attempting to bring the THP review process into further compliance with CEQA, ruled that THPs are government projects as defined under CEQA, thus requiring that full environmental impact reports (EIRs) be filed for each THP. A completed

EIR would represent a practical assessment of the cumulative impacts of all previous human activity on the site in question and require the California Department of Fish and Game and the Regional Water Quality Control Boards to participate in an extensive review of previous science. But the new requirement threatened to overburden the agencies involved because at the time more than one thousand THPs were being filed each year. Both industry and agency reacted by pressuring the state to mitigate this regulatory burden, and the California Legislature responded by amending CEQA with a provision that recognized the CDF-administered THP process as "functionally equivalent" to the EIR. Governor Jerry Brown then made this compromise into law by way of an executive order issued to the secretary of the Resources Agency. The cumulative-impact provision of CEQA and Z'berg-Nejedly had thus effectively been scuttled by the power of big-timber capital, which thereby continued to dominate the regulatory process.

This sequence of events set the task of forest defenders for the next twenty-five years; if they wanted to use the legal concept of cumulative impacts to halt the destruction of ancient forest and related public-trust values by industrial forestry corporations, they would have to sue the relevant agencies, particularly CDF, on a case-by-case basis. Retrospectively the problem appears simple; the THP approval process considered only the effects of individual logging plans, one at a time, without attempting to gauge the impact of consecutive THPs filed in the same watershed over a period of years—years that stretched into decades and regularly produced 100 percent removal of forest canopy. The irrational result was that, under the regime of law in place by 1976, whole watersheds were being completely deforested with legal approval by the resource agencies.

But the corporations ran into trouble in the redwood forest. Environmental consciousness was on the rise and beginning to organize. Resident subjects of environmental resistance formed the Environmental Protection Information Center of Garberville just as this legal environment was emerging, and over the years EPIC became California's leading watchdog litigator of the THP process.[55] The success of *EPIC v. Johnson* in 1983 established that CDF must consider cumulative impacts and consult the Native American Heritage Commission if it is known that the harvest will affect historic indigenous sites. The case appeared closed when the California Court of Appeals ruled that the THP was inadequate. But two years later, in 1985, Georgia Pacific filed exactly the same seventy-five-acre THP, another attempt to clear-cut Sally Bell Grove. EPIC sued again, but this time the case never went to trial. By publicizing the threat, defenders of Sally Bell Grove drew the attention of the Trust for Public Land, which purchased the grove from Georgia-Pacific as part of a larger preservation package.

The case of Sally Bell Grove set a historic precedent. Local grass-roots networks learned to form up and defend specific groves from specific corporations and individual THPs while simultaneously using the courts to force the regulatory agencies to fulfill the mandates of the new laws by considering the effects of the cut in relation to numerous other cuts and the history of cuts on each parcel, that is, the cumulative impacts. Publication of each particular conflict raises collective awareness of each contested site while legal challenges hold the chain saws at bay, buying time for collective determination of their fate.

Soon after saving Sally Bell Grove, EPIC shifted its attention to Maxxam. In 1987 it sued over two old-growth THPs filed in the Headwaters Forest Complex and won. This time the judge ruled that CDF had rubber-stamped the THPs in May and June 1987 and that CDF had intimidated the staffs of both the Regional Water Quality Control Board and the Department of Fish and Game, effectively dissuading them from participating fully and openly in the process. In 1988 EPIC sued Maxxam again over another THP in Headwaters, asserting that the plan did not fully account for impact on old-growth-dependent endangered species. This case went all the way to the California Supreme Court, which ruled in favor of EPIC and established that CDF must comply with both the Forest Practice Act and CEQA—once again reaffirming that cumulative impacts must be considered.

EPIC went on to sue Maxxam numerous times—too many, in fact, to give even a brief treatment here. But it is important to note these early suits because they embody the basic dynamic, which is not difficult to extract from this chronicle of deforestation, local resistance, and compromise formation of forestry legislation: Property holders in the redwood forest, financed by dollars extracted from the forests themselves, bought the political leverage they needed to control the regulatory process and fend off the rising challenge of environmentalists, whose primary resource was emergent local knowledge of environments under siege. Large timberland owners had several things in their favor: the tradition of logging with its mythical history, a solid foundation in the cultural authority of the concept of individual property, and the relative weakness of public-trust doctrines still under construction by ecological challengers. Individual citizens hoping to exercise their right to engage in the process—a right institutionalized by NEPA, CEQA, ESA, and Z'berg-Nejedly—faced the accumulated power of landed capital wielding highly paid corporate lawyers in a legal system that gives corporations the same rights as individuals. The entire redwood imaginary—a historical apparatus permeated by the ethos of property and corporate law—had in fact largely been built out of big corporate timber.

But forest defenders, many of whom became environmentalists only through witnessing the destructive industrial forestry in their own communities, were developing a deep local knowledge of ecology. They set about challenging the dominant culture, publicizing images of ecodestruction that amplified their claims enormously. Ecological conditions provided them with the grim specter of deforestation to use as a tool and weapon—and they put this landscape of memory to use. They banded together in local groups and campaigns and drew the attention of the large environmental groups, widening the public of each contested THP, ecosystem, species, or regulation. Only by collectivizing resources and legal expertise did they counter the aggregate force of accumulated timber capital and achieved timber culture. In so doing, their labor too entered the living, symbolic, and built place of the redwood imaginary, as surely as had decades of redwood labor and timber capital management. Their desire ensured that the second contradiction in capital culture was embodied not just in deforested mountains, extinct salmon runs, declining bird populations, an increasingly disempowered labor force, and the long downward trend in the lumber economy but also in stands of newly preserved ancient redwoods, salmon stream restoration, a local culture of environmental resistance with institutional structure, and a growing body of forestry law at least marginally more responsive to citizen review.

Takeover

I began this discussion by approaching the redwood timber wars as I did the labor trouble and the Indian wars—by allowing my fieldwork encounter with Humboldt's contemporary historical consciousness and landscape of memory to guide my study back to the signature event of this epochal struggle. Listening in on the timber war field of cultural production on the eve of Redwood Summer, we discovered a heterogeneous local movement of forest defenders rising first against threats to Sally Bell Grove in Mendocino and then against Maxxam in Humboldt when the takeover compromised Headwaters Grove. The Endangered Species Act had been invoked, first in the name of the spotted owl but then in the name of the red tree vole, the goshawk, the marbled murrelet, and the salmon as well. At the same time, the state voter initiative that would have saved all of California's ancient trees had grown popular and appeared to be passable. But the companies combined to defeat these concerns. And we saw how the Right to Work Rally was, as the ad for the event put it, "Sponsored by business and industries that have made the northcoast great!" But that event convened more than just local timber culture; property rights activists based outside California and working at the national level were concerned enough about the convergence of labor and environmental interests in Humboldt to help finance the spectacular rally and send in their representatives. Their money, their efforts, and especially their rhetoric presented to Humboldt a unified image of patriotism, work, responsibility, masculinity, and legitimate ownership—a narrative image, in other words, through which the community could reaffirm its collective obligations to the hegemonic order of achieved timber culture. The rally was a social ritual designed to supplement cultural identification with this order and its industrial base, the necessary psychical energies of which were flagging in a time of relentless environmental critique. Remember that the grassroots collective subject of holistic ecological consciousness did not rise up to challenge the capital culture and the legitimate authority of its feeling for property until the industrial forestry machine had almost exhausted its available redwood old-growth timber values—only a few thousand acres of an original two million were still privately owned and unprotected when the timber wars broke out in the 1980s, cued by the entry of Maxxam on the scene.

In 1985 the Houston businessman Charles Hurwitz used Maxxam to take control of Pacific Lumber Company. The hotly contested $900 million leveraged buyout left the company with more than $575 million in debt.[56] Before the takeover, Pacific Lumber had been operating on the basis of a thirty-year-old "timber cruise," planning to harvest all its remaining old-growth coastal redwoods in about twenty-five years. Maxxam's new management hired Hammon, Jensen, Wallen and Associates to conduct a new, more thorough inventory—the first exhaustive cruise since 1956. They found 31 percent more timber volume than Pacific Lumber had on record—both young and old growth—

an extraordinary windfall and grounds for developing a new business plan. Maxxam had acquired 194,000 acres, including approximately 75 percent of all unprotected old growth.[57]

Two years after the acquisition, Maxxam described its new plan in a public relations paper titled "The Pacific Lumber Company and the Issues of Forest Management," the tone of which suggests a new degree of image consciousness and a recognition that some answer must be made to the public challenges coming from the media-savvy protest movement (2–3). Maxxam was clearly on the defensive and framing the takeover in the best possible terms.

"An acknowledged objective," Maxxam wrote, "is to increase cash flow in order to repay more than $575 million of debt incurred in the merger, plus the cost of completing a $42 million co-generation power plant to serve the company's expanded Scotia facilities, and an $8 million expenditure to purchase the nearby Carlotta lumber mill." The first steps taken toward higher cash flow were asset liquidation. The San Francisco office building sold for $31 million dollars. The cutting and welding equipment subsidiary sold next for $325 million. Then four thousand acres of redwood land in San Mateo County, south of San Francisco, were unloaded for $250 million (13–15). But the big payoff would come from liquidating the forest.

According to the company, most of its old-growth inventory was residual in character—basically leftover, isolated trees on cutover lands, not so-called virgin timber aggregated in unentered, biologically important blocks. But several large groves did exist, and "on the basis of the new cruise and the accelerated schedule, the company now believes that it will take 20 years to harvest them, although the future rate of harvesting may vary from year to year according to market conditions." Company foresters estimated that "a doubling of the 1985 harvesting level can be sustained for that period without harm to the long-term viability of Pacific Lumber's timber resource." But the public need not worry, Maxxam continued, because "removing the old growth redwoods, which grow very little, if at all, will allow the land on which they stand to regenerate forests of faster growing new trees. So new forests will be growing where old growth stood to renew the company's basic resource in perpetuity" (14–15). Maxxam had thus publicly announced that it would liquidate most of the last unprotected old-growth groves in existence. According to the timber cruise inventory presented in this document, that meant 553,445 old-growth redwoods and 721,230 old-growth Douglas fir trees, spread over 194,00 acres in Humboldt. The places that forest defenders named Headwaters, Owl Creek, Allen Creek, Elk Head Springs, Shaw Creek, and All Species Grove would perish (4).

Although the plan's stated intent was to pay debt heaped on the company by the takeover itself, Maxxam assured the community that it would benefit greatly, pointing out that three hundred employees had already been added to the payroll to handle the extra work required for the speedup, expanding the workforce from 1,100 to 1,400 (8). Scotia's 272 houses were renting for $200 to $275 per month, it reported, a continuation of what it called a "conservative, paternalistic, [and] dependable" management style for the workforce, community, and land. As further evidence of good corporate stewardship, Maxxam reported that the town's Protestant and Catholic churches still paid just one dollar per year in rent. The kindergarten paid five dollars per year, and the town's health clinics, inn, bank, post office, clubhouse, civic auditorium, supermarket, drug store, hardware store, gas station, and coffee shop all similarly benefited from company generosity (8–9).

Then Maxxam described the forest defenders: "The company's announcement in 1986 that it would begin harvesting its old growth redwoods on an accelerated schedule—principally the 'residuals,' but some of the few remaining 'virgin' old growth stands as well— brought an explosive response. It ranged from angry opposition by a small group of self-proclaimed militants calling itself Earth First!, to threats of violence against forest workers. Another self-appointed environmentalist group filed lawsuits accusing the State Department of Forestry of improperly approving timber harvest plans" (5). The company's phrasing here is crucial. It completely elides the mainstream resistance embodied in organizations like the Salmon Group, Mattole Restoration Council, and EPIC. It does not precisely claim that Earth First! is threatening violence against timber workers but instead labels them militant in a sentence claiming that workers have been threatened. The implication is clear—Maxxam is telling its workers that they are under attack and physically threatened by forest defenders. The technique allows them to associate forest defenders with violence without having to provide any evidence. Is this the old charge of tree-spiking again? We cannot say, for no examples are given or incidents mentioned. In this hostile public-sphere struggle of claim and counterclaim, where appealing to publics and mobilizing attitudes are essential tactics, if there had been threats of violence against workers, the public would have been treated to continuous repetition of the details in the corporation's conscious use of the media spectacle for the shaping of public consciousness. But no mention is made of any specific threat or event because there was none—the forest defense in Humboldt was devotedly nonviolent.

In the following year, Maxxam's new managers ramped up their rhetoric of corporate paternalism, property rights, and American liberty. Perhaps the oratory peaked when *Leaders,* a business magazine, quoted Charles Hurwitz, CEO and controlling stockholder of Maxxam, in an article titled "Value Visionary":

> Our feeling is that this is America: there are private property rights; we have a Constitution. . . . Our greater concern with the extremists . . . is that they seem to have no regard for private property rights. That disregard is disturbing because, of course, private property rights are at the foundation of the political and economic systems in this country. . . . We know we're helping the environment.[58]

Hurwitz employs a reified concept of property rights here, saying that the so-called extremists have no regard (we can read "no respect") for property rights (we can read "authority"). But what are the rights to which he is referring? If no distinction is made, it is implied that they are all the same. He is drawing an equivalence between his own large, absentee ownership and citizens' and workers' ownership of their personal homes and cars. The problem is that the power represented by large corporate aggregation of permanent, fixed, and immovable so-called real property, such as land, is veiled by equating it to small-scale, more-movable personal property.

But contrary to Hurwitz's charge, while the environmental challengers tend not to respect the authority of property they see as having been unjustly accumulated and subsequently mismanaged, they do exhibit high regard for the process by which property rights are established. They see through the veil. They know exactly where property rights are constructed and understand the powers they both represent and exercise. And they know where to go to change them. Rights are living institutions that have been transformed continuously throughout U.S. history and will continue to be, if forest defenders have their way. They know that Hurwitz's financial power over labor and environment in Humboldt and elsewhere resulted from the legal deregulation of financial markets and the savings and loan industry in the early 1980s, when Republicans captured first the White House and then Congress. Forest defense literature is replete with this narrative and precise in its documentation of the history and its understanding of the takeover process. The junk bond debt that Hurwitz created to buy the company is seen as what it is, a new technology of finance capital—that is, of property—and thus a new name for juridical domination by property.

Maxxam's public proclamations of "property right" and "we have a Constitution" were rhetoric for the masses, whose identifications with or against the company shaped the political climate of its operations. There can be no doubt that if redwood labor had collectively determined that it was in its best interest to exercise its power to assemble and unionize, it could have transformed the flow of value it was producing, channeling it back into laboring communities and away from the Maxxam debt machine that was vacuuming value out of the region. But Maxxam fought to prevent that at any cost, telling its workers they were under personal attack by forest defenders and labeling environmentalists terrorists. In this context, redwood labor rolled over, letting their accumulated product—the Pacific Lumber Company—be sold off and liquidated. And so, in a sense, to a certain degree, they themselves were liquidated. The company's rhetoric of property rights held up a timeless, ethereal concept of individual liberty that effectively bolstered the sense of authority and respect with which the working populace recognized the company's property, helping to prevent the mass identification of labor folk and their lumber-dependent economic communities with the forest defenders, which alone could have countered the harmful effects of the takeover. In the process, the libidinal dimensions of this struggle were laid bare. Which path of accumulation would the objectified and alienated bodily energy of redwood labor take: accumulation to the restoration of the communities of labor and environment that originate those values, or accumulation to previous accumulations of corporate capital for even more power over labor and environment?

The stakes were thus the same as they had been in the labor struggle. The company feared that regulatory change would rob it of legitimate property, so it intervened in the lawmaking process that maintained its existence. The attitudes and life energies of the workers were principal objects of corporate interest. The company's history of paternal domination in its labor relations reflects this investment, and we find it manifest in the architectural register of Scotia, the company town whose architecture doubles as social control.

In place of a separate volume that might be written on the Scotia program alone, I offer instead one small image from the field of power. When the Scotia town center was destroyed by fire, incidentally at the height of the Headwaters campaign, the company rebuilt an impressive new building and blazoned the town's Code of Conduct on a prominent corner of the general store. "The town of Scotia" is private property, it reads, "and as a visitor . . . you are prohibited from . . . engaging in non-commercial expressive activity without the proper written permission of the management of Scotia." Other proscriptions posted included disturbing the peace, defacing private property, sitting around, annoying others, littering, begging, drinking, loitering, and lawbreaking—in other words, a comprehensive warning not to protest here.

The success of the Scotia program can be measured against the attitudes of people from Scotia and all of Humboldt toward Pacific Lumber before Maxxam's arrival. Pacific Lumber was famous for providing extensive benefits for families and nurturing generations of work from a single bloodline. An affectionate shorthand for the company name was the gently suggestive Palco, and "Palco Pride" became a widely recognized slogan. Into the 1990s, it was common to hear Pacific Lumber Company employees publicly testify to their second- and third-generation family commitment and pride in the Palco way of life. It was this complex amalgam of values, habits, and naturalized interests—the Palco *way of life*—that started coming apart after the takeover.

The symbolic effects of Maxxam's arrival were felt throughout the region. If even Palco could fall to global capital, then what could survive? Though forestry practices had been a matter of concern for decades, before Maxxam the basic legitimacy of this culture of private ownership for timber production had been fundamentally unchallenged. The system of unlimited redwood production under free competition that had for so long stood with the silent authority of an established, unquestioned reality—a hegemonic structure of feeling toward the industry—had now to be explicitly backed up with public displays of authority and law, for example, the Code of Conduct.

Forest defenders argued that the concept of private exploitation of forests is fundamentally flawed because forests are not and cannot be private entities. On the contrary, their existence is deeply embedded in the intricate web of life that is the basic object of ecological knowledge. Forests are intrinsically public.[59] Ecology, the forest defenders held, being the science and thus the knowledge of the totality of in-terrelations between organisms and their environment, represents a fundamental challenge to the culture of unhindered private ownership and unlimited resource exploitation. And whereas the culture of prop-erty functions by truncating relations between segments, sectors, and actors in the world, positing and championing what is private and self-organizing, the concept of ecology functions by describing their inter-relation, positing what is common and mutually dependent. Private property works by excluding the other, and capital accumulates by externalizing cost. Everything that goes into the product that can be signified otherwise accrues to the producer as a mythical excess—surplus value—which is, of course, the definition of profit.

The key term here is *signification*. Who will decide what will be signified and treated as inside the cost of production and what will be cast aside, outside the pricing mechanism, that is, thrown outside the price? The Wiyot? The workers? The owls, salmon, and salamanders? Who will take the power to name these prices? The case of Humboldt has shown how this question is held permanently open by the deep culture drive of American constitutionalism, whose spectacular in-vitation to speak up and challenge the law and remake the law by recognizing its authority—or not—addresses everyone in the idiom of universal rights. Here again we encounter the modern social imagi-nary in perhaps its most singular field of effects—the call that it places on everyone, all the time, to enter the zone of perpetual struggle for the power to name and therefore to channel the flow of accumulation. In the redwood timber wars, whose total field of cultural production embodies this modern social imaginary in a unique, historical instan-tiation, both Maxxam and the forest defense answered this call to step up and fight.

In a long-deferred and pyrrhic victory for the forest defense, that struggle ended and another one began in 2007, when Maxxam de-clared Pacific Lumber bankrupt. The judge ruled against Hurwitz's plan to subdivide the property for elite miniranches, choosing in-stead to throw him out of Humboldt and turn the remnants of Pacific Lumber over to the Mendocino Redwood Company (MRC). MRC agreed to create the Humboldt Redwood Company (HRC), which now controls the timberland and the mills. As part of the ruling, MRC/HRC agreed to run the operation as it runs the parent company MRC in Mendocino County—according to the principles of restoration tree farming under the auspices of the Forest Stewardship Council. MRC's partner in the new venture is the Marathon Structure Finance Fund of New York, an entity to whom Maxxam's Pacific Lumber owed $160 million on a debt for which it used the town of Scotia as collateral. Under the ruling, Marathon is handling the privatization of the town. This may take time, because the town was never surveyed. There are only two legal lots recorded. Selling the individual houses and businesses is turning out to be a messy process.[60] Headwaters had been saved, and forest practices had been improved on what had been Palco land, but much had been lost over twenty-three years of Hurwitz in Humboldt channeling the values that redwood labor blasted out of nature into the debt machine he created in 1985.

The deal to set up HRC and liquidate the company town cost MRC $530 million in cash and started the new company bearing $325 million in debt, substantially less of a burden for Humboldt's communities of labor and environment in what everyone is hoping will be a new era of restoration forestry. But it is still a burden. The debt must still be paid, with interest. Combined, MRC and HRC now control about 450,000 acres of redwood forestland. It is a massive accumulation of capital, but it is capital that has been forced to internalize at least some of the social and environmental costs it might otherwise have externalized and thrown off as profit, for example, the costs of stream restoration, old growth preservation, and selective cutting. But the unions have not been invited in. No provisions for organized labor have been made. Capital still rules in the redwoods.

Spectacular Forestry

By way of closing this account of hegemonic social order and insurgent social movement in the redwood imaginary, I turn one last time to psychoanalytic social theory and the role of advertising in the public-sphere struggle for power in the timber wars.[61] Who will dominate the regulatory environment that channels the flows of accumulating value variously into private accumulations or public-trust resources, like restored salmon runs, biodiversity, and healthier forests? One thing we find are capitalist enterprises—the big timber companies—who, in fighting for the power to name what is inside and outside the price, and therefore the market, reach directly for that resource of which both dominant and challenging identificatory publics always already need most: people's attention. Their attention is the conduit of their libidinal investment of life's passionate energies in the labor of self-identification, for example, at work in the economic sphere, at home in the family sphere, at the marketplace in the social sphere, and maybe even in the streets, so to speak, with social movements in the public sphere. And insofar as control over forestry regulation hinges on establishing and maintaining favorable public attitudes and political consciousness, their efforts must concentrate in the prized social space of state-sanctioned, rights-driven mass-media spectacle.

In the years between the takeover and Redwood Summer (1986–90), big timber worked hard to maintain its trembling monopoly on established reality. Advertising in the *Humboldt Visitor* and the *Eureka Times-Standard,* the companies tried to shore up their authority by equating family property, working-class community, and large commercial ownership using public narratives of paternal labor control and well-meaning corporate environmental stewardship. The industry was using its accumulated capital as power over media to prepare mass attitudes favorable to themselves, producing a running spectacle of corporate pedagogy meant to secure such practical libidinal-economic commitments as are necessary to keep the industry not just moving but moving profits in its direction.

For example, one of the world's big extraction corporations, with land, sawmills, and factories throughout the Pacific Northwest, including several in Humboldt, entered the advertising field of cultural production to reassure Humboldt visitors and citizens of their care and their place in the established order of things. "WE LIVE HERE, TOO," ran the ad's banner headline, published in the *Humboldt Visitor* in 1987: "We're the men and women of Louisiana-Pacific. We work here. More important, we live here. It's our home. And we're dedicated to making it a better place to live. That's why we're active in the life of the community. In schools. In churches. In government. In organizations large and small. We're your neighbors. We're Louisiana-Pacific. And we like it here." These words form two columns wrapped tightly around a graphic of a mill worker wearing a hard hat. The company must perceive that its commitment to home, community, labor, and environment is dangerously in doubt—but for readers informed about the history of antiunionist activities of combined redwood capital, such doubts are redoubled by the ad and then doubly exposed by the question it raises. Why is this smiling logger giving me the thumbs-up sign? What is wrong in the redwoods?

Being and living in the bay redwood region, just going to the store or talking to people in the streets, is always already immersion in this

WE LIVE HERE, TOO

We're the men and women of Louisiana-Pacific. We work here. More important, we live here. It's our home. And we're dedicated to making it a better place to live. That's why we're active in the life of the community. In schools. In churches. In government. In organizations large and small. We're your neighbors. We're Louisiana-Pacific. And we like it here.

LP *Louisiana-Pacific*

Coastal Division, P.O. Box 158, Samoa, CA 95564 (707) 443-7511

This advertisement by the Louisiana-Pacific Corporation was printed in the *Humboldt Visitor* in 1987. This publication is for the benefit of tourists, with a single issue continuously distributed throughout the year.

kind of imagery. It amounts to a dense network of power-laden and signifying social relations alive with the narratives, images, architectures, and built environments that embody the history of colonial, industrial, and ecological labor, the effects of which continuously exert their semiotic pressures, shaping everyday consciousness and life.

Similarly, in an advertisement by Simpson Timber in the *Eureka Times-Standard* on the eve of Redwood Summer in 1990, what lies outside the frame really constitutes the message. At the time Simpson's pulp mill on the shore of Humboldt Bay was under siege, facing lawsuits and new regulations, because boiling wood chips in chlorine to make paper produces cancer-causing dioxins that exit the plant via wastewater into the bay and the ocean and via smokestack into the prevailing onshore winds, creating a noxious and malodorous plume that had hung for decades over the county seat of coastal Eureka. It was the company's hundredth year of operations, which it commemorated in the advertisement. The words "SIMPSON PAPER COMPANY HUMBOLDT PULP MILL" appeared over a line drawing of the plant and its smokestack with text reading: "At Humboldt Pulp Mill over 250 experienced professionals are working together to produce high quality grades of bleached softwood kraft pulp. These highly trained people do their jobs with a sense of pride and fulfillment. They are also active participants in community groups and special projects. Our employees are a real asset to Humboldt County. Simpson Paper Company is proud to have them." Here again, in a context of collective oppositional outrage at the company, management is calling its workers by name, supplementing their identification with work and the company. As it turned out, the new regulations passed, and the company chose to shut down rather than upgrade the plant and maintain its investment in Humboldt. When the towers were toppled in the spring of 2001, I stood by and watched, one among hundreds of Humboldt citizens gathered to see this enduring symbol of timber hegemony come crashing down.

When the flotilla that had gathered on the bay blew their whistles simultaneously as the tower tipped over and disappeared in a cloud of debris, it was a powerful aural and visual symbol of change that punctuated the long decline of industrial timber culture. It was also a spectacle of environmental resistance, a victory that in this particular instance had literally and dramatically visually transformed the landscape of memory.

Also in the spring of 1990, with Redwood Summer looming, Forests Forever on the ballot, and the spotted owl about to be listed as an endangered species, big timber combined their efforts and published another ad in the *Times-Standard:* "WE'RE CREATING WILDLIFE HABITAT," read the prominent tag line, beneath which ran a simple but tellingly structured graphic of forest animals over a bit of text.

The image showed simple scenes of wildlife—but their placement on the page conveyed a concept of the natural order of things, with the dominant bear in lead position on the left, the eagle flying over the cougar at center, and the noble elk and cute but lowly rabbit coming in last. This order instructs us that the authors are in the business of correctly perceiving and representing the world and that they are vested in order and hierarchy—and it prepares the viewer to receive an authoritative message from a registered scientific expert. The mother bear, elk, and rabbit each look directly into the reader's eyes,

Simpson pulp mill stack implosion,
Humboldt Bay, Eureka, May 10, 2001.

capturing attention, establishing identification, and interpolating energies into the hierarchy of species by putting them in the position of the messenger, a registered professional forester: "I see more wildlife now, more diverse species, healthier individuals and larger populations of mammals and birds than I saw ten years ago and this is because of our management. It's a direct result of creating more wildlife habitat." Across the bottom, the industry group signed its name—NCFI (North Coast Forest Industry)—next to a slogan that establishes a chain of equivalence between the forest, its animals, the public, and the companies: "Our roots run deep in the community."

Whereas for years the forest defenders had claimed to stand for the rights of animals and identified themselves with the animals by suing in their names (*Marbled Murrelet and EPIC v. Pacific Lumber; Marbled Murrelet, Northern Spotted Owl, and EPIC v. Bruce Babbitt et al.; Coho Salmon, EPIC, et al. v. Pacific Lumber et al.;* etc.) and adopting their names for use as aliases when captured by police during trespassing, blockading, and protesting actions, in this advertisement the companies try to seize back and co-opt the symbolic power of these gestures. Who can most successfully claim the remnant community of endangered animals as part of their community? The companies advertise their paternal position—owners, stewards, even *creators*—while forest defenders claim solidarity and respect, often in the familiar terms of "our mother earth," and place their names, side by side, on lawsuits filed in defense of nature. This is a real turf war— a struggle for the psychical attentions and identifications that animate the place of redwood timber wars with opposing cultures of industrial extraction and ecological resistance.

Such advertising supports timber hegemony to the extent that it helps bind the flow of its subjects' life energies to the industrial forestry machine, the effect of which might be expressed, for example, in commitment to work, to the company, or in resistance to environmental initiatives. While exuding confidence at the manifest or denotative level, a closer look at these typical timber ads quickly reveals their common concern with the faltering status of achieved timber culture, if not their outright desperation. Their speech is plaintive and pleading for support, offering interpretations of the industry that favor itself in its struggle over ancient redwood forests, ever hopeful of building its identificatory publics precisely by mediating the timber community—that is, by bringing its elements, its people, into closer contact, establishing a space of desire and exchange across time and space, and effectively multiplying the symbolic sites and the intensity of symbolic channels through which their affective, identificatory collective life can be formed.[62] In the process, they add to the archive of public discourse with which all future mass-cultural identifications and political projects, whether for or against the forestry machine, will be constructed.

In the case study at hand, the perpetual struggles by all interested parties to create public cultures have accumulated as history in the field of effects I have called the living, symbolic, and built place of Humboldt—the redwood imaginary. This is the imaginatively performed and constructed social space that every subject of Humboldt encounters as an objective symbolic order, an archive of social memory available for use in the meaning-making project of living a life and participating in politics. It is that unconscious signifying structure and system in and through which they will act, for example, in the interest of defending this or that remnant community of labor or environment, or both, or perhaps in their own interests more narrowly defined, or in league with the corporations, or in line with the mythical and expertly advertised traditions of achieved timber culture.

Conclusion
Living in the Archive of
the Redwood Imaginary

Consciousness is created by certain symbols.
—Wlodzimierz Suleja, director, Wroklaw branch,
Polish Institute for National Memory

What we teach the subject to recognize as his
unconscious is his history.

—Jacques Lacan

When we shiver in witness of untimely death, or flush in the warm celebration of
extraordinary life, or shudder at the gravity of syndicated forces towering over us and acting on the world as if free of constraint, the thick
veil of routine that obscures so much of the social world grows thinner
for a moment or is cast aside completely. The routines of life, for an
instant no longer inured to the urgency of historical forces, quicken to
their pace. Exemplary figures identify us and bind us to the unfolding
story. We find ourselves attached to the plot. David "Gypsy" Chain,
Julia "Butterfly" Hill, and the company town of Scotia are elements
like this in the timber war story. They take us over the threshold of
daily life, attracting our attentions away from habitual concerns and
channeling them into Humboldt's public culture of environmental
conflict. They are living, symbolic, and built mediators—messengers,
in other words—if we choose to listen. Their particular transmissions are concrete expressions of the modern social imaginary in the
redwoods.[1]

But what we find being shared in this particular place—this living, symbolic, and built place we are calling the redwood imaginary—can hardly be called a simple consensus. A cultural space of perpetual conflict that people here cannot avoid is a better portrayal of what binds them together. Decades of labor energy—channeled through Indian trouble, labor trouble, and trouble in the forest—have entered the region's institutional, symbolic, and physical geography, building a spatial grammar and a material lexicon that subjects here use for self-understanding and narrative self-projection into the timber war story. In this linguistic metaphor, society is construed as discourse in action, and the timber wars are one particular discursive formation—an idiom of conflict in which Chain, Hill, and Scotia are contested figures that incite participation and implicate all in the history of struggles.

I am speaking from experience. It was they who attached my interest to the conflict and convinced me of its significance not just for the place of Humboldt and the struggle to preserve the ancient forest but for understanding the real challenge that converging First Peoples', labor, and environmental movements present to imperial capital in its present extended historical moment of globalization. We know a great deal about this process. It is, for example, driven by its own internal structural contradictions to expand exponentially, generating revolutions in transportation, communications, and information technologies that are breaking down spatial and temporal limits to the system's infinite expansion, ultimately and fearfully ensuring a future of social unrest, denuded resources, extreme climate change, and ecological collapse on a planetary scale.

But more can be learned from listening to local movements that set themselves against this powerful tendency. For example, by taking the totality of global political-economic and ecological relations as their desired object of transformation, while taking action to protect their immediate locale, Humboldt's forest defenders are challenging citizens of redwood country to engage that totality and start shaping its future. The movement thus serves a pedagogical function as its discourse permeates the region, but these effects also reach upward and outward through the national environmental movement and make themselves felt in global civil society, for example, when forest defenders take their stories and participate in antiglobalization politics and appear at social forums and protests around the world.

Scotia, on the other hand, as a symbol of industrial capital accumulation and everything that made Humboldt great, a bastion of hardworking, patriotic men whose labor moved mountains and made life here possible, incites feelings of respect and identification with the dominant national culture but also indignation at seeing the land and the company demeaned by the global market forces embodied in Maxxam. The spectacle of Maxxam in Humboldt also served a consequential pedagogical function. By acquiring Pacific Lumber first, then Kaiser Aluminum, and then leveling the forest while locking out striking steelworkers, Maxxam gave labor and environmental leaders a lesson in the growing power of networking movements. And when steelworkers marched together with Sierra Club turtles at the Battle of Seattle, with the Hurwitz street puppet carried by Humboldt forest defenders marching at the head of the column, imperial capital was put on notice that labor-environment alliances grounded in concrete local grievances are gaining the knowledge and skills they need to understand their common interests and master the technologies necessary to organize resistance. Maxxam was but one symbolic channel for this convergence. The expanding world system is producing more and more like it all the time.

As opposing collective subjects of public culture in the bay redwood region coalesced around Chain, Hill, and Scotia in the late twentieth century, they provided symbols around which imaginations could congeal and seek access to the totality of social and ecological relations. Chain, Hill, and Scotia were used to channel desire into the global struggle by people taking stands on local issues. It was a time in which complex totalizing concepts like planetary ecology and the global economy were becoming increasingly well understood and discussed as objects of collective historical agency. Yes, the timber wars have been fought over trees—but also over planetary survival and the direction of neoliberal globalization.

Such grand concepts themselves have become, as did the timber wars, a means for local citizens wherever they are to channel local grievances into the wide-flung oppositional networks of emergent global civil society. We see this in the rise of forum-style politics of the European and World Social Forums; in the debt relief, anti-globalization, and anti-WTO trade agreement movements; and even in the recent creation of a global union called Workers Uniting, in which the United Steelworkers of America merged with Britain's largest labor organization to form a transatlantic alliance of 2.8 million steel, paper, oil, health care, and transportation workers. Steelworkers president Leo W. Gerard remarked of Workers Uniting, "This union is crucial for challenging the growing power of global capital."[2] Such global imaginations and ambitions are on the lips of social movers in every sector these days, as they were on the lips of the coalition of forest defenders who saved Headwaters Grove from the Maxxam chain saws. This imagination helped the forest defenders to combine with the Steelworkers first against Maxxam and then at the 1999 Battle of Seattle, helping set off the antiglobalization movement that contributed to sinking first the incipient Free Trade Agreement of the Americas (FTAA) and then the Doha round of WTO talks. In this regard, it should be noted that if anyone in the future ever visits the primeval forests of Humboldt and feels a sublime call to environmental consciousness, the credit will fall to the forest defenders and their labor movement allies who took converging counterpublic stands against colonizing capital and helped take this remnant of redwood ecology out of the market. In the case of Humboldt, the symbolic life of Chain, Hill, and Scotia teaches this about globalization—it brings the local more dearly into play.

But the public life of Chain, Hill, and Scotia as popularly invested symbols did more. It carried the traces of redwood industrialization and colonization that led my study back through the series of epochal struggles for property: property first in land, then in labor, and finally in environment, in ecological relations, each of which periods produced a signature event of extraordinary violence that continues to channel attention down through the ages and transmits to the present an image of social relations prevailing at that historical moment. Scenes of the Wiyot massacre, the labor massacre, and the car bombing of environmentalists punctuate the modern redwood imaginary.

Each signature event did for its time what Chain, Hill, and Scotia have done for the bay redwood region at the end of the millennium. In a way that reveals the psychohistorical and geographic implications of colonization by the modern social imaginary, defined as the rights-based moral order of liberal institutions embodied in interwoven capitalist markets, media publics, and republican democracy, each event drew masses to the struggle at hand, attracting attentions and interests away from established investments in habitual concerns and providing a means of participating in emergent historical processes. They too created public culture. They too occasioned archives of comments and narratives that became enduring structures for psychological investment and the making of meanings that would accumulate over time. They are still working today, focusing people's attentions and thus providing the means for action-orienting memory and for shaping projects. And that is what I mean by archive culture: the modern social imaginary, in every new place where it digs in and takes root, in each of its differentiated and ecologically infused local instantiations, is a cultural dynamo that comes to function as a local *archon*. In ancient Greek city-states, the *archon* was a magistrate who had power over records and in whose house, the *arkheion,* were kept the records themselves, the archive *(arkhe).* Thus, under cultural colonization by the modern social imaginary, the place of the redwood imaginary becomes a producer, ruler, and keeper of accumulating documents, words, testimonies, images, stories, monuments, grievances, and their physical, architectural, geographic, and ecological equivalents, in short, an objective material symbolic order that future subjects of history perforce will encounter and necessarily use to assemble new identities and originate new projects.

From this perspective, we can start to see the implications of our modern social imaginaries—like the redwood imaginary they are made to remember. And because the modern social imaginary is internally structured to proliferate instances of its forms on its margins, it ensures that the places it colonizes will similarly gather available psychophysical labor energies into emergent public-sphere struggles over property that culminate in violence that new local archival social imaginaries will remember and transmit to the future.

I have used the signature events of redwood social history to trace the development of a modern social imaginary in this crucial ecological location, showing how its precolonial inhabitants and its biophysical resources shaped the redwood imaginary into a world historical structure for the environmental movement and its convergence with labor and First Peoples' struggles in opposing imperial capital. The case study presented suggests how we might best learn to study other such places that have similarly been made. Cultural sociologists, geographers, and environmental theorists should consider the archival character of modern social imaginaries as they seek to eradicate what Avery Gordon plainly calls "the injurious and dehumanizing conditions of modern life," which, she explains, have everything to do with the haunting "complexities of modern power and personhood."[3] The case of Humboldt should compel us to emphasize how, in every locale into which it reaches, the rights-based free-speech public spheres and property law of our modern social imaginaries tend to colonize the complexities of power and personhood and make them their own.

They enter into the embodied and performed identities of the people we study.

In the stories I have told of David Chain, Julia Hill, Scotia, Judi Bari, the takeover, the campaign to save Headwaters, the Great Lumber Strike, the Scotia Redmen, the redwood unions, the big timber companies, the Wiyot tribe, Indian Island, Captain Ottinger, and E. H. Howard, among others, I hope to have honored the complexities of power and personhood while still showing what critical theorists of every stripe might learn from them about our modern local archival imaginaries, about where the past lies and what it might promise to the present under future conditions in which new political subjects will necessarily respond to the growing contradictions of the capitalist expansion currently under way. Constructing alternatives to the possible futures of human and ecological disaster that are already immanent in globalization may very well depend on learning to use this archival culture and its symbolic power to structure new movements of desert, lake, river, ocean, forest, and species defense, and to bring them into increasing contact with the movements by unions, First Peoples, poor peoples, peasants, and small farmers that are already everywhere opposing imperial capital.

The case of Humboldt shows how the bay redwood region accumulated meaning-making potential over decades of continuous conflict, differentiating itself as it archived local history and memorialized events in its living institutions, symbolic texts, and physical geography, and especially its signature events of extraordinary violence. Culture does indeed enter nature through labor, as the environmental theorist James O'Connor put it—but out here on the nation's western frontier, on the cutting edge of modern free-spoken democratic capitalism, that means more than just commodification of the natural world: it means that social memories of the Wiyot holocaust, the labor massacre, and the car bombing of redwood forest defenders are objectifications of the labors of capital culture that structure its possible future by dint of the archival dynamo. Through such signature events, precapitalist peoples, laboring desires, and physical ecologies enter into and make the colonizing archival culture a thing of their own.

Both critical environmental sociology and cultural theory have been necessary for studying the case at hand. Environmental theories of capitalist development helped distinguish successive epochs of colonization, industrial unionism, and environmentalism as times that embody, respectively, the human stories of primitive accumulation, first contradictions, and second contradictions in capital culture. Taken together and examined as contiguous moments in a tale of local modernity in the making, each epochal conflict reveals itself to be inconceivable apart from its predecessor. It is a story of capital accumulation in general, 500 years in the making of its one global history, and 150 years in the making of its local redwood variation. These three concepts are just names for the troubled labors by which capital culture made a troubled place for itself in the redwoods—the place of the redwood imaginary, a concept I have extrapolated from cultural theories that take the social formation of psychological imagination seriously.

Let me be clear about the first and second contradictions. They are integrally related and determinative, largely, of the whole archival process of development. Capitalist enterprise as such, once its conditions of possibility are secured by primitive accumulation, is geared to accumulation of capital and succeeds only to the extent that it externalizes cost on communities of labor. Driven to despair by capital, increasingly self-conscious labor organizes itself and forces systemic transformation. And as unionization grows, capital culture accepts it; it has to. But it maintains profitability by expanding in scale to accommodate declining prices and rising wages, and henceforth the system is driven to expand both in space and time, seeking new places to mine cheaper labor and nature's resource inputs. This is a program for infinite expansion in space as well as in production. Resources become scarce, and laboring communities are degraded, raising the costs of both, such that capital begins having trouble maintaining profitability because of supply-side costs, as previously it did maintaining demand. In this way the second contradictions absorb and contain the first rather than surpass them, and capital comes to be doubly driven to find ever-newer sources of labor and nature to maintain its growth. These are the engines of globalization that Marx began describing in 1848 and whose continued relevance is sharply evident today in Humboldt and everywhere else that converging peoples, labor, and environmentalist movements attest to the process firsthand.

To briefly summarize this local tale of making the redwoods modern, the constitutional nation's law, embodied in and driving the claims of its land-hungry pioneers, made possible the coming feelings for property that drove the industrialization of the redwood markets. The result was wages and working conditions that roused labor struggle and led to the programmatic expansion in scale and mechanization that produced deforestation and finally stoked the redwood forest defense. This nation's law and the social memory of its violent history have been and remain essential for writing redwood history and for understanding and conducting redwood politics.

Emergent media technologies deserve special attention in this story. By facilitating public culture in each of the epochal struggles, rights-driven media spectacle transformed their spasms of violence into signature events with ostensibly permanent archival powers. Today redwood politics take place only in and through this text- and image-laden media archive. Newspaper reporting and visual documentation of Indian genocide, union repression, and attacks on union-friendly environmental activists are now lasting symbolic and material structures for all future meaning making and collective subject-forming identifications with, and narration of, timber war politics. This can be seen, for example, in the public slide presentations of the Humboldt Watershed Council during the Maxxam bankruptcy proceedings of 2007–8, which included photos of Judi Bari's bombed-out car; sinister video images of police violence, with uniformed officers applying pepper spray directly to the eyeballs of forest defenders engaged in nonviolent civil disobedience; and devastated landscapes that contextualize Maxxam's transformation of Pacific Lumber from respected corporate land steward to just another citadel of miscreant capital towering over labor and ravaging the earth. When the Watershed Council's Mark Lovelace told me in 2008 that he would be happy if his presentations had contributed to the lack of surprise with which the peoples of Humboldt greeted Maxxam's retreat into bankruptcy and legal exile from Humboldt, the force of this archive spoke through his words.[4]

Looking forward, my study suggests that if scholars and social movers alike are to grasp what is essential in emergent struggles over remnant First Peoples and their ecophysical places, over the qualities of labor and its time, and over the relics of species and their habitats for living, they must study places like Humboldt. They will find, I think, what I found in Humboldt—the place-bound conditions for converging movements of First Peoples, labor, and environmentalists. With that said, a final gloss on the archival redwood imaginary might do well to suggest what its power of return to the present might mean.

The Nation's Law

In colonial times, mass media telegraphed the nascent redwood public's prevailing concern with acquiring land according to the preemption and homestead laws by which the nation-state sought to create an expanding republic of free, propertied white men insulated from threats of majority faction. The trouble that early capital culture had, or rather made, was not yet with unionism and still less with environmentalism but with Native resistance to incursion, territorial appropriation, collective punishment, organized murder, and finally removal by force and concentration on reservations.

The mass murder of Wiyot in 1860? What justification did we hear for total violence? Supposedly they took cattle from the private property herds. In the media spectacle, editors refused to name the men who did it. No counterpublic succeeded in publicizing names or exacting justice: no one was charged for the crime. But if the word of future redwood historians can be taken as a measure of how the event yields symbolic material for identificatory self-understanding in the present, we see that in not naming the individuals, the public named itself. As evidence we have the work of Bledsoe, Irvine, Hittel, Bancroft, Loud, Coy, Hoopes, Norton, Carranco, Lowry, and finally Raphael and House—the massacre haunts all their accounts. It does more: it dominates them. First by its act and then by its spectacular public-sphere mediation, Humboldt's dominant collective colonial subject called into existence this catastrophic event's historical public, distant in time but anchored in place—an enduring effect of the archive. The traumatic landscape of memory remembers this historical public as a community of lawmaking violence. The Wiyot commons were taken, their voices silenced, at least for the time being, and now their killers, unnamed, forever make the name of the power that names.

There is something more primitive operating here than mere violence for economic accumulation. It was not just that a local performance of the nation's law set the culture system in motion. It was the failure of that law to prevent the white colonizers from acting out total violence in the name of the law that helped make the law and its culture system what they are today—a failure that determined not only whose material interest the law would institutionalize but the symbolic context in which its nominally legitimate reproduction and slow transformation would thereafter be performed.

Here is what that failure accomplished: among those who came later, the law is constructed and administered not merely from interpretation of the letter of the law, its denotative content, but also from the archival structure of the public, collective spirit and context of its original installation, its connotative content, a context in which the system was set up as legitimate authority, as the law, in a historical moment that failed that law's own philosophical content. The moral history of accumulation by force permanently entered the meaning-making structure of the archive. Consciously or not, contemporary citizens of Humboldt—agents of hegemony or resistance in the timber war field of cultural production—construct their politics using this archive as a deep cultural structure. It holds the language they are bound to speak. And in the sphere of its power, the silence of actors—for example, on the Indian question, the labor question, the environmental question—will speak as loud as, if not louder than, words. That is the logic of archive culture, American style; modern, legal, and rational, yes, but haunted and largely unconscious.

Work in Common

Once the Indians were down and the modern culture system was up, land was capital wired for power over workers, and unionization was imminent. Wage labor and capital were opposing forces engaging each other through legitimate politics as well as through not yet legitimized worker combination and corporate repression of organizing efforts. This conflict, too, seized people's attention in public-sphere discourse, and in the signature moment of violence during the Great Lumber Strike of 1935, it archived an image of new social relations coming into play in the next prevailing contest for accumulating value—the struggle for right of control over property in labor.

At this time of ascendant Fordist relations, in which mechanization drove expanding social productivity to unparalleled levels, capital culture entered into what Guy Debord would later call, perhaps prematurely, its final phase—the society of the spectacle. "The spectacle," Debord wrote, "corresponds to the historical moment at which the commodity completes its colonization of social life. It is not just that the relationship to commodities is now plain to see—commodities are now all there is to see. The world we see is the world of commodity."[5] Spectacular society runs on a state-sanctioned fusion of commercial and media interests, creating an ethico-political-libidinal-economic display that tends to reproduce the socio-psychological conditions of production. When redwood capital experienced the need to expand and produce consuming desire for its accumulating inventories, it embodied this fusion of government, commerce, and media and entered the advertising age of spectacular capital, signaling how the redwood imaginary had entered the time in which the production of signs—and therefore of consciousness—was becoming synonymous with production itself.

The growing value of the sign can thus be derived from the first contradiction in capitalism and connected to the second contradiction described by environmental theorists like James O'Connor and John Bellamy Foster. It is the mechanism through which accumulated capital continues to adjust its operating style upon recognizing that it must begin preparing mass attitudes for ramped-up consumption and opening the horizon for infinite growth.

After the machine-gunning of redwood strikers at the Holmes-Eureka gate in 1935, defenders of the timber barons and their empires explained that the tragic outcome was necessary so that law and order should prevail. The American juridical body politic, whose sacred constitutional value of liberty had been assuming the name of property for decades, had to be defended at any cost. But the plaque at the mall that marks the site reminds us that no one was ever charged for the killings. It reminds us of death and of what was not done, what was not said, and how that other sacred constitutional value—equality—had still to be substantively constructed. When the shots rang out at the gates of the Holmes-Eureka mill, another deep structure for symbolic identification was built into the redwood imaginary. Once again, something about the failure of justice uniquely inscribed this event in the archive and guaranteed it a future of symbolic power to generate and shape the construction of future historical consciousness. It too created public culture. It still sustains a historical public.

Was it legitimate violence? To the victors went the spoils of naming the world. They defined legitimacy for themselves, but not under conditions of their own choosing. Though the unions were shackled, employers raised wages to halt the strike action, and workers won better conditions and shorter hours. Later, when the long strike of 1946–48 finally broke the redwood timber monopolies and set the stage for the post–World War II proliferation of small operators, that outcome reflected the broad social power of a union movement finally being realized. Working folk had transformed the libidinal-economic circuits of timber culture, values flowed more fully to labor, and postwar timber towns achieved a semblance of working-middle-class affluence. But Scotia kept the unions out—it held on to power and delivered up Palco to global finance capital without much opposition because workers were completely unorganized.

So went the age of first contradictions in capital culture in the bay redwood region—the time of internal contradictions in which competitive firms exercising great power over unorganized labor externalized cost on laboring communities and treated laborers' bodily and psychical energies as market commodities even though, in a real sense, those energies are not commodities at all. They had been made not by the market imperatives of supply and demand but, for entirely different reasons, by the families and communities on which the market was learning to feed for values that might otherwise have accumulated to households and lives in some other fashion, as we know they did among First Peoples and farmers under previous regimes and other social imaginaries in other times and other places.[6] But here we find modern, rational, scientific capitalist enterprises continuously revolutionizing the means of production, dividing up labor, expanding in scale, and substituting machinery for labor, trying to eke out efficiencies however possible. In so doing, Humboldt's pioneer lumber barons built the great redwood timber empires up to their legendary proportions. The result, of course, was workers' revolt, for their consciousness had thresholds at which they could suffer no additional extraction of their laboring energies or absorb any more externalized costs in their deteriorating lives. They entered into historical consciousness and action, taking matters into their own hands and disrupting the machine of accumulation just enough to force some concessions that channeled some of the value they were producing back into their lives. But not without provoking violence—not without inciting the hegemonic powers to murderous actions that would work through the spectacle to durably mark the archival redwood imaginary and therefore come to structure the future, insofar, at least, as the future remains a domain of political possibility.

Globalization

Over the ensuing decades of redwood industrialization, before the trouble with radical forest defenders began, external costs of the nation's land privatization and economic development project accrued locally and precisely externally—that is, outside the price, not to the firms but to their sources of value in the nested communities and animate bodies of physical labor and nature, creating the conditions of both social and environmental decline that would eventually form the basis of environmental revolt. The social relations of labor to land that produced these conditions, having long been neglected in the discourse of power, were inserted there by social movers. By the early 1980s, they had built powerful grass roots for this knowledge-based resistance. And as an era of conservatism dawned in 1980, ecological consciousness gathered its potential around Sally Bell Grove—named for the last full-blooded Sinkyone Indian by environmentalists who were allied with Native Americans. Forest defenders saved Sally Bell Grove and set a new precedent for legal struggles over forestry regulations: they made cumulative impacts and species preservation into new limits on industrial timber.

When Maxxam rode the wave of globalization into Humboldt in 1985, powered by deregulated finance capital, it ran headlong into these pent-up forces, which apparently were ready for a struggle to the death for the power to name the local physical and social ecology. And so it began: the timber wars were on. Maxxam named the largest and most pristine grove of uncut and unprotected ancient redwoods timber harvest plans 87-240 and 87-241, announcing that "there are property rights; we have a Constitution," and "these trees are ours, we have a right to cut them." But apropos of the open invitation presented by the universal, cosmopolitan Enlightenment philosophy enumerated in the Constitution, forest defenders renamed these trees the Headwaters forest, making the grove a sacred place of resistance and launching the court cases and direct-action campaigns that would ultimately rename them again: Headwaters Forest Reserve is now legally defined as part of the public domain, no longer part of the private market.

In the course of the struggle, the forest defenders made Humboldt into a global spectacle of resistance against corporate excess and the place of an organic culture of environmental consciousness. They made Headwaters into a psycho-geographic engine of future politics that is permanently built into the traumatic landscape of social memory, right alongside the labor plaque at the mall, the Holmes-Eureka gate at Sequoia Park, the museum of Indian trouble at Fort Humboldt, and Indian Island itself.

The bombing of Judi Bari in 1990? When the would-be assassins struck this feminist union organizer in the very moment of forging an alliance between unionist timber workers and radical environmentalists, what function was served in the redwood imaginary—or rather, what system or mechanism for social memory was set up? The redistribution of power and property in the world she imagined in her visionary writings was a lightning rod for violence, for it directly challenged the body politic of legitimate property in land and in labor. She was labeled a terrorist—an image purveyed by established powers in media, industry, and the state. It stuck to the forest defense movement at least long enough to ensure the defeat of the Forests Forever ballot initiative and guarantee the continued destruction of ancient forest.

So went the age of second contradictions in capital culture's ongoing colonization and occupation of the North Coast redwoods, the time of external contradictions in which firms ran up against environmental limits to accumulation. Maxxam, for example, was chasing fat old-growth logs deeper and deeper into its acreage, farther and farther up the hills from which they are more expensive to haul or fly out by helicopter.

Extralegal violence again made the law, just as it had in the labor trouble and Indian wars, wherein the defensive blow struck against radical workers helped keep unionization under control, the mills working with open shops, and the timber barons' monopolies intact, and the defensive blow struck against radical Indians helped accumulate their land and set the system up for capital investment in a legal environment of secure property rights. In each case, the prevailing economic conditions were made by illicit violence against collective forces labeled as violent outsiders, events that helped seal outcomes in the domain of modern, rational, and open legislative conflict.

In each epoch of struggle, resistance rose up to proclaim something incommensurable with the colonizing property culture. Natives, workers, and forest defenders each charged that something of life should not be alienated, privatized, and reduced to merely a means of potential commodification. Something must remain as an end in itself. Something must escape diminution to mere means to an alienated economic end. Something must be set aside from market culture—sacred and unprofaned. Natives fought for their lives and their place of production. Strikers fought for the quality of life and more time for its living. Forest defenders fought to extend both life and its quality to the forest and its nonhuman living beings, and they did so in the name of all the remnant communities of life, labor, and environment that have always been cast outside the price, so to speak, by the prevailing culture of capital.

Resistance to colonization by capital changed the balance of power in each of these contests; in each case it set new conditions that rechanneled the flows of property accumulation. This history repudiates the reification of the concept of property evidenced not just in the actions but in the rhetoric of various players in the timber wars, most notably by representatives of the dominant order of achieved timber culture. Property is never just a thing but is rather a right made by law that is itself forged in perpetual social conflict. It is an arbitrary, historical signification of power: it is the name of power performatively uttered, delegated, and distributed in the field of the law—a power delegated by the state through which the nation speaks when it says that such and such an entity has this or that right. Property and right and power thus tend to lose their distinction, though they remain analytically separable for purposes of explanation. In practice they become different names for the same type of hierarchical social relations. This is an effect of the deep culture drive that set up the constitutional republic as a machinic producer of modern social order that articulates rights-based property markets with rights-based free-speech public spheres and the whole democratic republican political apparatus of universal franchise. As it happens, changes in one institutional sphere register in each of the others and so ultimately rechannel the flow of accumulation, at least to some extent. The result is that, in the American modern imagined by the framers and enumerated in the Constitution, every law has its price. This is a stunning achievement—providing a tangible measure of the Enlightenment philosophy of limited government. But its profound future effects in the ecological register are only now becoming clear. The growing autonomy of property has become difficult to challenge, especially *before* the damage is done. Recall that the timber wars revolve around the final 1 percent of uncut and unprotected ancient trees. Ninety-six percent had already been taken. Three percent had been preserved.

Speaking of damages, looking back, we must ask ourselves not merely what the nation's preemption laws achieved by naming only white male citizens as legitimate customers in the nation's land privatization program, but what the law accomplished by ignoring prior claims on the land. Preemptions were not merely applications of a law already established—they were performative announcements of the arrival of a law being made on the spot. But the archive of social memory shows that that law was not able to make itself in Humboldt ex nihilo. It took violent, historical agency to install it. Violence ensured that legal, rational accumulation of property entered predominantly white male formations that still prevail today. The redwood imaginary is a racial and gender formation.

In those early days of the formation of the redwood imaginary, individual actions had the potential to make major differences, in moral, historical, and archival terms. We remember how townsfolk ran Bret Harte out of Humboldt for daring to question the genocidal impulse. The people who did this were victorious in defining what future publics in Humboldt would see. But Harte's outspoken support of the Indians shows us again that the frontier public sphere was not consensus but conflict: it was not just a hegemonic national project carried out by free heroic individuals. Also at work were the hesitant utterances of more or less potent individuals acting out counterpublic dreams in an uncertain world full of powerful compulsions. In that dismal historical moment, Harte's words are a balm, though they sting in the wounds of this weary reader. But what stands out in the archive is the trace of the dominant sentiments that rose up against Harte and rejected his identification with the victims of the massacre. Henceforth he was unwelcome, un-American, a threat to the community, just as would be the radical strikers shot down at the gates of the Holmes-Eureka mill seven decades later, when the AFL broke its own redwood strike and purged the union leadership of radicals and communists. And in the same American vein, the law of which expresses itself in the propertied modern, Humboldt's dominant timber public stood by in silence when the bomb hit Judi Bari. Forest defenders, too, are unwelcome, situated outside the hegemonic order by that order itself—dismissed as radicals who stand outside the law.

Having been through this history, we are no longer mystified when we encounter white male accumulation exercising excessive power over labor and environment in the timber wars. The authority and freedom associated with property ownership was produced by an organized state program of land privatization that racialized and gendered the material concentration of power. That power was legislated, certainly, but violence beyond the law signed its name to the fact. White men have always dominated Humboldt with state-delegated power to name, and they have done so from the beginning in the name of all the freedoms that are promised by the universalizing language—the cultural code—of rights enumerated in the Constitution. This history is the cultural unconscious of collective agency operating in Humboldt today—a history of ambiguous success in fulfilling the twin promises of freedom and equality, but unambiguous exertion of the will to demand them. It was not the Wiyot who sold Headwaters to Maxxam. Nor was it redwood laborers, or even a board of directors including a labor representative. But Native Americans, timber workers, and environmental movers today do stand up before the law and make their demands, which invariably take the form of a claim on those promises, as they are contained in the founding speech that institutionalized the law.

This shows how capital accumulations in the bay redwood region are really compromise formations between opposing sociopolitical forces represented by collective subjects, especially the First Peoples, capitalists, laboring folk, and forest defenders who have struggled to take history into their own hands. The forest defense, for example, is a hotbed of ideas for living outside the consumer system, not totally, but by degrees, and their ideas about sustainable forestry are just one domain of alternative knowledge within the movement. If this idea of sustainable production were seized by the masses, if its public could capture more attention and continue to grow, it might achieve the material force that it needs to institute changes in the land, its people, and their conditions of possibility for an ecological future. But that, of course, would change the bottom line. Sustainability is a value that must struggle for attention with the established authority of the value of property in generating profit.

For Alberto Melucci, new social movements like the forest defense essentially work by projecting challenging codes: as movements their labor is "reversing the symbolic order."[7] But when we discover in Humboldt the modern culture of nation, race, gender, and property embodied in the mental architecture of institutionalized authority, by which socially structured psychology the nation mediates its citizens' encounters with their world, binding their desire in pleasurable circuits that tend overwhelmingly to follow the law, however exploitative, we begin to see the depth of the problem facing forest defenders. To fight against property in Humboldt is to challenge the embodied unconscious of the American form of fictive ethnicity and its dynamic institutional engines of racial and gender capitalism—the body politic of modernity itself, in other words, as it has been achieved in this very successful local articulation. But they do try—and to some degree they succeed. By reversing the symbolic order, living differently, building a network of grassroots movements, and projecting a global vision, forest defenders are trying to disembody the mentality of modern capital culture and establish a new way of life.

I hope to have shown how interested scholars need both critical cultural and environmental theory to grasp what is at stake in these politics and to grapple with the evidence of converging First Peoples, labor, and environmental movements.[8] They suggest, and the case of Humboldt confirms, that the cultural conditions for solidarity between the movements have always been present, in the same way and for the same reasons that Karl Polanyi explained why, in the early to mid-twentieth century, national competitive capitals developed into their opposite, namely, state-managed and class-dominated capitalism: the internal tendency of market liberalism to fail, and of new movements born of such failures to morph into one or another form of state interventionism and control, democratic or otherwise.[9] The work of environmental theorists on the first and second contradictions in capital culture, especially James O'Connor and John Bellamy Foster, suggests how deeply emergent labor and environmental consciousness are linked to this same internal tendency, only this time on the scale of globalization. The rise of transnational movements and First Peoples' alliances in the late 1990s must also be seen in this light. And the culturalist concept of social imaginaries, developed by Charles Taylor and Craig Calhoun, among others, best describes these cultural engines of perpetual conflict because it keeps theory focused on the coarticulation of rights-driven institutions of markets, publics, and politics. But still more has been needed than environment and culture: psychoanalytic social theory has provided the tools for describing how the psychical attentions and energies of subjects are called up to labor in these institutional spheres, how they answer the spectacular call of the nation form that public-sphere spectacle puts on constant display, and how they accumulate in archival events that always invite, if not ensure, their historic return to the present.

Spectacular Violence

Unlike the Wiyot massacre and the killing of redwood strikers at the Holmes-Eureka gate, the bombing of Bari and Cherney is closer in time to this writing, complicating my argument for its status as a signature event around which collective memory and future politics will continue to orbit, calling up its archive to help narrate memories, identities, and conscious projects for change. Will its impression on the redwood imaginary survive over decades and centuries? I believe that it will.

The bombing of Bari and Cherney allowed FBI covert operations to publicly tarnish the movement and contributed to the defeat of the Forests Forever voter initiative by creating a media climate in which forest defenders were labeled as violent and associated with terrorism.

Like redwood strikers and Indian peoples, they were charged with disrespecting property, violently attacked, and then, in spectacular media display, publicly situated at the center of antisocial violence, terror, and threat to *the law.*

The disruption of Redwood Summer and the failure of the initiative meant a double defeat for progressive democracy. It amplified the power that accumulated redwood capital exercises over labor in the making of redwood commodity circuits. Specifically, it helped shape local and regional consciousness in ways that aided timber capital in escaping the immediate regulation of ancient forest conversion, with the result being that additional hundreds of millions of dollars have been blasted into private accumulation, values that might otherwise have come to rest in less-quantifiable public-trust values, for example, in restoring salmon streams, recovering endangered bird populations, decreasing risk of flooding, improving water supplies, and lowering sediment transport rates into the clogged shipping channels of Humboldt Bay.

Once again it was violence that made the law, not all of it, maybe not even most of it, but at the crucial moment in which the forest defense public was poised to take the power to name into its own hands, the weapon was detonated, the charge of terrorism rang out, and the identificatory public of forest defense lost ground, so to speak, to the established authority of property culture in the body politic. It was not just rational discourse, not just open dialogue, but spectacular violence against the body of a woman whose idea of organizing unions together with forest defenders against corporate power provoked the attack that set the movement back. Nothing close to Forests Forever was ever achieved, but the successful movement to save Headwaters registered a significant, if partial, victory, and it would be difficult to argue that the symbolic power of Judi Bari's body did not aid in constructing the wide base of support on which the Headwaters Forest Reserve has been erected. For seven years after the bombing, Bari took her case before the law, a rallying point and identificatory challenge to the unconscious political body of property.

On May 24, 2000, I attended a rally outside the Federal Courthouse in San Francisco, joining for the first time what had become a yearly ritual. Representatives of the American Indian Movement took the stage briefly and called for an alliance between all progressive movements that have suffered police repression. Radical union labor bard Utah Phillips, who had driven with Bari in her booby-trapped car

Judi Bari gives the signature Earth First! salute—arm raised, clenched fist facing forward—in front of the Federal Courthouse in San Francisco, March 3, 1995. A broad spectrum of unionists, Black Power activists (including Black Panthers), antiapartheid activists, communists, anarchists, and socialists have historically adopted this gesture as an expression of resistance to oppressive regimes. Photograph by Xiang Xing Zhou.

down the coast to Oakland on the day before it exploded, sang songs and spoke intimately about Bari's philosophy. "I was particularly interested in her nonviolence," he said, "so I was really moved by what she was trying to do with regard to the Mississippi Summer and translating that into action in the forests of northern California with the Redwood Summer. . . . Judi understood that nonviolence is not a tactic; it is a way of life, you understand, the people with all of the guns have to be absolutely convinced, absolutely convinced, that you mean it, that you're nonviolent, and that you mean it, and that you're not going to jump this way and jump that way, which is what tactics mean, doesn't it? . . . The man owns the gun. I mean these folks here [nodding over his shoulder at the courthouse], they can escalate from a hand gun on the block all the way up to a hydrogen bomb, and they're always saying to you, 'How far up the road do you want to go? how far up? follow us up that road,' and you get so far up that road, and pretty soon they grease you, and you're done for, so no, you pick another road, you pick a road that the man doesn't know anything about, he doesn't know how to deal with peace, he doesn't know how to deal with nonviolence, so you use your strongest weapon—moral courage—against his weakest weapon, because he hasn't got any."

Labor bard Utah Phillips speaks in front of the Judi Bari street puppet at a rally for the Redwood Justice Fund, Federal Courthouse, San Francisco, May 24, 2000.

While Phillips was talking, a street puppet likeness of Bari stood by, its long fabric body whipping in the San Francisco wind. Darryl Cherney sang songs, introduced the speakers, sold T-shirts, and updated the crowd of about one hundred people on the progress of their case, which at that time had not yet reached trial. It was a lively show that reflected the gathering energies of revolt that erupted on the global stage at the Battle of Seattle in 1999. It presaged the coming epoch of new social movements allying for world social justice—the movement of movements, as it has been called by the antiglobalization brigades, antiwar movers, and World Social Forum activists, among others on the gathering progressive front, harbingers all of an emergent planetary civil society newly enabled by emergent media technologies to engage the powers that be in what promises to be the greatest struggle of all—the struggle for power over communications, information, and images. The case of Humboldt has shown how emergent media made possible the colonizing culture by projecting its power, transmitting its call to labor on the land, putting its sacred motivating and subjectifying symbols on display, and obliterating the obstacles of time and space, therein making a national imaginary possible. Now new media are helping form a global imaginary, similarly marshaling the world's available psychical energies and channeling them through new publics and counterpublics. To the extent that the global imaginary models itself on the modern, Western, and specifically American imaginary, setting up rights-driven institutions of free property markets, free-speech public spheres, and free democratic-republican polity, the world is facing a spectacular future of perpetual public-sphere struggles over property that will determine the interwoven futures of its remnant precapitalist peoples, its laboring multitudes, and their common environment—just as the arrival of national capital did, and still does, on a local scale in the bay redwood region. By that account, the archival redwood imaginary might be seen as pregnant with the future.

Notes

Entry Point

1. The fullest physical description of Humboldt Bay is Roger A. Barnhardt, Milton J. Boyd, and John E. Pequegnat, *The Ecology of Humboldt Bay, California: An Estuarian Profile,* U.S. Fish and Wildlife Service Biological Report (Washington, D.C.: U.S. Department of the Interior, 1992).

2. Reed F. Noss, *The Redwood Forest* (Washington, D.C.: Island Press, 2000), 10, 46–47.

3. Heather A. Enloe, Robert C. Graham, and Stephen C. Sillett, "Arboreal Histosols in Old-Growth Redwood Forest Canopies, Northern California," *Soil Science Society of America Journal* 70 (2006); also R. M. Burns and B. H. Honkala, *Silvics of North America,* vol. 1, *Conifers* (Washington, D.C.: USDA Forest Service Agriculture Handbook, 1990). On canopy salamanders and crablike crustaceans, see Sillett et al., "Evidence of a New Niche for a North American Salamander," *Herpetological Conservation and Biology* 1, no. 1 (2006).

4. Stephen C. Sillett, "Tree Crown Structure and Vascular Epiphyte Distribution in Sequoia Sempervirens Rainforest Canopies," *Selbyana* 20, no. 1 (1999).

5. David Anderson, "Salmon Come First on the Eel, Agencies Say," *Eureka Times-Standard,* May 6, 1999. Friends of the Eel River maintain an online archive at eelriver.org.

6. David Anderson, "Battle over Trinity Flow Anticipated," *Eureka Times-Standard,* May 25, 1999. See Dane J. Durham, "How the Trinity Lost Its Water," Friends of the Trinity River, fotr.org (2005).

7. John Driscoll, "State Warns Klamath Dam Owner over Delays," *Eureka Times-Standard,* August 8, 2007.

8. Thomas M. Mahony and John D. Stuart, "Status of Vegetation Classification in Redwood Ecosystems," USDA Forest Service Gen. Tech. Rep. PSW-GTR-194 (2007).

9. Noss, *The Redwood Forest,* 153–54.

10. On cultural geography and the concept of capitalist culture systems, see Joseph E. Spencer, "The Growth of Cultural Geography," *American Behavioral Scientist* 22, no. 79 (1978).

11. See Ray Raphael and Freeman House, *Two Peoples, One Place,* vol. 1 of *Humboldt History* (Eureka, Calif.: Humboldt County Historical Society, 2007), 39–90, for the best account of European exploration and discovery; see also Owen C. Coy, *The Humboldt Bay Region, 1850–1875* (Los Angeles: California State Historical Association, 1929; reprinted by Humboldt County Historical Society, 1982), esp. 27–32.

12. See L. K. Wood's narrative of the Gregg expedition in Oscar Lewis, ed., *The Quest for Qual-a-wa-loo: Humboldt Bay; A Collection of Diaries and Historical Accounts of the Area Now Known as Humboldt County, California* (San Francisco: College Publishing Company, 1943).

13. Letter of H., *Humboldt Times,* September 9, 1954 (italics mine). Here and throughout the book, when primary sources are cited, the original irregular wording and punctuation has been retained for historical accuracy.

14. Letter of Citizen, *Humboldt Times,* October 21, 1854; Citizen also appeared in the *Humboldt Times* on November 7, 1854.

15. *Humboldt Times,* October 21, 1854.

16. "Redwood trees grow in an interrupted 724-km belt along the Pacific Coast from the southwestern tip of Oregon (42°09' N. latitude) to southern Monterey County in California (35°41' N. latitude), once covering some 647,500–770000 ha." Noss, *The Redwood Forest,* 39.

17. Kimberly Wear, "Humboldt County Sees Its Share of Poverty," *Eureka Times-Standard,* August 29, 2007.

Introduction

1. "Pacific Lumber Cited for Illegal Practices," *Eureka Times-Standard,* September 25, 1998. On the history, culture, and politics of the international Earth First! movement, see Susan Zakin, *Coyotes and Towndogs: Earth First! and the Environmental Movement* (New York: Penguin, 1993); John Opie, *Nature's Nation: An Environmental History of the United States* (Fort Worth, Tex.: Harcourt Brace College Publishers, 1998); Christopher Manes, *Green Rage* (Boston: Little, Brown, 1990); Derek Wall, *Earth First! and the Anti-roads Movement: Radical Environmentalism and Comparative Social Movements* (New York: Routledge, 1999); Rik Scarce, *Eco-Warriors: Understanding the Radical*

Environmental Movement (Chicago: Noble Press, 1990); and Timothy Luke, "Ecological Politics and Local Struggles: Earth First! as an Environmental Resistance Movement," *Current Perspectives in Social Theory* 14 (1994).

2. Suzanne Zalev, "Death in the Forest," and "Activists Weep for Comrade," *Eureka Times-Standard,* September 18, 1998. Bullwinkel, quoted in "PL Officials 'Saddened,'" *Eureka Times-Standard,* September 18, 1998.

3. Zalev, "Death in the Forest."

4. Suzanne Zalev, "PL Blamed for Forest Death," *Eureka Times-Standard,* September 19, 1998.

5. Farmer published "How Gypsy Really Died: An Eyewitness Account," on October 1, 1998, at envirolink.org. Jordan's account appeared at envirolink.org before the end of October.

6. Greg Magnus, "PL Workers Shocked, Not Surprised at Death," *Eureka Times-Standard,* September 19, 1998.

7. *Eureka Times-Standard,* September 23, 1998.

8. The timber wars are a particular case of the general configuration of social struggles in the age of media spectacle, consumer society, and image politics. I use the terms *spectacle* and *spectacular* to analyze the transition from industrial to consumer society that is driving the contradiction between capitalism and ecology, in other words what John Bellamy Foster writes about in "The Absolute General Law of Ecological Degradation under Capitalism," *Capitalism Nature Socialism* 3, no. 3 (1992). The vast technological apparatus of display that is mass consumer culture grows to be so thoroughly captured by capital, devoted to its imperative of defending current profits at any cost, that it tends to prevent or delay public culture from recognizing and acting in its own social and environmental interest. On the spectacular society, see Guy Debord, *Comments on the Society of the Spectacle* (New York: Verso, 1988), and *The Society of the Spectacle* (1967; New York: Zone Books, 1994); as well as Henri Giroux, *Beyond the Spectacle of Terror: Global Uncertainty and the Challenge of New Media* (Boulder, Colo.: Paradigm, 2006); Michael Hardt and Antonio Negri, *Empire* (Cambridge, Mass.: Harvard University Press, 2000), xiv, 8–9, 47–49, 186–87, 321–23, 347, 458n17, and 458n18; Giorgio Agamben, *The Coming Community* (Minneapolis: University of Minnesota Press, 1993), 79–82, and *Homo Sacer* (Stanford, Calif.: Stanford University Press, 1998), 10–11; Kevin G. Barnhurst and John Nerone, *The Form of the News* (New York: Guilford Press, 2001), 298–310; and especially Anselm Jappe, *Guy Debord* (Berkeley: University of California Press, 1999). George Ritzer lucidly explains Debord's concept of the spectacular society in terms of consumer society. Ritzer, "The New Means of Consumption and the Situationist Perspective," in

Explorations in the Sociology of Consumption (London: Sage, 2002). On image politics, see Kevin Michael Deluca, *Image Politics: The New Rhetoric of Environmental Activism* (New York: Guilford Press, 1999).

9. By "discourse of free trade" I mean both the collective representation and performance of globalization—what Jeffrey Alexander calls its semantics and pragmatics. In "'Globalization' as Collective Representation: The New Dream of a Cosmopolitan Civil Sphere," *International Journal of Politics, Culture, and Society* 19 (2007), Alexander suggests that the emergent global civil sphere can be described as a discourse, a collective representation, an imaginary, and a dream. In a similar way, the Battle of Seattle was both symbolic and material. On the role of forest defenders from the Pacific Northwest in Seattle 1999, see Eddie Yuen, ed., *The Battle of Seattle* (New York: Soft Skull Press, 2001); on the role of redwood forest defenders in particular, see Alexander Cockburn, *Five Days That Shook the World: Seattle and Beyond* (New York: Verso, 2000). Cockburn's text is of special interest to me for having captured my anonymous image in the crowd photo chosen for the cover.

10. Gaye LeBaron, "Remembering Scotia, the Last of the Company Towns," *Santa Rosa Press Democrat,* October 12, 2008.

11. Alberto Melucci, "The Global Planet and the Internal Planet: New Frontiers for Collective Action and Individual Transformation," in *Cultural Politics and Social Movements,* ed. Marcy Darnovsky, Barbara Epstein, and Richard Flacks (Philadelphia: Temple University Press, 1995).

12. Here I draw on Marshall McLuhan's language in the preface to *The Mechanical Bride: Folklore of Industrial Man* (Boston: Beacon Press, 1951). See also Paul Willis and Mats Trondman, "Manifesto for Ethnography," *Ethnography* 1, no. 1 (2000): 5–16; Michael Burawoy, *Global Ethnography* (Berkeley: University of California Press, 2000), and "Manufacturing the Global," *Ethnography* 2, no. 2 (2001); and Zsuzsa Gille and Seán Ó Riain, "Global Ethnography," *Annual Review of Sociology* 28 (2002).

13. A social system, dialectically conceived, is a set of relationships between people, labor, its products, and the world; and concrete and particular events, processes, connections, and developments are viewed as inseparable from the systemic whole, but not reducible to it. For Theodor Adorno, each concrete and particular cultural artifact is an intersection between social structure (institutions, symbolic order) and social action; its field of possible meanings is a function the social totality. Adorno, "Sociology and Psychology," *New Left Review* 46 (1968). Adorno wrote admiringly that Walter Benjamin "never wavered in his fundamental conviction that the smallest cell of observed reality offsets the rest of the world." Adorno, "A Portrait of Walter Benjamin," in *Prisms* (Cambridge,

Mass.: MIT Press, 1967), 236. For concise primers on dialectical thinking and ecological dialectics, respectively, see Bertell Ollman, "Why Dialectics? Why Now?" *Science and Society* 62, no. 3 (1998): 338–57; and
David Harvey, *Justice, Nature, and the Geography of Difference* (Cambridge: Blackwell, 1996).

14. I see primitive accumulation as the constitutive moment of capital
and its precondition as such, without being a stage that is somehow sur-
passed but which rather enters into the form of capital culture. On the
continued relevance of the concept, see David Harvey, *The New Imperialism* (Oxford: Oxford University Press, 2003), esp. 145–46; Michael
Perelman, "Primitive Accumulation from Feudalism to Neoliberalism,"
Capitalism Nature Socialism 18, no. 2 (2007); and Werner Bonefeld,
"History and Social Constitution: Primitive Accumulation Is Not Primitive," thecommoner.org, March 2002. Bonefeld cites Marx, *Capital*,
vol. 1: primitive accumulation, defined as "the separation of labour from
its product, of subjective labour-power from the objective conditions of
labour, was therefore the real foundation in fact and the starting-point of
capitalist production. But that which at first was but a starting point, becomes, by the mere continuity of the process, by simple reproduction, the
peculiar result, constantly renewed and perpetuated, of capitalist production." He ends his discussion with a note to Walter Benjamin: "The violence of capital's original beginning is the formative element of the 'civilized' forms of equality, liberty, freedom and utility. These forms mystify
the real content of 'equality' as an equality in the inequality of property.
They are the constituted forms of the original violence—violence as civilized normality." Cf. Walter Benjamin, *Zur Kritik der Gewalt und andere
Aufsätze* (Frankfurt: Suhrkamp, 1965).

15. Environmental theorists have debated theories of the second contradiction in capitalism in the journals *Capitalism Nature Socialism
(CNS)* and *Monthly Review*. See especially John Bellamy Foster, "Capitalism and Ecology: The Nature of the Contradiction," *Monthly Review*
54, no. 4 (September 2002); and James O'Connor, "Capitalism, Nature,
Socialism: A Theoretical Introduction," *CNS* 1 (1988); "On the Two Contradictions of Capitalism," *CNS* 2, no. 3 (1991); *Natural Causes* (New
York: Guilford Press, 1998); and "What Is Environmental History? Why
Environmental History?" *CNS* 8, no. 2 (1997). See also John Bellamy
Foster, *Capitalism against Ecology* (New York: Monthly Review Press,
2002); "Marx's Ecological Value Analysis," *Monthly Review* 52, no. 4
(2000); "The Scale of Our Ecological Crisis," *Monthly Review* 49, no. 11
(1998); and "The Absolute General Law of Environmental Degradation
under Capitalism," *CNS* 3, no. 3 (1992). Further see Samir Amin, "A
Note on the Depreciation of the Future," *CNS* 3, no. 3 (1992): 21–22;

Victor Toledo, "The Ecological Crisis: A Second Contradiction of Capitalism," *CNS* 3, no. 3 (1992); and Michael A. Lebowitz, "Capitalism: How Many Contradictions?" *CNS* 3, no. 3 (1992).

16. See my remarks in chapter 6; also see Environmental Protection Information Center, "Litigation Summary," wildcalifornia.org.

17. Mark Lovelace, the Humboldt Watershed Council, "Palco Presentation," humboldtwatersheds.org, 2006.

18. The redwood timber wars have rarely been treated in a scholarly way, but see John Bellamy Foster, "The Limits of Environmentalism without Class: Lessons from the Ancient Forest Struggle in the Pacific Northwest," in *Capitalism against Ecology* (New York: Monthly Review Press, 2002); Andrew Rowell, *Green Backlash: Global Subversion of the Environment Movement* (New York: Routledge, 1996), 155, 157–81; Jacqueline Vaughn Switzer, *Green Backlash: The History and Politics of Environmental Opposition in the U.S.* (Boulder, Colo.: Lynne Rienner, 1997), 214–15; David Helvarg, *The War against the Greens: The Wise Use Movement, the New Right, and Anti-environmental Violence* (San Francisco: Sierra Club Books, 1994), 330–39. Susan R. Schrepfer's history of environmental reform in the redwoods, *The Fight to Save the Redwoods* (Madison: University of Wisconsin Press, 1983), was written before Maxxam moved in and blasted the struggle into the sphere of globalization. Several journalists tell a more up-to-date story: Susan Zakin, *Coyotes and Towndogs: Earth First! and the Environmental Movement* (New York: Penguin, 1993); David Harris, *The Last Stand: The War between Wall Street and Main Street over California's Ancient Redwoods* (New York: Times Books, 1995); Patrick Beach, *A Good Forest for Dying* (2003). See also Julia Hill, *The Legacy of Luna: The Story of a Tree, a Woman, and the Struggle to Save the Redwoods* (San Francisco: Harper Collins, 2000); and Judi Bari, *The Timber Wars* (Monroe, Maine: Common Courage Press, 1994).

19. Ethnographers have examined the timber conflicts and the social worlds of the loggers across the Pacific Northwest in the 1990s; see, for example, Terre Satterfield, *Anatomy of a Conflict: Identity, Knowledge, and Emotion in Old Growth Forests* (Vancouver: University of British Columbia Press, 2002); Fred Rose, *Coalitions across the Class Divide* (Ithaca, N.Y.: Cornell University Press, 2000); Matthew S. Carroll, *Community and the Northwestern Logger: Continuities and Changes in the Era of the Spotted Owl* (San Francisco: Westview Press, 1995); Kathie Durbin, *Tree Huggers: Victory, Defeat, and Renewal in the Northwest Ancient Forest Campaign* (Seattle: Mountaineers, 1996); Beverly Brown, *In Timber Country* (Philadelphia: Temple University Press, 1995); W. Scott Prudham, *Knock on Wood: Nature as a Commodity in Douglas Fir Country* (New York: Routledge, 2005); also James D. Proctor,

"Whose Nature? The Contested Moral Terrain of Ancient Forests," in *Uncommon Ground: Rethinking the Human Place in Nature,* ed. William Cronon (New York: W. W. Norton, 1996); Jake Kosek, *Understories: The Political Life of Forests in Northern New Mexico* (Durham, N.C.: Duke University Press, 2006).

20. Susan Beder, *Global Spin: The Corporate Assault on Environmentalism* (White River Junction, Vt.: Chelsea Green, 1997); Helvarg, *The War against the Greens.*

21. Environmental Protection Information Center of Garberville, newsletter, Spring 2001.

22. John Sterling, "Thousands Rally for Headwaters," *Earth Island Journal,* Fall 1995; Mary Lane, "2,400 Rally against PL," *Eureka Times-Standard,* September 16, 1995.

23. Kie Relyea, "897 Cited at Carlotta Protest," *Eureka Times-Standard,* September 16, 1996; Mike Geniella, "Record 6,000 at Headwaters Protest," *Santa Rosa Press Democrat,* September 15, 1997.

24. *Pacific Lumber Co. v. United States* no. 96-257L (Fed. Cls). Maxxam's "Complaint for Inverse Condemnation" was filed in the U.S. Court of Federal Claims on May 6, 1996.

25. See my discussion of the deal in chapter 1.

26. In the film *Tree-Sit* (James Ficklin, Headwaters Action Video Collective, earthfilms.org, 2001), members of the United Steelworkers are shown protesting on the streets in Scotia. Union members told the filmmakers the story of Maxxam's unfair labor practices after the takeover of Kaiser, including the company's active recruitment and use of laid-off Palco workers as maintenance personnel to scab at Kaiser's plant in Tacoma, Washington.

27. Alliance for Sustainable Jobs and the Environment, asje.org.

28. Richard White and John M. Findlay, eds., *Power and Place in the North American West* (Seattle: University of Washington Press, 1999).

29. O'Connor, "What Is Environmental History?" Without touching on the redwood region, the works of William G. Robbins are important for a general understanding of the timber industry's place in the history of the U.S. West; see his *Colony and Empire: The Capitalist Transformation of the West* (Lawrence: University Press of Kansas, 1994); "The Social Context of Forestry: The Pacific Northwest of the Twentieth Century," *Western Historical Quarterly* 16 (1985); "The Western Lumber Industry: A Twentieth-Century Perspective," in *The Twentieth Century West: Historical Interpretations,* ed. Gerald D. Nash and Richard Etulian (Albuquerque: University of New Mexico Press, 1989); and *Lumberjacks and Legislators: Political Economy of the U.S. Lumber Industry, 1890–1941* (College Station: Texas A&M University Press, 1982).

30. Noel Castree, "Geographies of Nature in the Making," in *Handbook*

of Cultural Geography, ed. Kay Anderson, Mona Domosh, Steve Pile, and Nigel Thrift (London: Sage, 2003), 179. Joseph E. Spencer, "The Growth of Cultural Geography," *American Behavioral Scientist* 22, no. 79 (1978), expertly surveys the discipline and its notions of culture systems, especially the culture system of capitalism. Essential reading in the contemporary cultural geography and the critical theory of nature-culture dialectics includes Bruce Braun and Noel Castree, eds., *Remaking Reality: Nature at the Millennium* (Routledge: New York, 1998); Bruce Braun and Noel Castree, eds., *Social Nature: Theory, Practice, and Politics* (Malden, Mass.: Blackwell, 2001); Steve Pile, *The Body and the City: Psychoanalysis, Space, and Subjectivity* (New York: Routledge, 1996); Donald Worster, "Doing Environmental History," in *The Ends of the Earth: Perspectives on Modern Environmental History,* ed. Donald Worster (Cambridge: Cambridge University Press, 1988); Walter Cronon, ed., *Uncommon Ground: Rethinking the Human Place in Nature* (New York: W. W. Norton, 1996), and *Changes in the Land: Indians, Colonists, and the Ecology of New England* (New York: Hill and Wang, 1983); Carolyn Merchant, *Ecological Revolutions: Nature, Gender, and Science in New England* (Chapel Hill: University of North Carolina Press, 1989); Robert Bunting, *The Pacific Raincoast: Environment and Culture in an American Eden, 1778–1900* (Lawrence: University Press of Kansas, 1997); Richard White, *Land Use, Environment, and Social Change: The Shaping of Island County* (Seattle: University of Washington Press, 1980); and Alexander Wilson, *The Culture of Nature* (Blackwell: Cambridge, 1992).

31. Louis Althusser, *Lenin and Philosophy* (New York: Monthly Review Press, 1971), 176, 174–81.

32. In what follows, my descriptive theory of the redwood imaginary draws much from Jacques Lacan's conceptualization of the symbolic and imaginary orders, especially in *Ecrits* (New York: W. W. Norton, 2002), 65–67, where he further elaborates the exemplary law concerning kinship structures: "This law, then, reveals itself clearly enough as identical to a language order." And finally, "Symbols in fact envelop the life of man with a network so total that they join together those who are going to engender him 'by bone and flesh' before he comes into the world; so total that they bring to his birth, along with the gifts of the stars, if not the gifts of the fairies, the shape of his destiny; so total that they provide the words that will make him faithful or renegade, the law of the acts that will follow him right to the very place where he is not yet and beyond his very death."

33. On uses of psychoanalysis for social and particularly cultural theory, see Richard Widick, "Flesh and the Free Market (On Taking Bourdieu to the Options Exchange)," *Theory and Society* 32 (2003), in which I build on Pierre Bourdieu's idea of "socioanalysis," as well as on the works of Judith Butler, Jacques Lacan, Herbert Marcuse, Theodor Adorno,

Max Horkheimer, Walter Benjamin, Eric Fromm, Leo Lowenthal, and Jürgen Habermas. In *An Invitation to Reflexive Sociology* (Chicago: University of Chicago Press, 1992), 136–37, Bourdieu described his program of socioanalysis in nearly psychoanalytic terms when he wrote that social "determinisms operate to their full only by the help of unconsciousness, with the complicity of the unconscious." Theodor Adorno's "Sociology and Psychology" is still a required introduction to sociological use of psychoanalysis for the critical, dialectical representation of the reciprocal constitution and institutional reproduction of structure and agency; but see Judith Butler, *Bodies That Matter* (New York: Routledge, 1993), for a lucid rendering in Lacanian terms. In an earlier effort, Herbert Marcuse anticipated volumes of emerging social theory and influenced a generation of youth counterculture with his revision of Freud's reality principle in the concept of a historically changing performance principle. Marcuse, *Eros and Civilization: A Philosophical Inquiry into Freud* (Boston: Beacon Press, 1955), 35. His *One Dimensional Man: Studies in the Ideology of Advanced Industrial Society* (Boston: Beacon Press, 1964) and *Five Lectures* (Boston: Beacon Press, 1970) add a great deal to psychoanalytic sociology. Jürgen Habermas's chapter "Psychoanalysis and Social Theory," in *Knowledge and Human Interests* (Boston: Beacon Press, 1971), esp. 274–75, grounds critical theory in psychoanalysis: "Freud conceived of sociology as applied psychology. . . . The superego, constructed on the basis of substitutive identifications with the expectations of primary reference persons, ensures that there is no immediate confrontation between an ego governed by wishes and the reality of external nature. The reality which the ego comes up against and which makes the instinctual impulses leading to conflict appear as a source of danger is the system of self-preservation, that is, society, whose institutional demands upon the emergent individual are represented by the parents." Joel Whitebook's *Perversion and Utopia: A Study in Psychoanalysis and Critical Theory* (Cambridge, Mass.: MIT Press, 1996) should also be required reading in psychoanalytic cultural theory; see especially his comments on Marcuse's performance principle, 24–41.

34. See, for example, Steven Pile, *The Body and the City: Psychoanalysis, Space, and Subjectivity* (New York, Routledge, 1996).

35. On psychoanalysis, psychical energies, and psychical publics, see Mustafa Emirbayer and Mimi Sheller, "Publics in History," *Theory and Society* 27 (1998).

36. In *Private Property and the Limits of American Constitutionalism* (Chicago: University of Chicago Press, 1990), Jennifer Nedelsky shows how the Constitution's dual initiatives of democracy and liberty program the schism between political rights and civil rights into U.S. political culture. On the role of enumerated rights in the performance of U.S. political

culture and sociological treatments of U.S. political culture that empha-
size the constitutional tension between liberty and democracy, see Knud
Haakonssen and Michael J. Lacey, eds., *Culture of Rights* (Cambridge:
Cambridge University Press, 1991); Richard Flacks, *Making History: The
American Left and the American Mind* (New York: Columbia University
Press, 1988); Robert Bellah et al., *Habits of the Heart: Individualism and
Commitment in American Life* (Berkeley: University of California Press,
1985); Jeffrey C. Alexander, *The Civil Sphere* (Oxford: Oxford University
Press, 2006), and "Citizen and Enemy as Symbolic Classification: On the
Polarizing Discourse of Civil Society," in *Cultivating Differences: Sym-
bolic Boundaries and the Making of Inequality,* ed. Michèle Lamont and
Marcel Fournier (Chicago: University of Chicago Press, 1992); Richard
Harvey Brown, *Society as Text: Essays on Rhetoric, Reason, and Re-
ality* (Chicago: University of Chicago Press, 1987); and Theodore Meyer
Greene, *Liberalism: Its Theory and Practice* (Austin: University of Texas
Press, 1957).

   37. While Hobbes might best be considered "the founder of liberalism" (Leo Strauss, *Natural Right and History* [Chicago: University of
Chicago Press, 1953], 182) and modern constitutionalism because he first
established the necessity of written law (Gary McDowell, "Private Conscience and Public Order: Hobbes and *The Federalist,*" *Polity* 25, no. 3
[Spring 1993]), most analyses of liberal political philosophy, constitutionalism, and the U.S. Constitution in particular ultimately turn to the
contributions of John Locke and his *The Second Treatise of Government*
(1690), ed. C. B. Macpherson (New York: Liberal Arts Press, 1952), especially the chapter "On Property," acknowledging how well he formulated
the philosophy and how widely influential it became as a justification for
revolution in the New World.

   38. George Mace, *Locke, Hobbes, and the Federalist Papers: An Essay
on the Genesis of the American Political Heritage* (Carbondale: Southern
Illinois University Press, 1979), 139.

   39. Clinton Rossiter, ed., *The Federalist Papers* (New York: Mentor,
1961), article 10, paragraph 19. Hereafter cited in the text.

   40. Harry N. Scheiber shows the importance of Supreme Court decisions on the promotion of commercial interests throughout the nineteenth
century in "Public Rights and the Rule of Law in American Legal History," *California Law Review* 72, no. 2 (March 1984), and "Law and
the Imperatives of Progress: Private Rights and Public Values in American Legal History," in *Ethics, Economics, and the Law,* ed. J. Roland
Pennock and John W. Chapman (New York: New York University Press,
1982). Alfred H. Kelly and Winfred A. Harbison point out in their chapter "The New Deal" that the National Industrial Relations Act, signed
in 1933, ushered the state into commerce on such a large scale that in

1935 the Supreme Court felt compelled to rule the act unconstitutional. Kelly and Harbison, *The American Constitution: Its Origins and Development* (New York: W. W. Norton, 1948). But the ruling could not alter the fact that the era of state capitalism had in fact already been launched. Irving Bernstein, *The New Deal Collective Bargaining Policy* (1950; New York: Da Capo Press, 1975).

41. "The facts about a man-made institution which creates and maintains certain relations between people—and that is what property is— are never simple." C. B. Macpherson, *Property: Mainstream and Critical Positions* (Toronto: University of Toronto Press, 1978), 1. See also Macpherson, *The Political Theory of Possessive Individualism: Hobbes to Locke* (Oxford: Clarendon Press, 1962); and Gordon J. Schochet, ed., *Life, Liberty, and Property: Essays on Locke's Political Ideas* (Belmont, Calif.: Wadsworth, 1971), 5–6.

42. "Violence haunts liberal political thought," write Candice Vogler and Patchen Merkell in "Introduction: Violence, Redemption, and the Liberal Imagination," *Public Culture* 15, no. 1 (2003). "The defining image of early modern European social contract theory—and an image that remains potent in contemporary contractarian moral and political theory—locates the possibility of civil society in a compact among men who are long accustomed to the use of force in the bloody business of self-assertion and self-preservation." The liberal state substitutes normalized, legitimate, monopolized, and patient violence for the pathological violence of unorganized life.

43. In *Ethics, Institutions, and the Right to Philosophy* (New York: Rowman and Littlefield, 2002), 3, Jacques Derrida described the international political institutions constructed after World War II, namely, the United Nations, in terms useful for the case of Humboldt: "These institutions are already *philosophemes,* as is the idea of international law or rights that they attempt to put into operation. They are philosophical acts and archives, philosophical productions and products, not only because the concepts that legitimate them have an assignable *philosophical history* and therefore a philosophical history that is inscribed in UNESCO's charter or constitution; but because, by the same token and for the very same reason, such institutions imply the sharing of a culture and a philosophical language."

44. Fred C. Alford examines this aspect of liberal political theory in *The Self in Social Theory: A Psychoanalytic Account of Its Construction in Plato, Hobbes, Locke, Rawls, and Rousseau* (New Haven, Conn.: Yale University Press, 1991).

45. Jürgen Habermas, *The Structural Transformation of the Public Sphere* (1962; Cambridge, Mass.: MIT Press, 1989), 83, especially note 60.

46. Ibid., 266n62.

47. Nicolas Garnham, "The Media and the Public Sphere," in *Habermas and the Public Sphere,* ed. Craig Calhoun (Cambridge, Mass.: MIT Press, 1992), 360–61.

48. On race, class, and gender particularism in the Constitution, see Michael Warner, "The Mass Public and the Mass Subject," in *Habermas and the Public Sphere,* ed. Craig Calhoun (Cambridge, Mass.: MIT Press, 1992), 383; as well as "Publics and Counter Publics," *Public Culture* 14, no. 1 (2002), and *Letters of the Republic: Publication and the Public Sphere in Eighteenth-Century America* (Cambridge, Mass.: Harvard University Press, 1990).

49. Étienne Balibar, "The Nation Form: History and Ideology," in *Becoming National: A Reader,* ed. Geoff Eley and Ronald Grigor Suny (Oxford: Oxford University Press, 1996), 140. Balibar draws on Louis Althusser's notion of ideological state apparatuses, the central function of which Althusser argued is the institutional reproduction of economic and cultural domination; see also Étienne Balibar and Immanuel Wallerstein, *Race, Nation, Class: Ambiguous Identities* (New York: Verso, 1991), 10, 49, 96, 102, 105, 223; and Balibar, *We the People of Europe: Reflections on Transnational Citizenship* (Princeton, N.J.: Princeton University Press, 2004), esp. 8, but also 9, 26, 29, 94.

50. Balibar, "The Nation Form," 138.

51. Cornelius Castoriadis, *The Imaginary Institution of Society* (Cambridge, Mass.: MIT Press, 1987); Benedict Anderson, *Imagined Communities* (New York: Verso, 1983). In the words of Manuel Castells, "If we mean by imaginary something that is symbolically communicated and expressed, all worlds are imaginary, as Baudrillard, Barthes and a number of other semiologists showed us long ago." Castells, *The Making of the Network Society* (London: Institute of Contemporary Arts, 2001), 19. In the words of Craig Calhoun, "To speak of the social imaginary is to assert that there are no fixed categories of external observation adequate to all history; that ways of thinking and structures of feeling make possible certain social forms and that such forms are thus products of action and historically variable. It follows that cultural creativity can be seen to be basic even to such seemingly 'material' forms as the corporation or the nation. These exist precisely because they are imagined; they are real because they are treated as real; and new, particular cases are produced through the recurrent exercise of the underlying social imaginary." Calhoun, "Imagining Solidarity: Cosmopolitanism, Constitutional Patriotism, and the Public Sphere," *Public Culture* 14, no. 1 (2002): 152.

52. Balibar, "The Nation Form," 143.

53. Anderson, *Imagined Communities,* 26. Thomas C. Leonard, *News for All: America's Coming of Age with the Press* (Oxford: Oxford University Press, 1995), 29–32; Richard D. Brown, *Knowledge Is Power: The Diffusion of Information in Early America, 1700–1865* (New York:

Oxford University Press, 1989), 13; John B. Thompson, *The Media and Modernity: A Social Theory of the Media* (Stanford, Calif.: Stanford University Press, 1995), 62–63.

54. On the problems of structure and agency in cultural theory, see Widick, "Flesh and the Free Market"; Jeffrey Alexander, "The Reality of Reduction: The Failed Synthesis of Pierre Bourdieu," in *Fin-de-Siècle Social Theory* (New York: Verso, 1995), 128–217; Pierre Bourdieu, *The Logic of Practice* (Stanford, Calif.: Stanford University Press, 1990); Anthony Giddens, *Central Problems in Social Theory* (Berkeley: University of California Press, 1979).

55. Charles Taylor, "Modern Social Imaginaries," *Public Culture* 14, no. 1 (2002): 106.

56. Ibid.

57. Emile Durkheim, *The Rules of Sociological Method* (New York: Free Press, 1966), 2–3.

58. Max Weber writes: "On the basis of this calculation, the American system of 'scientific management' enjoys the greatest triumphs in the rational conditioning and training of work performances. The final consequences are drawn from the mechanization and discipline of the plant, and the psycho-physical apparatus of man is completely adjusted to the demands of the outer world, the tools, the machines—in short, to an individual function. The individual is shorn of his natural rhythm as determined by the structure of his organism; his psycho-physical apparatus is attuned to a new rhythm through a methodical specialization of separately functioning muscles and an optimal economy of forces is established corresponding to the conditions of work." Weber, "The Meaning of Discipline," in *On Charisma and Institution Building,* ed. S. N. Eisenstadt (Chicago: University of Chicago Press, 1968), 38–39.

59. Karl Marx, "The Brumaire of Louis Bonaparte" (1852), in *The Marx-Engels Reader,* ed. Robert C. Tucker (New York: W. W. Norton, 1978), 594–95.

60. O'Connor, "What Is Environmental History?" 8.

61. Ibid., 18–19.

62. Ibid., 22.

63. On the transformation of nature by capitalism, see Karl Polanyi, *The Great Transformation* (1957; Boston: Beacon Press, 2001); also Prudham, *Knock on Wood,* 8.

64. George Ritzer, *Explorations in the Sociology of Consumption* (New York: Sage, 2001), 181–83.

1. Power and Resistance in Redwood Country

1. "Pacific Lumber Cited for Illegal Practices," *Eureka Times-Standard,* November 25, 1998.

2. Alexander Cockburn, "Why David Chain Died," *Nation,* October 26, 1998, 8.

3. Jeff Goodell, "Death in the Redwoods," *Rolling Stone,* January 1999, 60.

4. Ibid., 67. See also Patrick Beach, *A Good Forest for Dying: The Tragic Death of a Young Man on the Front Lines of the Environmental Wars* (New York: Doubleday, 2004).

5. Maxxam, "The Headwaters Forest," white paper, May 1997, palco.com.

6. Because the case was settled out of court there is no case decision on record, but see "The Pacific Lumber Company, a Delaware Corporation; Scotia Pacific Holding Company, a Delaware Corporation; and the Salmon Creek Corporation, a Delaware Corporation, Plaintiffs, vs. United States of America: Complaint for Inverse Condemnation," filed by Maxxam/Pacific Lumber on May 6, 1996, in the United States Court of Federal Claims, archived at judibari.org; see also Environmental Protection Information Center, "Memorandum of Points and Authorities in Support of Motion to Intervene (No. 96-257-C) to Judge James T. Turner," July 23, 1996. I received photocopies of these public legal documents directly from EPIC on August 29, 2001.

7. On "incidental take permits," see section 10(a) of the ESA, 16 U.S.C. § 1539[a] of the Endangered Species Act.

8. Maxxam/Pacific Lumber's Habitat Conservation Plan/Sustained Yield Plan can be accessed at pacificlumber.com.

9. EPIC, *Headwaters Forest Update,* newsletter, September 1998, 3; see also Trees Foundation, *Headwaters Forest Stewardship Plan Draft Summary,* 1997, treesfoundation.org.

10. The Headwaters Preservation Agreement of September 28, 1996, Bureau of Land Management, blm.gov.

11. Greg Magnus, "Headwaters in Public Hands: Wilson Signs Landmark Bill to Fund Purchase," *Eureka Times-Standard,* September 20, 1998.

12. Kevin Bundy, "Voodoo Ecology: The Pacific Lumber Company's Habitat Destruction Plan," Environmental Protection Information Center, *Headwaters Forest Update* (September 1998).

13. Bureau of Land Management, "Headwaters Forest Fact Sheet," http://www.blm.gov/ca/arcata/headwaters_factsheet.html.

14. On the Luna tree-sit, see the films of James Ficklin, Headwaters Action Video Collective (HAVC), earthfilms.org, especially *Luna: The Stafford Giant* (1998) and *Tree-Sit* (1998); also Julia Hill, *The Legacy of Luna: The Story of a Tree, a Woman, and the Struggle to Save the Redwoods* (San Francisco: Harper Collins, 2000).

15. Suzanne Zalev, "50 Enter PL Land, Honor Tree-Sitter," *Eureka Times-Standard,* March 21, 1998.

16. *Eureka Times-Standard,* September 4, 2003; "FBI Clears Humboldt Deputies in Pepper Spray Case," *Eureka Times-Standard,* December 18, 1998; Anderson, "Pepper Spray Ruling Overturned," *Eureka Times-Standard,* May 5, 2000. The trial documents are archived at nopepperspray.org and judibari.org.

2. Convoking the Opposition

1. In 2003, in a corporate image makeover and greening of Simpson's public image, the company adopted the name Green Diamond Resource Company and changed its color scheme and logo from red to green.

2. On the idea of psychical collectives, see my discussion of sociological use of psychoanalysis in the introduction. See also Theodor Adorno, "Sociology and Psychology," *New Left Review* 46 (1968); and Mustafa Emirbayer and Mimi Sheller, "Publics in History," *Theory and Society* 27 (1998).

3. *Timber Gap,* a film by Headwaters Action Video Collective (earthfilms.org, 2000); Judi Bari tells the story of Louisiana-Pacific and the Fisher family in *The Timber Wars* (Monroe, Maine: Common Courage Press, 1994). The Environmental Information Protection Center's (EPIC) groundbreaking legal case in ancient-forest defense sought protection for Sally Bell Grove in Mendocino County from LP; see *EPIC v. Johnson 1983–1985,* EPIC Litigation Summary, wildcalifornia.org.

4. From Cherney's self-published audiotape *Timber,* http://www.darrylcherney.com/timber.

5. Judi Bari, "Timber: An Interview with Judi Bari," in *Sounding Off,* ed. Ron Sakolsky and F. Wei-han Ho (Brooklyn: Autonomedia, 1995).

6. Headwaters Action Video Collective (1998).

7. Margot Hornblower, "Five Months at 180 Ft.," *Time,* May 11, 1998, 4; *People,* "Out on a Limb," April 6, 1998, 128; Erik Brazil, "Life atop a Forest Giant," *San Francisco Examiner,* February 13, 1998; "California Is Talking," *Newsweek,* April 6, 1998; "The 30th Annual Most Admired Women Poll," *Good Housekeeping,* September 1999; Gloria Mattioni, "Il richiamo della foresta," Italian *Elle,* n.d.; Susan Carpenter, "Fighting for Treedom," *Los Angeles Times,* February 20, 1998; Giles Whittell, "Tree Crusader Claims Moral High Ground," *Times (London),* March 7, 1998; Louise Parc, "Julia, sur un arbre haut perche'e," *Le Monde,* August 13, 1998; "Is Butterfly Insane?" *Jane,* May 1998, 136.

8. Marshall McLuhan, *Understanding Media: The Extensions of Man* (New York: Signet Books, 1964).

9. On the role of Indymedia in Seattle, see Eddie Yuen, ed., *The Battle of Seattle* (New York: Soft Skull Press, 2001); and Dorothy Kidd, "Indymedia.org: A New Communications Commons," in *Cyberactivism: Online Activism Theory and Practice,* ed. Martha McCaughey

and Michael D. Ayers (New York: Routledge, 2003); also see Alexander Cockburn, Jeffrey St. Clair, and Allan Sekula, *Five Days That Shook the World: Seattle and Beyond* (New York: Verso, 2000).

10. David Helvarg, *The War against the Greens: The Wise Use Movement, the New Right, and Anti-environmental Violence* (San Francisco: Sierra Club Books, 1994); and Sharon Beder, *Global Spin: The Corporate Assault on Environmentalism* (White River Junction, Vt.: Chelsea Green, 1997).

11. See Kevin Michael DeLuca, *Image Politics: The New Rhetoric of Environmental Activism* (New York: Guilford Press, 1999), for an analysis of the use of media by environmentalists, especially Earth First!

12. Glen Martin, "Clear-Cutting Blamed for Many Mudslides: Residents Say Storm Damage Could Have Been Avoided," *San Francisco Chronicle*, January 9, 1997.

13. Alberto Melucci, *Challenging Codes: Collective Action in the Information Age* (Cambridge: Cambridge University Press, 1996).

14. Michael Warner, "Publics and Counterpublics," *Public Culture* 14, no. 1 (2002): 86.

15. Sven Lütticken, "Secrecy and Publicity: Reactivating the Avant-Garde," *New Left Review* 17 (2002): 133.

16. Melucci, *Challenging Codes*, 357.

17. Ibid., 358.

18. Ernesto Laclau and Chantal Mouffe, in *Hegemony and Socialist Strategy: Towards a Radical Democratic Politics* (New York: Verso, 1985), describe the principal goal of democratic theory and practice as articulation, a process by which political subjects frame their claims and addresses to the public in ways that create equivalencies between disparate subject positions.

19. The Alliance for Sustainable Jobs and the Environment, http://www.asje.org.

3. Everybody Needs a Home

1. *Eureka Times-Standard*, October 16, 1999. This ad also appeared in *California Forests* 3, no. 4 (October–December 1999).

2. Thoreau's "Civil Disobedience," first published in May 1849, had earlier been delivered as a lecture on January 26, 1848.

3. Suzanne Zalev, "'Butterfly' Still Out on a Limb," *Eureka Times-Standard*, December 10, 1998.

4. Jim Smith, president of the Humboldt–Del Norte Central Labor Council, interview with the author, 2001. Smith explained that unions "are almost nonexistent" in Humboldt's timber industry.

5. "Community Disapproves, Stays Silent," *Eureka Times-Standard*, December 10, 1998.

6. Ibid.

7. The Center for Defense of Free Enterprise, cdfe.org. See Ron Arnold, *Ecoterror: The Violent Agenda to Save Nature; The World of the Unabomber* (Bellevue, Wash.: Free Enterprise Press, 1997). See my remarks on the CDFE and Ron Arnold in chapter 6.

8. James O'Connor, "What Is Environmental History? Why Environmental History?" *Capitalism Nature Socialism* 8, no. 2 (June 1997); see my discussion of O'Connor in the introduction.

9. Lynwood Carranco, *Redwood Lumber Industry* (San Marino, Calif.: Golden West Books, 1982), 149; Pacific Lumber, "The History of Pacific Lumber and Scotia," pacificlumber.com, October 1998.

10. Claudia Wood, "The History of the Pacific Lumber Company," unpublished paper (1956), Humboldt Room, Humboldt State University, Arcata, Calif.

11. Lowell S. Mengell II, "The Murphy Family and TPL Co.," *Humboldt Historian* 25, no. 4 (July–August 1977); Brett H. Melendy, "One Hundred Years of the Redwood Lumber Industry, 1850–1950" (Ph.D. diss., Stanford University, 1952), 196–97; Carranco, *Redwood Lumber Industry,* 149–52.

12. Melendy, "One Hundred Years," 196–97.

13. "Scotia Will Be One of the Largest Lumbering Towns on the Pacific Coast," *Humboldt Times,* December 19, 1912.

14. Wood, "History of the Pacific Lumber Company"; Daniel Cornford, *Workers and Dissent in the Redwood Empire* (Philadelphia: Temple University Press, 1987), 14.

15. Cornford, *Workers and Dissent,* 204. For a regional history of the International Workers of the World, see Robert L. Tyler, *Rebels in the Woods: The IWW in the Pacific Northwest* (Eugene: University of Oregon Books, 1967).

16. Wood, "History of the Pacific Lumber Company," 34.

17. Cornford, *Workers and Dissent,* 121.

18. Pacific Lumber Company, "Scotia: Home of the Pacific Lumber Company" (Pacific Lumber, 1957), Pamphlet File, Humboldt Room, Humboldt State University, Arcata, Calif.

19. Pacific Lumber Company, "Facts about Scotia: Home of PALCO Redwood Products" (Pacific Lumber, 1938), Pamphlet File, Humboldt Room, Humboldt State University, Arcata, Calif.

20. Letter of John H. Emmert to Mr. Donald Macdonald, May 6, 1925. Pacific Lumber Collection, Bancroft Library, Berkeley, Calif.; Wood, "History of the Pacific Lumber Company."

21. Andrew Genzoli, "Scotia Band, 1935–1985," Pamphlet File, Humboldt Room, Humboldt State University, Arcata, Calif.

22. Pacific Lumber Company, "Facts about Scotia."

23. Carranco, *Redwood Lumber Industry,* 152.

24. Pacific Lumber Company, "Scotia."

25. Held, "Scotia, the Town of Concern," *Pacific Historian* 16, no. 2 (Summer 1972): 86–87.

26. Pacific Lumber Company, "Scotia."

27. Held, "Scotia," 8.

28. Greater Eureka Chamber of Commerce and Humboldt County Board of Trade, "List of Lumber-Logging and Forest Products Plants of Humboldt County, California" (Eureka: Humboldt County Tax Collector, July 1961), Pamphlet File, Humboldt Room, Humboldt State University, Arcata, Calif.

29. Held, "Scotia," 89.

30. Cited in Held, "Scotia," 81.

31. Ibid.

32. Frank J. Taylor, "Paradise with a Waiting List," *Saturday Evening Post,* February 1951.

33. Carranco, *Redwood Lumber Industry,* 152.

34. Held, "Scotia," 90–91. See also Ray Raphael, *More Tree Talk: The People, Politics, and Economics of Timber* (Washington, D.C.: Island Press, 1994); and Hugh Wilkerson and John van Der Zee, *Life in the Peace Zone: An American Company Town* (New York: Macmillan, 1971).

35. Walter Wiley, "Company Town . . . Scotia Is Rooted In," *Sacramento Bee,* April 27, 1981.

36. Raphael, *More Tree Talk.*

37. PL Rescue Fund, *Takeback* 1, no. 1 (February 1989), a newsletter of the PL Rescue Fund, Pamphlet File, Folder "Pacific Lumber Co., 1988–9," Humboldt Room, Humboldt State University, Arcata, Calif. The takeback effort was still alive on January 22, 1992, when they summarized their investigation of the takeover in a full-page advertisement in the *Eureka Times-Standard.*

38. Pacific Lumber Company, "Seasons Greetings," *Eureka Times-Standard,* December 1989, Pamphlet File, Folder "Pacific Lumber, 1988–9," Humboldt Room, Humboldt State University, Arcata, Calif.

39. This Maxxam/Pacific Lumber employee requested anonymity as a precondition of our 2001 interview.

40. John Driscoll, "PL Closes Scotia's Last Mill; Latest Action Cuts 140 Jobs," *Eureka Times-Standard,* December 1, 2001. Before this announcement, Maxxam had already discharged 250 employees since the Headwaters deal in 1999.

41. Maxxam/Pacific Lumber's Scotia Development LLC, in the United States Bankruptcy Court for the Southern District of Texas, Corpus Christi Division, *Notice of Chapter 11 Bankruptcy Cases and Meeting of Creditors,* http://www.loganandco.com, January 18, 2007.

42. John Driscoll, "Judge Clears Mendocino Redwood Plan for Palco, Appeal to Follow," *Eureka Times-Standard,* July 9, 2008.

43. Michael Warner, "Publics and Counterpublics," *Public Culture* 14, no. 1 (2002): 82.

44. Ibid., 81.

45. Ibid., 82.

46. Mike Geniella, "Scramble to Save Redwood," *Eureka Times-Standard,* November 29, 2000.

47. John Driscoll, "Hill Returns to Grieve for Luna," *Eureka Times-Standard,* November 30, 2000.

48. Matt Morehouse, "A Monument to 'Lunacy,'" *Eureka Times-Standard,* November 31, 2000.

4. Indian Trouble

1. The definitive history of Humboldt's early years is Ray Raphael and Freeman House, *Humboldt History,* vol. 1, *Two Peoples, One Place* (Eureka, Calif.: Humboldt County Historical Society, 2007).

2. In *Tree-Sit: The Art of Resistance,* a documentary film by the Headwaters Action Video Collective (James Ficklin, earthfilms.org, 2001), Chris Peters, a Karuk/Yaruk Indian, is shown addressing a forest defense rally. "Headwaters forest is a sacred area," he says. "Those of you who have been in the area understand that it is the last pristine area in this part of the world."

3. See my discussion of primitive accumulation and the first and second contradictions of capital culture in the introduction.

4. *Northern Californian,* February 29, 1860.

5. Letter from "Exodus," *San Francisco Bulletin,* May 11, 1860.

6. Norris A. Bleyhl, ed., *Indian-White Relations in Northern California, 1849–1929: In the Congressional Set of United States Public Documents* (Chico: University of California, 1978), 7, entry 20. In 1853 President Millard Fillmore appointed Edward Beale as the superintendent of Indian Affairs covering Nevada and California; Beale served until 1856.

7. Owen C. Coy, *The Humboldt Bay Region, 1850–1875* (Los Angeles: California State Historical Society, 1929), 115.

8. Ibid., 135–36.

9. Ibid., 111–16.

10. Ibid., 135.

11. *Humboldt Times,* November 18, 1854; also Coy, *The Humboldt Bay Region,* 19–20.

12. It would be a mistake to misrecognize the pre-Columbian redwood commons as a space of nonownership. On social and legal customs of the redwood tribes, see Llewellyn L. Loud, *Ethnogeography and Archaeology of the Wiyot Territory* (University of California, 1918); Chad L. Hoopes, *Lure of Humboldt Bay Region: Early Discoveries, Exploration, and Foundations Establishing the Bay Region* (Dubuque, Iowa: Kendall/

Hunt, 1966); Ray Raphael, *Little White Father: Redick McKee on the California Frontier* (Eureka, Calif.: Humboldt County Historical Society, 1993); T. T. Waterman, "Yurok Geography," *American Archaeology and Ethnology* 16, no. 5 (1920); Lucy Thompson, *To the American Indian: Reminiscences of a Yurok Woman* (Berkeley, Calif.: Heyday Books, 1916); and Maureen Bell, *Karuk: The Upriver People* (Happy Camp, Calif.: Naturegraph, 1991), 83.

13. Letter of Exodus, *San Francisco Bulletin,* May 11, 1860.

14. Ibid.

15. Letter of Major G. J. Rains (May 22), *San Francisco Bulletin,* May 24, 1860.

16. Letter of "Eye-Witness," *San Francisco Bulletin,* March 13, 1860.

17. Bret Harte, *Northern Californian,* February 29, 1860.

18. Ibid.

19. *Humboldt Times,* March 3, 1860.

20. George R. Stewart Jr., *Bret Harte: Argonaut in Exile* (Cambridge, Mass.: Riverside Press, 1931), 88–89.

21. See Jack Norton, *When Our Worlds Cried: Genocide in Northwestern California* (San Francisco: Indian Historical Press, 1987), 105. On white representations of California's Native Americans, see James Rawls, *Indians of California: The Changing Image* (Norman: University of Oklahoma Press, 1984); and R. Berkhoffer, *The White Man's Indian: Images of the American Indian from Columbus to the Present* (New York: Vintage, 1979).

22. Étienne Balibar, "The Nation Form: History and Ideology," in *Becoming National: A Reader,* ed. Geoff Eley and Ronald Grigor Suny (Oxford: Oxford University Press, 1996). See my remarks on Balibar and fictive ethnicity in the introduction.

23. On white attitudes toward the Indians of California, especially in Humboldt County, see Rawls, *Indians of California;* as well as Norton, *When Our Worlds Cried,* 105.

24. *Humboldt Times,* October 28, 1854.

25. Charles Sellers, *The Market Revolution* (Oxford: Oxford University Press, 1991); Shawn Wilentz, "Society, Politics, and the Market Revolution, 1815–1848," in *The New American History,* ed. Eric Foner (Philadelphia: Temple University Press); Michael Rogin, *Fathers and Children: Andrew Jackson and the Subjugation of the American Indian* (New York: Alfred A. Knopf, 1975), esp. the chapter titled "Market Revolution and the Reconstruction of Paternal Authority"; Kevin G. Barnhurst and John Nerone, *The Form of News: A History* (New York: Guilford Press, 2001).

26. Sellers, *The Market Revolution;* Eli Zaretsky, *Capitalism, the Family, and Personal Life* (New York: Harper Colophon, 1973); and Mary Ryan, *Womanhood in America: From Colonial Times to the Pres-*

ent (New York: Franklin Watts, 1983), especially the chapter titled "Creating Woman's Sphere: Gender in the Making of American Capitalism, 1820–1985."

27. Rogin, *Fathers and Children.*

28. Ryan, *Womanhood in America,* 9. The average American family size fell from six in the seventeenth century to three and one-half by the mid-twentieth century.

29. Sellers, *The Market Revolution,* 9.

30. Michael Omi and Howard Winant, *Racial Formation in the United States: From the 1960s to the 1990s* (New York: Routledge, 1994), 56.

31. Sellers, *The Market Revolution,* 29–30.

32. Ibid. See also Ryan, *Womanhood in America,* 10.

33. Roger Friedland and Richard Hecht, *To Rule Jerusalem* (Cambridge: Cambridge University Press, 1996), 5.

34. Coy, *The Humboldt Bay Region,* 107–9; U.S. Government, "Population of the United States in 1860; Compiled from the Original Returns of the Eighth Census" (1864).

35. Coy, *The Humboldt Bay Region,* 110.

36. David Dary, *Red Blood and Black Ink: Journalism in the Old West* (New York: Knopf, 1998), 15. The punctuation, spelling, and capitalization of these sentences are retained from the original.

37. Sellers, *The Market Revolution,* 369–70. The Postal Act compelled paper exchange for the cut-up method of circulating news, which remained strong until the telegraph went transcontinental in 1861. Newspaper postage was free up to thirty miles and one penny up to one hundred miles. By 1850 2,526 weekly and 254 daily newspapers were operating in the United States.

38. *Alta California* 1, no. 1 (January 4, 1849).

39. Dary, *Red Blood and Black Ink,* 127.

40. Thomas C. Leonard, *News for All: America's Coming of Age with the Press* (Oxford: Oxford University Press, 1995), 118–24.

41. On the Humboldt and Mendocino Telegraphy Company, see Coy, *The Humboldt Bay Region,* 286, 300, 61, 63n, 299–302, 304, 305, 321; also Charles Strope, "Times Printing Company," *Humboldt Historian,* Summer 1997.

42. Dary, *Red Blood and Black Ink.*

43. A. J. Bledsoe, *Indian Wars of the Northwest: A California Sketch* (1865; San Francisco: Bacon and Co., 1956), 18; Coy, *The Humboldt Bay Region,* 46, 297.

44. Bledsoe, *Indian Wars of the Northwest,* 14–18, quotes E. H. Howard from the official minutes of the annual gathering of the Society of Humboldt Pioneers, organized January 22, 1876, incorporated May 12, 1881, here meeting on May 12 to commemorate the twenty-third anniversary of

the legislative act incorporating the County of Humboldt (May 12, 1853). The pioneer lumber baron John Vance called the meeting to order at Russ Hall in Eureka.

45. Oscar Lewis, *The Quest for Qual-a-wa-loo: Humboldt Bay; A Collection of Diaries and Historical Accounts of the Area Now Known as Humboldt County, California* (San Francisco: College Publishing Company, 1943), 180–82.

46. On the second Bank of the United States and its role in land distribution, see Rogin, *Fathers and Children.* On land law, see Paul Wallace Gates, *History of Public Land Law Development,* written for the Public Land Law Review Commission, with a chapter by Robert W. Swenson (Washington, D.C.: Government Printing Office, 1968); "The Homestead Law in an Incongruous Land System," *American Historical Review* 41, no. 4 (July 1936).

47. Gates, *History of Public Land Law,* 122.

48. Jefferson's letter of October 28, 1785, quoted in Peter Barnes, *The People's Lands: A Reader on Land Reform in the United States* (Emmaus, Pa.: Rodale Press, 1975).

49. Marion Clawson, *Man and Land in the United States* (Lincoln: University of Nebraska Press, 1964), 41–55; also Malcolm J. Rohrbough, *The Land Office Business: The Settlement and Administration of American Public Lands, 1789–1837* (New York: Oxford University Press, 1968); Rogin, *Fathers and Children;* Sellers, *The Market Revolution.*

50. Gates, *History of Public Land Law,* 145–77; Samuel Trask Dana and Myron Krueger, *California Lands: Ownership, Use, and Management* (New York: Arno Press, 1958), 243. Coy, *The Humboldt Bay Region,* 81, refers to the "Act of April 28th."

51. On the extent and formative role of title disputes in early American history, see the chapter "Nature, Property, and Title," in Rogin, *Fathers and Children,* 85.

52. Coy, *The Humboldt Bay Region,* 79.

53. Ibid., 78–94; Sellers, *The Market Revolution;* and especially Rogin, *Fathers and Children,* 79.

54. Coy, *The Humboldt Bay Region,* 78–94.

55. Ibid. The General Allotment Act of 1887 (the Dawes Severalty Act) reflected the same type of thinking. It simultaneously legislated Christianization of the Indians and their subjectification by the property system. The bill's author, congressman Henry L. Dawes of Massachusetts, in the constitutional tradition of Jefferson, defended the bill, saying that to be civilized was "to wear civilized clothes . . . cultivate the ground, live in houses, ride in Studebaker wagons, send children to school, drink whiskey [and] own property."

56. Coy, *The Humboldt Bay Region,* 98; I have kept the original spelling

and format that Coy presented, presumably that of the original document. See *U.S. Statutes at Large,* vol. 24, 388–91; and *House and Senate Journals,* 49th Cong., 1st and 2nd sess.; also see the *Congressional Record.*

57. Coy, *The Humboldt Bay Region,* 83.

58. Ibid., 85.

59. Dana and Kreuger, *California Lands,* 244.

60. Coy, *The Humboldt Bay Region,* 89.

61. Ibid., 2.

62. Ibid., 305, citing *History of Humboldt County, California* (San Francisco: Mid-Cal Publishers, 1881), 213–14; Charles Volney Anthony, *Fifty Years of Methodism: A History of the Methodist Episcopal Church, within the Bounds of the California Annual Conference from 1847 to 1897* (San Francisco: Methodist Book Concern, 1901); Gail Karshner, *A Bell Rang in Uniontown: The First Hundred Years of Arcata and Its Methodist Church, 1850–1950* (Virginia Beach, Va.: Donning, 1994).

63. The inaugural speeches of California's governors, including Bigler (1852–56), are archived by the state at http://www.californiagovernors .ca.gov.

64. Coy, *The Humboldt Bay Region,* 201–2, 239, 238; on the College Land Act of 1862, see *Statutes at Large,* vol. 12 (Washington, D.C.: Government Printing Office), 503.

65. The information in this paragraph was compiled from the various tribes' Web sites.

66. Black Elk, Counting Coup, and Lame Deer sat down with professional writers and recorded what they could of their nations' traditional wisdom and history. Lucy Thompson's *To the American Indian: Reminiscences of a Yurok Woman* (1916; Berkeley, Calif.: Heyday Books, 1991) was written by a literate Native woman without the help of a professional writer.

67. Bob Doran and Greg McVicar, "The Return of Indian Island: Restoring the Center of the Wiyot World," *North Coast Journal,* July 1, 2004. See also wiyot.com.

68. The Sacred Sites Fund art auction was held at the Eureka Concert and Film Center, formerly the Eureka Theatre, on April 7, 2001.

5. Labor Trouble

1. On the presence of the Earth First! forest defender Alicia Littletree at the dedication ceremony, see "Democracy Is Not a Spectator Sport," radio interview with Dave Chism (former sawmill union representative) and Bob Cramer (member, Taxpayers for Headwaters), by Dan Fortson, November 27, 1997, KMUD, FM Radio 91.1. Transcript archived at http://www.iww.org/unions/iu120/local-1/PALCO/DFortson1.shtml.

2. See my discussion in the Introduction of capital culture's first and second contradictions.

3. *Humboldt Times,* June 22, 1935.

4. M. Graham, ed., *Man! An Anthology of Anarchist Ideas, Essays, Poetry, and Commentaries* (London: Cienfuegos Press, 1974). *Man!* was an internationally minded anarchist journal published in San Francisco from its first issue of January 1933 until its last issue of May 1940. The Romanian anarchist Marcus Graham edited the journal and in 1974 published this compendium of excerpts arranged by subject and not by date.

5. *Humboldt Times,* June 22, 1935.

6. Graham, *Man!,* citing vol. 3, no. 5, 226–27.

7. For literature and case histories showing how nineteenth-century legal doctrine and practice continuously elevated the value of increased productivity over both vested rights and public rights, see Harry N. Scheiber, "Public Rights and the Rule of Law in American Legal History," *California Law Review* 72, no. 2 (March 1984); and "Law and the Imperatives of Progress: Private Rights and Public Values in American Legal History," in *Ethics, Economics, and the Law,* ed. J. Roland Pennock and John W. Chapman (New York: New York University Press, 1982).

8. "Union Claims Gain; Mills Operate," *Humboldt-Standard,* May 16, 1935.

9. *Humboldt-Standard,* June 6, 1935; Brett H. Melendy, "One Hundred Years of the Redwood Lumber Industry, 1850–1950" (Ph.D. diss., Stanford University, microfilm title B26, Stanford, 1952).

10. Lumber Code Authority, *Lumber Code Authority Bulletin* 1, no. 1 (August 21, 1933): 6.

11. David Lawrence, "From the Battlefield in Washington," *Humboldt Times,* June 11, 1935.

12. *Humboldt-Standard,* June 21, 1935.

13. *Humboldt-Standard,* June 21, 1935.

14. *Polk's Eureka City Directory, 1935.*

15. Federated Trades and Labor Council, *Official Yearbook of Organized Labor: Humboldt County, 1929* (Eureka, Calif.: Federated Trades and Labor Council, 1929), Redwood District Council Collection, Onstine Papers, Humboldt Room, Humboldt State University, Arcata, Calif.

16. Ibid.

17. Harry C. Breit, "Hotel and Restaurant Employee's International Alliance," in *Official Yearbook of Organized Labor: Humboldt County, 1929* (Eureka, Calif.: Federated Trades and Labor Council, 1929), Redwood District Council Collection, Onstine Papers, Humboldt Room, Humboldt State University, Arcata, Calif.

18. Knud Haakonssen and Michael J. Lacey, eds., *Culture of Rights* (Cambridge: Cambridge University Press, 1991).

19. Lynwood Carranco, *Redwood Country* (Belmont, Calif.: Star, 1986), 35–70; Daniel Cornford, *Workers and Dissent in the Redwood Empire* (Philadelphia: Temple University Press, 1987). Humboldt's Chinese population increased from 38 in 1870 to 242 in 1880.

20. Melendy, "One Hundred Years," 337, citing *Humboldt Times,* March 27, 1903.

21. Cornford, *Workers and Dissent,* 195. E. A. Blockinger, general manager of the Pacific Lumber Company, writing in a 1911 edition of the Pacific Coast timber trade journal *Pioneer Western Lumberman,* exhorted lumber employers to diversify their workforce as a means of containing labor agitation. "Don't have too great a percentage of any one nationality," he wrote. "For your own good and theirs mix them up and obliterate clannishness and selfish social prejudice." On the other side of the issue, John Ericksen, vice president of the Eureka Trades Council, in the same year told the annual convention that "since the woodmen's strike here in 1907, men speaking many languages have been imported to make harmony among the workers hard to obtain." Cornford (*Workers and Dissent,* 194) described the ethnic constitution of Pacific Lumber's workforce as representative of the whole in 1911: 33 percent were native-born; 16 percent Italian; 25 percent foreign; Finnish, Swedish, and Norwegian were the second largest in proportion; Austrians were next. His data were taken from the 1910 Manuscript Census.

22. Melendy, "One Hundred Years," 339.

23. See Cornford, *Workers and Dissent,* 199; and Melendy, "One Hundred Years." See also Lizbeth Cohen's remarks on racial hiring strategies in Chicago's steel industry in *The Making of a New Deal: Industrial Workers in Chicago, 1919–1939* (New York: Cambridge University Press, 1990).

24. *Humboldt Times,* May 14, 1935.

25. Arthur Preuss, *A Dictionary of Secret and Other Societies* (St. Louis, Mo.: B. Herder, 1924), 180, quoting the *Fellowship Forum* 2, no. 41 (March 31, 1923): 7.

26. Carl Lemke, ed., *The Official History of the Improved Order of Redmen* (Waco, Tex.: Davis Bros., 1964), 3–4.

27. Preuss, *Dictionary of Secret and Other Societies,* 180–81.

28. Cornford, *Workers and Dissent;* Melendy, "One Hundred Years"; Ray Raphael, *Edges* (New York: Alfred Knopf, 1976), 43–53; Frank Taylor, "Paradise with a Waiting List," *Saturday Evening Post,* February 24, 1951.

29. Cornford, *Workers and Dissent,* 11–12.

30. *Polk's Eureka City Directory, 1935,* 41.

31. See my discussion of Balibar and the nation form in the introduction.

32. Étienne Balibar, "The Nation Form: History and Ideology," in *Becoming National: A Reader,* ed. Geoff Eley and Ronald Grigor Suny (New York: Oxford University Press, 1996), 137–38.

33. See the introduction for my discussion of the first and second contradictions.

34. Here I allude to Cornelius Castoriadis, *The Imaginary Institution of Society,* trans. Kathleen Blamey (Cambridge, Mass.: MIT Press, 1987).

35. Ferruccio Gambino, "A Critique of the Fordism of the Regulation School," *Wildcat-Zirkular,* nos. 28–29 (October 1996). See Paolo Virno, *A Grammar of the Multitude* (New York: Semiotext[e], 2004), on the transition of Fordism to post-Fordism and its relevance to new movements.

36. Irving Bernstein, *The New Deal Collective Bargaining Policy* (1950; New York: Da Capo Press, 1975), 133, citing *Investigation of the Concentration of Economic Power,* 77th Cong., 1st sess., Sen., Final Rep. and Recommendations of the TNEC (March 31, 1941).

37. "The Boston Journeymen Bootmakers' Society was charged with being an unlawful conspiracy since it practiced the closed shop. The court ruled that the combination itself was not conspiracy and that the tests of legality were the purposes and the means employed. The maintenance of labor conditions through the closed shop met these standards." Bernstein, *New Deal,* 18, citing *Commonwealth v. Hunt* (1842), 4 Metcalf, 111.

38. Bernstein, *New Deal,* 18, citing *U.S. Stat. at Large,* XXX, 424. Bernstein reports that this provision was declared unconstitutional in *Adair v. United States* (1908), 208 U.S. 161.

39. On the Supreme Court's so-called Incorporation Doctrine, see Alfred H. Kelly and Winfred A. Harbison, *The American Constitution: Its Origins and Development* (New York: W. W. Norton, 1948). Jennifer Nedelsky criticizes the tendency of progressive historians (like Charles Beard and Howard Zinn) to view the Madisonian framework that triumphed in the Constitution as merely the rich protecting their property interests. Nedelsky, *Private Property and the Limits of American Constitutionalism* (Chicago: University of Chicago Press, 1990), 279n2.

40. *Santa Clara County v. Southern Pacific Railroad Company,* error to the Court of the United States for the District of California. Argued January 26, 27, 28, 29, 1886; filed May 10, 1886.

41. See H. Graham, *Everyman's Constitution: Historical Essays on the Fourteenth Amendment, the "Conspiracy Theory," and American Constitutionalism* (Madison: State Historical Society of Wisconsin, 1968), 566–84. In *Santa Clara County v. Southern Pac. R.R.,* 118 U.S. 394, 396 (1886), Chief Justice Waite "announced from the bench that the Court would not hear argument on the question whether the equal

protection clause applied to corporations. 'We are all of the opinion that it does.'"

42. Bernstein, *New Deal*, 21–22, citing *U.S. Stat. at Large*, XLVII, 70.

43. Ibid., 40, 57–62. Section 7[a] is the first hesitant step of the New Deal, according to Bernstein; the NIRA's National Labor Board, set up in 1934, made interpretations of section 7[a] that paved the way for acceptance of the language pertaining to the Wagner Act.

44. Ibid., 37.

45. Ibid., 148.

46. Joseph G. Rayback, *A History of American Labor* (New York: Free Press, 1959), 341–42.

47. Bernstein, *New Deal*, 137.

48. Rayback, *History of American Labor*, 398–400; Bernstein, *New Deal*, 137–49.

49. C. B. Macpherson, *Property: Mainstream and Critical Positions* (Toronto: University of Toronto Press, 1978), and *The Political Theory of Possessive Individualism: Hobbes to Locke* (Oxford: Clarendon Press, 1962); Carol M. Rose, *Property and Persuasion: Essays on the History, Theory, and Rhetoric of Ownership* (Boulder, Colo.: Westview Press, 1994).

50. Cornford, *Workers and Dissent*, 14.

51. A. J. Doolittle, "The Official Township Map of Humboldt Co., Cal." (San Francisco: Grafton T. Brown Lith., 1985). In the University Archives, Bancroft Library, University of California, Berkeley.

52. Ibid., 14.

53. "Strike Vote Ordered by Union Here," *Humboldt Standard*, May 11, 1935.

54. "The Pacific Lumber Company: Calendar of Events" (1925), Bancroft Library, Pacific Lumber Collection, Box 4. See also Lowell S. Mengell II, "The Murphy Family and TPL Co.," *Humboldt Historian* 25, no. 4 (July–August 1977); and Melendy, "One Hundred Years," 196–97.

55. See my discussion of Scotia in chapter 3.

56. Cornford, *Workers and Dissent*, 14.

57. Ibid., 205.

58. Ibid., 221.

59. Mengell, "Murphy Family."

60. Lynwood Carranco, *The Redwood Lumber Industry* (San Marino, Calif.: Golden West Books, 1982), 155–67; Melendy, "One Hundred Years," 106–7, 187–89.

61. Melendy, "One Hundred Years," 201.

62. Cornford, *Workers and Dissent*, 16.

63. Melendy, "One Hundred Years."

64. Ibid., 311–13.

65. Ibid., 313.

66. Ibid., 314.

67. Cornford, *Workers and Dissent*, 1–3.

68. Melendy, "One Hundred Years," 325.

69. Cornford, *Workers and Dissent*, 1.

70. International Workingmen's Association, *To the Laboring Men of Humboldt County* (San Francisco: International Workingmen's Association, n.d. [c. 1880]).

71. Ibid.

72. Cornford, *Workers and Dissent*, 85.

73. Redwood District Council, *Redwood District Council of Lumber and Sawmill Workers Collection, Inventory, 1990–1991* (1990–91), 3, Humboldt Room, Humboldt State University, Arcata, Calif.

74. The Woodmen of the World was a social organization of lumbermen that the owners mandated; see Melendy, "One Hundred Years," 329, citing *Humboldt Times,* January 8, 1910, 3.

75. Melendy, "One Hundred Years," 329, citing *Humboldt Times,* January 8, 1910.

76. *Humboldt Times,* April 28, 1907.

77. On the strike of May 1, 1907, see Cornford, *Workers and Dissent,* 166.

78. Ibid., 166–73.

79. Ibid., 121–22.

80. Melendy, "One Hundred Years," 330–31, citing *Humboldt Times,* December 30, 1919, 1.4; see also *Humboldt Times,* April 28, 1927.

81. On the IWW in Humboldt, see Charlotte Todes's partisan *Labor and Lumber* (New York: International Publishers, 1931). On the IWW in the Pacific Northwest, see Robert L. Tyler, *Rebels of the Woods: The IWW in the Pacific Northwest* (Eugene: University of Oregon Press, 1967).

82. Cornford, *Workers and Dissent*, 190–26.

83. Melendy, "One Hundred Years," 212.

84. Ibid., 213.

85. *Cornford, Workers and Dissent,* 213, citing *Industrial Pioneer* 4 (April 1924): 19–26.

86. Melendy, "One Hundred Years," 209.

87. Lumber Code Authority, *Lumber Code Authority Bulletin* 1, no. 1 (August 21, 1933); Melendy, "One Hundred Years," 209–10.

88. Melendy, "One Hundred Years," 110–14.

89. Ibid.

90. Guy Debord, *The Society of the Spectacle* (New York: Zone Books, 1994), 30. See also Anselm Jappe, *Guy Debord* (Berkeley: University of California Press, 1999).

91. *Humboldt Times,* May 15, 1935.

92. The American slang term "coolie" referred to Asian workers transported to the United States in the nineteenth century.

6. Trouble in the Forest

1. Cecilia Lanman, interview with the author, 2001. Lanman recalled the vote being swung by the violence. The campaign began in 1988, and as soon as the initiative was announced, the companies started mowing down the residual groves, while the defenders struggled to protect the main groves.

2. The official title of Proposition 138 on the ballot was "Forestry Programs. Timber Harvesting Practices. Initiative Statute." See State of California, *California Ballot Pamphlet* (California Ballot Propositions Database, Hastings College of Law, Berkeley, Calif., 1990). In the end, 5,201,891 people voted against the proposition (71.16 percent), and 2,108,389 voted for it (28.84 percent).

3. Lanman interview, 2001.

4. Forests Forever failed to pass, with 3,842,733 (52.13 percent) voting no and 3,528,887 (47.87 percent) voting yes. The official title of Proposition 130, popularly called Forests Forever, was the Forest and Wildlife Protection and Bond Act of 1990. See State of California, *California Ballot Pamphlet* (California Ballot Propositions Database, Hastings College of Law, Berkeley, Calif., 1990).

5. The Redwood Summer Justice Project maintains an online archive of court documents, news reports, and audiovisual media documenting the project's long struggle against political repression by the FBI at judibari.org. I attended the Bari rally of May 24, 2000, at the Federal Courthouse in San Francisco, as well as the rally on the opening day of the trial in Oakland. May 24 has since been designated Judi Bari Day by the city of Oakland.

6. In July 2003, the FBI and the Oakland Police Department announced that they would not appeal. On the verdict, see "United States District Court for Northern California Official. Third [final] Verdict Form, No. C 91-01057. June 11, 2002," facsimile, unsigned and undated, judibari.org, June 12, 2002. The Cherney quote is from Mike Geniella, "FBI, Police Ordered to Pay $4.4 Million to Activists. Federal Jury: Car-Bombing Arrests Violated Rights of Earth First's Bari, Cherney," *Santa Rosa Press Democrat,* June 12, 2002. While the trial received little daily reportage by local, regional, and national media, the verdict received national notice. The verdict came on June 11. Associated Press, "Environmental Activists Arrested in Car Bombing Win Case," *New York Times,* June 11, 2002; see also Evelyn Nieves, "Environmentalists Win Bombing

Lawsuit," *New York Times,* June 12, 2002; Jim Herron Zamora and Henry K. Lee, "Earth First Activists Win Case; FBI, Cops Must Pay $4.4 million for Actions after Car Bombing," *San Francisco Chronicle,* June 12, 2002; Josh Richman, "Jury Awards Earth First Activists $4.4 Million," *Eureka Times-Standard,* June 12, 2002.

7. See especially *Eureka Times-Standard,* May 25–31, 1990; Judi Bari, *The Timber Wars* (Monroe, Maine: Common Courage Press, 1994); Redwood Summer Justice Project, "Who Bombed Judi Bari," compact audio disc, judibari.org, 1997; Judi Bari, *Earth First! and Timber Workers: Alliance for Sustainable Communities,* ed. Tanya Brannan (Willits, Calif.: Redwood Summer Justice Fund, Rainy Day Women Press, 2000); "Earth First! in Northern California: Interview with Judi Bari," *Capitalism Nature Socialism* 4, no. 4 (1993); "Timber: An Interview with Judi Bari," in *Sounding Off,* ed. Ron Sakolsky and F. Wei-han Ho (Brooklyn, N.Y.: Autonomedia, 1995); "Revolutionary Ecology," *Capitalism Nature Socialism* 8, no. 2 (1993); Beth Bosk, *"New Settler* Interview with Judi Bari," *New Settler* (1995), reprinted in *Albion Monitor,* January 13, 1997.

8. See my discussion in the introduction of the first and second contradictions in capital culture.

9. *Eureka Times-Standard,* May 26, 1990.

10. David Harris, *The Last Stand: The War between Wall Street and Main Street over California's Ancient Redwoods* (New York: Times Books, 1995).

11. Ibid., 163–69.

12. Tim Abate, "Timber Wars: Can a Virgin Forest Resist Wall Street?" *California Journal,* August 1998; also Ray Raphael, *More Tree Talk: The People, Politics, and Economics of Timber* (Washington, D.C.: Island Press, 1994).

13. Bari, "Earth First! in Northern California."

14. Pacific Lumber Company, "The Pacific Lumber Company and the Issues of Forest Management," public relations handout/position paper, 1987, Folder "Pacific Lumber," Humboldt Room, Humboldt State University, Arcata, Calif.

15. Bari, *Earth First! and Timber Workers,* 9–10.

16. Bari, *The Timber Wars,* 11, 67.

17. Raphael, *More Tree Talk,* and *Edges* (New York: Alfred Knopf, 1976). On the so-called back-to-the-land migration, see *New Settler Interview,* a magazine published by Beth Bosk (P.O. Box 72, Mendocino, Calif.); and Chris Bowman, "The Redwood Wars," *Sacramento Bee Magazine,* February 12, 1989.

18. Bari, *Earth First! and Timber Workers.*

19. The writings of Freeman House (*Totem Salmon: Life Lessons from Another Species* [Boston: Beacon Press, 1999]), the literature of the

Mattole Restoration Council, and Beth Bosk's *New Settler Interview* are perhaps the best sources of the back-to-the-land folks. Their indigenous bioregionalism grew out of a creative fusion of urban migrants, rural culture, Native American history, and resistance to corporate industrial forestry. See also Ray Raphael's *Edges* and *More Tree Talk*. Journalistic sources include Cliff Slaughter, "Humboldt's Rift," *California* magazine, October 1987; Abate, "Timber Wars"; and Bowman, "The Redwood Wars."

20. On Bari's union efforts in the redwoods, see Jeffrey A. Shantz and Barry D. Adam, "Ecology and Class: The Green Syndicalism of IWW/Earth First Local 1, Earth First! and IWW Local 1," *International Journal of Sociology and Social Policy* 19, nos. 7–8 (1999). In *Coyotes and Towndogs: Earth First! and the Environmental Movement* (New York: Penguin, 1993), Susan Zakin described Bari as a pink-diaper baby and an antiwar activist at the University of Maryland and Cherney as a child actor in TV commercials hailing from New York City (344). Bari and Cherney met in 1987 during Cherney's run for a California congressional seat, when he asked Bari to help with the redwood defense. More on Bari, Cherney, and North Coast Earth First! can be found in Jonathan K. London, "Common Roots and Entangled Limbs: Earth First! and the Growth of Post-wilderness Environmentalism on California's North Coast," *Antipode* 30, no. 2 (1998); Harris, *The Last Stand;* Rik Scarce, *Eco-Warriors: Understanding the Radical Environmental Movement* (Chicago: Noble Press, 1990); Christopher Manes, *Green Rage* (Boston: Little, Brown, 1990); and Jeff Shantz, *Clearcut the Bosses: Radical Ecology and Class Struggle* (San Francisco: Free Press, 2006). But the best sources are the writings of Bari herself and the weblog of the successful lawsuit against the FBI at judibari.org.

21. *Eureka Times-Standard,* May 26, 1990.

22. Bari conscientiously distinguished her biocentric and deep ecological analysis from the one espoused by Dave Foreman, for example, in his *Confessions of an Eco-Warrior* (New York: Harmony Books, 1991) and *Ecodefense: A Field Guide to Monkeywrenching* (Tucson, Ariz.: Ned Ludd Books, 1983), and by Christopher Manes in *Green Rage*. She disagreed with any "nature first, people second" approach to the common problem addressed by environmentalists, although she would not reduce their contributions to that diagnosis (and neither would I). Her positive contribution was to seek the social middle ground without giving up any ground to capitalist assumptions. Proactively seeking to avoid racist and white-masculinist overtones, she steered the movement away from environmental misanthropy and espoused an inclusive, socialist, internationalist, and feminist ecology that contradicted capitalism (for its untenable first principle of private property in public goods, like the redwoods),

communism (for offering no alternative other than an equally destructive centralized industrialism), and patriarchy (for unreconstructed masculinist values of conquest and domination directed at women, environment, and people). See especially Bari, "Revolutionary Ecology."

23. Bari, *The Timber Wars.*

24. Ibid. As the timber wars were heating up in Humboldt, Louisiana-Pacific in Mendocino County was shutting down factories, laying off workers, and shipping production to Mexico under the direction of CEO Harry Merlo. See Mike Geniella, "Impoverished Town Opens Arms to L-P"; and "Harvest for Export," *Santa Rosa Press Democrat,* December 10, 1989; also "Environmentalists, Lawmakers Battle Marketplace Mind-Set" and "Few Openly Criticize L-P Mexico Plan," *Santa Rosa Press Democrat,* December 11, 1989.

25. Cited in Bari, "Revolutionary Ecology," 10.

26. John Opie, *Nature's Nation: An Environmental History of the United States* (Fort Worth, Tex.: Harcourt Brace College Publishers, 1998), 429–31.

27. Western States Center, "Inside the 1993 Wise Use Conference," *Western Horizons* newsletter (Wise Use Public Exposure Project, a joint effort of the Western States Center and the Montana State AFL-CIO, Portland, Oregon, 1993), obtained from clearproject.org, 2001.

28. Ron Arnold and Alan Gottlieb, *Trashing the Economy: How Runaway Environmentalism Is Wrecking America* (Bellevue, Wash.: Free Enterprise Press, 1993).

29. The Mountain States Legal Foundation (MSLF) is on the Internet at http://www.mountainstateslegal.com.

30. See, for example, Peter Roper, "Study Links Spotted Owl, Social Woes," *Eureka Times-Standard,* May 5, 1990.

31. Bob Egelko, "Ex–Pacific Lumber Chief Must Face Fraud Trial," *San Francisco Chronicle,* March 18, 2009. In a whistle-blower lawsuit, former state Department of Forestry director Richard Wilson and former state forester Chris Maranto charged that Hurwitz falsified key forestry data on which the Sustained Yield Plan rested. The case was settled out of court in 2009.

32. Clifford Geertz, *Negara: The Theatre State in Nineteenth-Century Bali* (Princeton, N.J.: Princeton University Press, 1980), 135.

33. Barney Dowdle, "Perspectives on the Sagebrush Rebellion," *Policy Studies Journal* 12, no. 3 (March 1984); Howard E. McCurdy, "Public Ownership of Land and the 'Sagebrush Rebellion,'" *Policy Studies Journal* 12, no. 3 (March 1984).

34. For example, membership in the Sierra Club grew from approximately 147,000 in the mid-1970s to 550,000 in 1989 (Scarce, *Eco-Warriors,* 16).

35. Opie, *Nature's Nation,* 429–31, summarizes the sagebrush rebellion and its relation to the wise-use movement, calling the latter "a classic backlash movement" (431).

36. Dowdle, "Perspectives on the Sagebrush Rebellion," 474.

37. Ann Riesch, "Conservation under Franklin D. Roosevelt" (Ph.D. diss., University of Wisconsin, 1952), 478–79.

38. Edith Woodridge, "Humboldt County Should Replant Her Depopulated Forests," *Humboldt Beacon* (Fortuna), 1909.

39. Samuel G. Blythe, "The Last Stand of the Giants," *Saturday Evening Post,* December 6, 1919.

40. For the definitive account of the early redwood struggles, see Susan R. Schrepfer, *The Fight to Save the Redwoods: A History of Environmental Reform, 1917–1978* (Madison: University of Wisconsin Press, 1983).

41. John B. Dewitt, *California Redwood Parks and Preserves* (San Francisco: Save the Redwoods League, 1993).

42. Pacific Lumber Company, "The Pacific Lumber Company and the Issues of Forest Management."

43. Ibid.

44. Henry J. Vaux, *Timber in Humboldt County,* California Agricultural Experiment Station Bulletin 748 (University of California, 1955).

45. Slaughter, "Humboldt's Rift."

46. William D. Pine, "Humboldt's Timber: A Present and Future Problem," pamphlet (Eureka, Calif.: Humboldt County Board of Supervisors, 1952), Bancroft Library, Berkeley, Calif.; Vaux, *Timber in Humboldt County.*

47. Opie, *Nature's Nation.*

48. Ibid., 429.

49. Ibid., 7.

50. Ibid., 101.

51. London, "Common Roots and Entangled Limbs," 157.

52. Zakin, *Coyotes and Towndogs.*

53. See Kevin Michael Deluca, *Image Politics: The New Rhetoric of Environmental Activism* (New York: Guilford Press, 1999).

54. For the 1973 act, see California Public Resources Code, § 4511 et seq. On the Simpson plan, see "We Have Big Stake in Forest Debate," *Eureka Times-Standard,* January 2, 1972. See also Reed F. Noss, ed., *The Redwood Forest* (Washington, D.C.: Island Press, 2000); Kim Rodriguez, "The History of Conflict over Managing Coast Redwoods," www.cnr .berkeley.edu, accessed April 30, 2002; and *Redwood Coast Watersheds Alliance v. California State Board of Forestry et al.,* City and County of S.F. Super. Ct. No. 932123, 1999.

55. Cecilia Lanman, interview with the author, October 2001. For

documentation of *EPIC v. Johnson,* see Environmental Protection Information Center, "Litigation Summary," wildcalifornia.org.

56. See Robert Lindsey, "Ancient Forests Fall to a Wall Street Takeover," *New York Times,* March 2, 1988, for a summary of the takeover's business dimension. See also Pacific Lumber, "The Pacific Lumber Company and the Issues of Forest Management," 13. Gina Bentzley reported that the price paid was $872 million ("Scotia Rally Protests PL Harvest Plan," *Eureka Times-Standard,* December 4, 1986); see also David Abramson, "The Take-Over," *Image* magazine, July 1986; David Harris, *The Last Stand;* Lisa Newton, "The Chainsaws of Greed: The Case of Pacific Lumber," *Business and Professional Ethics Journal* 8, no. 3 (1989); Bill McKibben, "Milken, Junk Bonds, and Raping Redwoods," *Rolling Stone,* August 10, 1989; Stephen Taub, "Hurwitz's Bid to Maxxam-ize Value," *Financial World,* August 20, 1991; Reed McManus, "Maxxamizing Profits," *Sierra,* July–August 1994; Charles Hurwitz, "The Value Visionary: An Interview with Charles E. Hurwitz, Chairman, President, and Chief Executive Officer, Maxxam Inc. and Maxxam Group Inc., Houston, Texas," *Leaders* 17, no. 2 (April–June 1994); and London, "Common Roots and Entangled Limbs." For an analysis of the media treatment of the takeover that concludes that Maxxam was treated unfairly and emotionally and that "the threat to PL's old-growth is *not* attributable to junk bonds, high leverage, or hostile takeovers," see Harry D'Angelo and Linda D'Angelo, "Ancient Redwoods, Junk Bonds, and the Politics of Finance: A Study of the Hostile Takeover of the Pacific Lumber Company," unpublished paper, Humboldt Room, Humboldt State University, Arcata, Calif., 1996.

57. Pacific Lumber, "The Pacific Lumber Company and the Issues of Forest Management," 2–3. In 1987 there were 37 billion board feet of redwood trees in existence; of these, 13 percent were old-growth trees not protected in parks or other preservation agreements, approximately 4.7 billion board feet. Maxxam acquired approximately 75 percent of this available old growth (approximately 3.5 billion board feet).

58. Hurwitz, "The Value Visionary."

59. Joseph L. Sax, "The Public Trust Doctrine in Natural Resource Law: Effective Judicial Intervention," *Michigan Law Review* 68, no. 471 (January 1970); Harry N. Scheiber, "Public Rights and the Rule of Law in American Legal History," *California Law Review* 72, no. 2 (1984).

60. See the Mendocino Redwood Company LLC's pamphlet *After Nine Years* (2008), http://www.mrc.org. Also see Gaye LeBaron, "Remembering Scotia, the Last of the Company Towns," *Santa Rosa Press Democrat,* October 12, 2008; Derek J. Moore, "Timber's New Dawn," *Santa Rosa Press Democrat,* October 12, 2009.

61. See my discussion of the uses of psychoanalytic theory for social and cultural analysis in the introduction.

62. On the relevance of psychoanalysis and psychical energies and functions to the formation of publics, see Mustafa Emirbayer and Mimi Sheller, "Publics in History," *Theory and Society* 27 (1998).

Conclusion

1. See my discussion of the modern social imaginary in the introduction.

2. Steven Greenhouse, "Steelworkers Merge with British Union," *New York Times,* July 3, 2008.

3. Avery Gordon, *Ghostly Matters: Haunting and the Sociological Imagination* (Minneapolis: University of Minnesota Press, 1997).

4. Mark Lovelace, Humboldt Watershed Council, personal communication, January 2008.

5. Guy Debord, *The Society of the Spectacle* (1967; New York: Zone Books, 1994), 29; Giorgio Agamben, *The Coming Community* (Minneapolis: University of Minnesota Press, 1991), 78–79.

6. On fictitious commodities, see Karl Polanyi, *The Great Transformation: The Political and Economic Origins of Our Time* (1957; Boston: Beacon Press, 2001); and John Bellamy Foster's use of Polanyi to explain the first and second contradictions of capital culture in "Capitalism and Ecology: The Nature of the Contradiction," *Monthly Review* 54, no. 4 (September 2002).

7. Alberto Melucci, *Challenging Codes: Collective Action in the Information Age* (Cambridge: Cambridge University Press, 1996), 357.

8. On movement convergence, see, for example, Gloria Muñoz Ramirez, *The Fire and the Word: A History of the Zapatista Movement* (San Francisco: City Lights, 2007).

9. Polanyi, *The Great Transformation.*

Index

Richard Widick is a visiting scholar at the Orfalea Center for Global and International Studies at the University of California, Santa Barbara.

www.ingramcontent.com/pod-product-compliance
Lightning Source LLC
Chambersburg PA
CBHW020821270326
41928CB00006B/389